W0245904

N. D. Chatterjee

Applied Mineralogical Thermodynamics

Selected Topics

With 79 Figures

Springer-Verlag Berlin Heidelberg GmbH

Prof. Dr. NIRANJAN D. CHATTERJEE
Ruhr-Universität Bochum
Institut für Mineralogie
Postfach 10 21 48
D-4630 Bochum 1

ISBN 978-3-540-53215-6

Library of Congress Cataloging-in-Publication Data. Chatterjee, N. D. (Niranjan D.) Applied
mineralogical thermodynamics: selected topics/N. D. Chatterjee.   p.   cm. Includes biblio-
graphical references (p.   ) Includes index.
ISBN 978-3-540-53215-6   ISBN 978-3-662-02716-5 (eBook)
DOI 10.1007/978-3-662-02716-5
                         1. Mineralogy. 2. Thermodynamics. I. Title.   QE364.2.T45C49
1990   549–dc20   90-10394   CIP

© Springer-Verlag Berlin Heidelberg 1991
Originally published by Springer-Verlag Berlin Heidelberg New York in 1991

Typesetting (media-conversion)

32/3145-543210 – Printed on acid-free paper

Preface

Thermodynamic treatment of mineral equilibria, a topic central to mineralogical thermodynamics, can be traced back to the turn of the century, when J.H.Van't Hoff and his associates pioneered in applying thermodynamics to the mineral assemblages observed in the Stassfurt salt deposit. Although other renowned researchers joined forces to develop the subject - H.E.Boeke even tried to popularize it by giving an overview of the early developments in his *"Grundlagen der physikalisch-chemischen Petrographie"*, Berlin, 1915 - it remained, on the whole, an esoteric subject for the majority of the contemporary geological community.

Seen that way, mineralogical thermodynamics came of age during the last four decades, and evolved very rapidly into a mainstream discipline of geochemistry. It has contributed enormously to our understanding of the phase equilibria of mineral systems, and has helped put mineralogy and petrology on a firm quantitative basis. In the wake of these developments, academic curricula now require the students of geology to take a course in basic thermodynamics, traditionally offered by the departments of chemistry. Building on that foundation, a supplementary course is generally offered to familiarize the students with diverse mineralogical applications of thermo-dynamics. This book draws from the author's experience in giving such a course, and has been tailored to cater to those who have had a previous exposure to the basic concepts of chemical thermodynamics. The presentation has much benefited through the feedback received from the students at the Universities of Bochum, FRG, and Allahabad, India. In response to their needs, I may have occasionally gone into length, may be even to the point of tedium, to reiterate certain concepts.

Applications of thermodynamics to mineral systems are too many to be done justice to within the framework of one course. That has led to my selecting a set of topics. Those included cover numerous aspects of heterogeneous equilibria between solids and fluids, the glaring omissions being ionic equilibria and the thermodynamics of silicate melt; the title of the book emphasizes this. The approach is to handle the selected topics explicitly, and in a step-by-step manner, such that the beginners have little difficulty in working on their own.

The book has been structured, and held at a level appropriate for use as course material for the first year graduate class. Its purpose is to serve as a practical guide toward the application of thermodynamics to solid-solid and solid-fluid phase relations. The emphasis is mostly on the methods of mineralogical thermodynamics. Consequently, the worked examples have been chosen to demonstrate the methods, less so because of their geological importance. It is believed that the knowledge of the methods involved may help raise the critical awareness with which the currently published papers are absorbed by the students. Numerous worked examples are given with supportive details, some of which can be solved by hand-held calculators.

I thank Klaus-Dieter Grevel and Helmut Rimbach for assistance with some worked examples. Ms. Almut Fischer and Renate Lehmann's help in drafting the figures is gratefully acknowledged. Thanks are due to the Institute of Mineralogy, Ruhr University, Bochum, for supporting my effort. Critical reviews of the earlier drafts of this manuscript by a few friends and colleagues including Bernard W. Evans (Seattle, USA), Alok K. Gupta (Allahabad, India), Darrell J. Henry, (Baton Rouge, USA), Walter V. Maresch (Münster, FRG), Yastami Oka (Yokohama, Japan), and Konstantin K. Podlesskii (Chernogolovka, USSR), were most helpful. I express my sincere appreciation for their constructive comments and at the same time exonerate them for any error or oversight. And last but not least, I thank Springer-Verlag for patient cooperation in producing this book.

February 1990   NIRANJAN D. CHATTERJEE

Contents

Symbols, Abbreviations, and Constants

A	Helmholtz energy as an extensive quantity
A	A Redlich-Kister parameter, unless otherwise stated
A, A'	Coefficients of virial or corresponding state equations
a	A constant of a vdW, RK, or MRK equation related to the attractive intermolecular interactions in a fluid
$\mathbf{a}_{mix}$	**a** for a fluid mixture
a	A coefficient of the $C°_p(T)$ function
a_i	Activity of the component i in a given phase
a^{id}_i	Ideal activity of the component i in a given phase
B	Redlich-Kister parameter, unless otherwise stated
B, B'	Coefficients of virial or corresponding state equations
b	A constant of a vdW, RK, or MRK equation related to the repulsive intermolecular interactions in a fluid
$\mathbf{b}_{mix}$	**b** for a fluid mixture
b	A coefficient of the $C°_p(T)$ function
C	A Redlich-Kister parameter, unless otherwise stated
C, C'	Coefficients of virial or corresponding state equations
°C	Degree Celsius
c	A coefficient of the $C°_p(T)$ function
d	A coefficient of the $C°_p(T)$ function
E	Electromotive force
e	A coefficient of the $C°_p(T)$ function
F	Faraday constant, 96487 J Volt^{-1} mol^{-1}
$f°_i$	Standard state [1 bar, T] fugacity of the species i
f_i	Fugacity of the pure species i at specified T and P
f^*_i	A dimensionless quantity defined as $f_i/f°_i$
f'_i	Fugacity of the component i in a fluid mixture at a given set of T, P, and X_i
f^∇_i	A dimensionless quantity defined as $f'_i/f°_i$
f	A coefficient of the $C°_p(T)$ function

G	Gibbs energy as an extensive quantity
G	Gibbs energy as an intensive (molar) quantity
$G^{\circ}_{f,T,i}$	Molar Gibbs energy of formation of i at 1 bar and T
G^{*}_{i}	Molar Gibbs energy of i at specified P and T
$\Delta_a G^{P,T}_{i}$	Apparent molar Gibbs energy of formation of i at P and T
G_m	Integral molar Gibbs energy of mixing
G^{id}_m	Integral molar ideal Gibbs energy of mixing
G^{ex}_m	Integral molar excess Gibbs energy of mixing
g_i	Partial molar Gibbs energy of i
$g_{m,i}$	Partial molar Gibbs energy of mixing of i
$g^{id}_{m,i}$	Partial molar ideal Gibbs energy of mixing of i
$g^{ex}_{m,i}$	Partial molar excess Gibbs energy of mixing of i
H	Enthalpy as an extensive quantity
H	Enthalpy as an intensive (molar) quantity
H°_{soln}	Enthalpy of solution
$H^{\circ}_{f,T,i}$	Molar enthalpy of formation of i at 1 bar and T
H^{*}_{i}	Molar enthalpy of i at specified P and T
$\Delta_a H^{P,T}_{i}$	Apparent molar enthalpy of formation of i at P and T
H_m	Integral molar enthalpy of mixing
H^{id}_m	Integral molar ideal enthalpy of mixing
H^{ex}_m	Integral molar excess enthalpy of mixing
h_i	Partial molar enthalpy of i
$h_{m,i}$	Partial molar enthalpy of mixing of i
$h^{id}_{m,i}$	Partial molar ideal enthalpy of mixing of i
$h^{ex}_{m,i}$	Partial molar excess enthalpy of mixing of i
K	Kelvin
K	Equilibrium constant
K_D	Distribution coefficient of species between a set of phases in a specified reaction
K_X	Product of mole fractions of species in a given reaction
K_{γ}	Product of activity coefficients of species for a given reaction
n	Number of moles
P	Pressure (bar or kbar, as specified)
P_c	Critical pressure
P_r	Reduced pressure ($\equiv P/P_c$)
q	Site multiplicity of a specified sublattice of a crystalline solution
R	Gas constant, 8.3143 J K^{-1} mol^{-1}

S	Entropy as an extensive quantity
S	Entropy as an intensive (molar) quantity
S°_T	Standard molar entropy at 1 bar and T
$S^\circ_{f,T,i}$	Molar entropy of formation of i at 1 bar and T
S^*_i	Molar entropy of i at specified P and T
S_m	Integral molar entropy of mixing
S^{id}_m	Integral molar ideal entropy of mixing
S^{ex}_m	Integral molar excess entropy of mixing
s_i	Partial molar entropy of i
$s_{m,i}$	Partial molar entropy of mixing of i
$s^{id}_{m,i}$	Partial molar ideal entropy of mixing of i
$s^{ex}_{m,i}$	Partial molar excess entropy of mixing of i
T	Temperature in Kelvin (K), unless indicated as degree Celcius (°C)
T_c	Critical temperature
T_r	Reduced temperature ($\equiv T/T_c$)
U	Internal energy as an extensive quantity
U	Internal energy as an intensive (molar) quantity
V	Volume as an extensive quantity
V	Volume as an intensive (molar) quantity
V°_T	Standard molar volume at 1 bar and T
V_m	Integral molar volume of mixing
V^{id}_m	Integral molar ideal volume of mixing
V^{ex}_m	Integral molar excess volume of mixing
v_i	Partial molar volume of i
$v_{m,i}$	Partial molar volume of mixing of i
$v^{id}_{m,i}$	Partial molar ideal volume of mixing of i
$v^{ex}_{m,i}$	Partial molar excess volume of mixing of i
$W_{G,ij}$	Margules Gibbs energy parameter for i in i-j solution
$W_{U,ij}$	Margules internal energy parameter for i in i-j solution
$W_{H,ij}$	Margules enthalpy parameter for i in i-j solution
$W_{S,ij}$	Margules entropy parameter for i in i-j solution
$W_{V,ij}$	Margules volume parameter for i in i-j solution
$W_{CP,ij}$	Margules heat capacity parameter for i in i-j solution
X_i	Mole fraction of i
Y	Any extensive quantity
Y	Any intensive (molar) quantity
Y_m	Any integral molar quantity of mixing
Y^{id}_m	Any integral molar ideal quantity of mixing
Y^{ex}_m	Any integral molar excess quantity of mixing

y_i	Any partial molar quantity of i
$y_{m,i}$	Any partial molar quantity of mixing of i
$y^{id}_{m,i}$	Any partial molar ideal quantity of mixing of i
$y^{ex}_{m,i}$	Any partial molar excess quantity of mixing of i
Z	Compressibility factor
Z_c	Critical compressibility factor
α	Isobaric thermal expansion
β	Isothermal compressibility
γ_i	Activity coefficient of component i in a given phase
ϕ_i	Fugacity coefficient of a pure fluid species i
ϕ''_i	Fugacity coefficient of i in a fluid mixture
μ_i	Chemical potential of component i at P,T, and $X_i(\neq 1)$
μ^*_i	Standard chemical potential of i at P and T
μ°_i	Standard chemical potential of i at 1 bar and T
v_i	Stoichiometric coefficient of i in a reaction
σ_i	Standard deviation of i
σ^2_i	Variance of i
ρ_{ij}	Correlation coefficient of i and j

Chapter 1

Summary of Basic Thermodynamic Concepts:
A Refresher

1.1 Introduction

One goal of geological research is to understand the physical and chemical processes responsible for the origin and evolution of the earth. In terms of time and space, most such processes are not accessible to direct observation, but they can be reconstructed by deciphering the physicochemical frameworks of formation of rocks and mineral deposits which formed in response to them. Supposing the observed mineral assemblage preserved in a rock records the state of chemical equilibrium, thermodynamics can help translate the compositional data on the coexisting minerals to temperature and pressure of their equilibration. What thermodynamics cannot do is to tell us how long a quenched mineral assemblage is likely to survive once it is outside its field of stability, or what the mechanism of a certain mineral reaction will be. Such questions - geologically of equal importance - can be answered by kinetics, a subject not treated in this book.

Classical thermodynamics, which we shall be concerned with here, operates on a phenomenological basis. It sets up functional relationships between some measurable macroscopic properties of a system. Such properties are known as the state variables or state functions. For a system in equilibrium, the mutually interrelated state functions remain constant over an indefinite period of time. To describe the state of a system, it thus suffices to specify the numerical values of the state functions. Besides pressure (P) and volume (V), known from mechanics, thermodynamics defines three new state functions. These are temperature (T), internal energy (U), and entropy (S), introduced on the basis of the so-called zeroth, first and second laws of thermodynamics[1], respectively. Depending upon the types of problem to be handled, appropriate sets of state functions are chosen to describe the state of the system. Although the five quantities noted above suffice in principle, treatment of certain problems are facilitated by introducing three additional state functions. These are enthalpy (H), Helmholtz energy (A), and Gibbs

---

[1] A knowledge of the laws of thermodynamics is assumed in this book.

energy (G), defined as

$$H \equiv U + PV, \tag{1.1}$$

$$A \equiv U - TS, \text{ and} \tag{1.2}$$

$$G \equiv U + PV - TS = H - TS. \tag{1.3}$$

Because pressure and temperature are the prime variables within the earth, Gibbs energy, G, with its variables T and P, is central to mineralogical thermodynamics. In any system, G, T, and P are not independent of each other; they are mathematically related. The functional relationship, written explicitly as $G = f(T,P)$, is an example of an equation of state; G being the state function, T and P its characteristic variables. Note that the macroscopic description of the equilibrium state of a system is not concerned with questions like ways in which G of a phase (mineral) could be affected by its microscopic properties such as atomic environment, nature of bonding etc. The microscopic description of the topic is handled by statistical thermodynamics, a subject beyond the scope of this book. Any conclusion based on microscopic arguments must, however, agree with those derived from the macroscopic properties.

For our subsequent treatment, we need to distinguish between two categories of state variables - intensive and extensive. An intensive variable is independent of mass; T and P are examples of intensive variables. An extensive variable, like V, U, S, H, A, or G, depends on the mass of substance; its numerical value doubles when the number of moles of the substance is doubled. However, any extensive quantity may be converted to an intensive (e.g. molar) quantity simply by dividing it by the number of moles of substance involved (see Sect. 1.2.2). The extensive variables are printed in boldface italics to distinguish them from the intensive variables, which appear in normal upper case letters.

1.2 Gibbs Energy as a State Function

1.2.1 *Temperature and Pressure Dependence of Gibbs Energy, and the Choice of a Standard State*

Because of its pivotal role in mineralogical thermodynamics, we continue by exploring the properties of Gibbs energy in detail. The variation of G with respect to T and P for a given amount of a homogeneous pure substance (be it a mineral end-member or a pure fluid species) is obtained by differentiating Eq.(1.3),

$$dG = dU + PdV + VdP - TdS - SdT. \tag{1.4}$$

From a combined expression for the first and the second laws of thermodynamics (Denbigh 1971, pp.44-45), it is known that

$$dU = TdS - PdV. \tag{1.5}$$

Substituting (1.5) into (1.4), we derive what is known as a Gibbs' fundamental equation

$$dG = -SdT + VdP. \tag{1.6}$$

Being a state function of the characteristic variables T and P, G is single-valued and continuously differentiable (unless there is a phase transition), and as such, an exact differential[2]. Thus

$$dG = \left(\frac{\partial G}{\partial T}\right)_P dT + \left(\frac{\partial G}{\partial P}\right)_T dP. \tag{1.7}$$

Equating (1.6) and (1.7), it follows that

$$\left(\frac{\partial G}{\partial T}\right)_P = -S, \quad \text{and} \tag{1.8}$$

$$\left(\frac{\partial G}{\partial P}\right)_T = V. \tag{1.9}$$

Measurement of S and V are thus of interest to us; they help derive the temperature and pressure dependences of Gibbs energy. The change in G, ΔG, between an initial state $T°,P°$, and the final state T,P, may be obtained by integrating (1.6) between the limits $T°$ to T and $P°$ to P. The Gibbs energy difference, ΔG [identical to $G(T,P) - G(T°,P°)$], is independent of the path followed during the change of state. Therefore, the Gibbs energy of the system in the final state, $G(T,P)$, is given by an expression of the general form

$$G(T,P) = G(T°,P°) - \int_{T°}^{T} S_{P°}(T)\, dT + \int_{P°}^{P} V_T(P)\, dP. \tag{1.10}$$

An unambiguous calculation of $G(T,P)$ evidently requires that two conditions be met:
1. The lower limit of integration in (1.10), denoted henceforth as the *standard state*, is properly chosen and uniquely specified, and
2. $G(T°,P°)$, the integration constant in (1.10), is known.

---

[2] For a treatment of the properties of an exact differential, see Klotz and Rosenberg (1972 pp.10-13).

It is emphasized that the choice of the standard state is an arbitrary affair; it is chosen in a manner that is best suited to handle a problem. In the following, we choose P° as 1 bar and T° as any T. This is a convenient choice, compatible with the tables of the thermodynamic data of minerals (Robie et al. 1979), which will be extensively used later on. To calculate $G(T,P)$, the integration starts at P° = 1 bar and a T°, for which $G(T°,P°)$ is known. If T° is not identical to T, the upper temperature limit of integration, both the integrals in Eq.(1.10) are sequentially executed. However, given $G(T°,P°)$ data at 1 bar and T, the final temperature, only the $_{P°}\int^P V_T(P)dP$ integral must be evaluated. But regardless of the T° at which the integration begins, we always end up with a unique value for $G(T,P)$. From the above expositions it is evident that whenever we deal with a change of state, we *must* specify the standard state. The problem of choosing the appropriate standard state will be repeatedly encountered in later sections. The 1 bar, T (K) standard state chosen here will be abbreviated as the 1 bar standard state throughout this book.

We also need to know $G(T°,P°)$ to obtain $G(T,P)$. Because only the relative value for $G(T°,P°)$ can be measured, but not its absolute value, $G(T°,P°)$ is arbitrarily set equal to the Gibbs energy difference of the formation reaction of a given amount of a phase from its constituent stable elements at the standard state. The stable elements themselves are assigned a zero energy datum *by convention*. The $G(T°,P°)$ for 1 mol of any phase is its standard molar Gibbs energy of formation, $G°_{f,T}$. In this notation, the superscript (°) indicates 1 bar pressure, whereas the subscripts $(_{f,T})$ denote the *formation* from the stable elements at a given T. It should be emphasized, however, that $G°_{f,T}$ need not necessarily be referred back to the constituent elements. For example, for the oxygen-base minerals, referring $G°_{f,T}$ to the constituent oxides is an equally viable alternative. In fact, Robie et al. (1979) give two sets of values for $G°_{f,T}$ (and other quantities of formation) of the oxygen-base minerals; one refers to the elements, the other to the oxides at their standard state. For the latter set, the $G°_{f,T}$ of the oxides is zero by convention. In this treatise, we shall invariably stick with the values referred to the elements, because they apply to all minerals, oxygen-based or not.

An alternative way of looking at the problem of the T- and P-dependences of G is to start by splitting the Gibbs energy into the enthalpy and the entropy terms according to Eq. (1.3):

$$G(T,P) = H(T,P) - TS(T,P). \tag{1.11}$$

This approach is very relevant, because G is only seldom measured directly. The most common source of our knowledge of G is from calorimetric measurements of H and S and combining them according to (1.11). This is tantamount to saying that we need to know H and S as functions of T and P.

Recalling that both S and H are state functions, we may write

$$dS = \left(\frac{\partial S}{\partial T}\right)_P dT + \left(\frac{\partial S}{\partial P}\right)_T dP, \quad \text{and} \tag{1.12}$$

$$dH = \left(\frac{\partial H}{\partial T}\right)_P dT + \left(\frac{\partial H}{\partial P}\right)_T dP .$$
(1.13)

The temperature dependences of S and H can be calculated from the heat capacity at constant pressure, C_P according to the relations

$$\left(\frac{\partial S}{\partial T}\right)_P = \frac{C_P}{T} , \quad \text{and}$$
(1.14)

$$\left(\frac{\partial H}{\partial T}\right)_P = C_P .$$
(1.15)

To derive $(\partial S/\partial P)_T$ and $(\partial H/\partial P)_T$ of Eqs.(1.12) and (1.13), let us start by considering (1.6). Because G is a state function, and as such, an exact differential, its crossed partial derivatives are equal (Klotz and Rosenberg 1972, p.12),

$$\left(\frac{\partial^2 G}{\partial T \partial P}\right) = \left(\frac{\partial^2 G}{\partial P \partial T}\right) .$$
(1.16)

Evaluating each side of Eq.(1.16) individually,

$$\left(\frac{\partial^2 G}{\partial T \partial P}\right) = \frac{\partial}{\partial T}\left(\frac{\partial G}{\partial P}\right)_T = \left(\frac{\partial V}{\partial T}\right)_P , \quad \text{and}$$
(1.17)

$$\left(\frac{\partial^2 G}{\partial P \partial T}\right) = \frac{\partial}{\partial P}\left(\frac{\partial G}{\partial T}\right)_P = -\left(\frac{\partial S}{\partial P}\right)_T .$$
(1.18)

Equating (1.17) and (1.18), we have

$$\left(\frac{\partial S}{\partial P}\right)_T = -\left(\frac{\partial V}{\partial T}\right)_P ,$$
(1.19)

which is one of Maxwell's equations. And finally, $(\partial H/\partial P)_T$ may be derived from

$$H = G + TS.$$
(1.3a)

Differentiating (1.3a) with respect to P,

$$\left(\frac{\partial H}{\partial P}\right)_T = \left(\frac{\partial G}{\partial P}\right)_T + T\left(\frac{\partial S}{\partial P}\right)_T .$$
(1.20)

Substituting (1.9) and (1.19) into (1.20), we have

$$\left(\frac{\partial H}{\partial P}\right)_T = V - T\left(\frac{\partial V}{\partial T}\right)_P.$$

(1.21)

Introducing Eqs.(1.14) and (1.19) into (1.12), and Eqs.(1.15) and (1.21) into (1.13), the total differentials of S and H are recovered

$$dS = \frac{C_P}{T}dT - \left(\frac{\partial V}{\partial T}\right)_P dP, \quad \text{and}$$

(1.22)

$$dH = C_P\,dT + \left[V - T\left(\frac{\partial V}{\partial T}\right)_P\right]dP.$$

(1.23)

Derivation of $G(T,P)$ from Eq.(1.11), therefore, demands that $H(T,P)$ and $S(T,P)$ be evaluated first. From (1.22) and (1.23) it is evident that these goals may be accomplished by integrating those equations between $T°,P°$ and T,P, $H(T°,P°)$ and $S(T°,P°)$ being known independently. Now, like Gibbs energy, only relative changes in enthalpy are amenable to measurement. Therefore, for the 1 bar and $T(K)$ standard state chosen earlier, $H(T°,P°)$ for 1 mol of any given substance will be identical to its standard molar enthalpy of formation, $H°_{f,T}$. Contrary to Gibbs energy and enthalpy, the absolute value of entropy *can* be measured. Consequently, $S(T°,P°)$ need not necessarily be defined with reference to formation from the elements; for 1 mol of a substance, it is designated as the standard molar entropy, $S°_T$. It must be emphasized, however, that occasions may arise requiring us to manipulate entropy in terms of the standard molar entropy of formation, $S°_{f,T}$ (see Sect. 1.2.2.1). For this purpose, it is defined in a manner analogous to the $H°_{f,T}$; it is the entropy difference of the reaction leading to the formation of the phase from its constituent stable elements at the standard state. Therefore, $S°_{f,T}$ follows from $S°_T$ of the compound and those of the elements.

From the foregoing discussions, it is apparent that $C_P(T)$ and the volume equation of state,

$$V = f(T,P),$$

(1.24)

are also required for thermodynamic calculations. Volume being a function of state, we have

$$dV = \left(\frac{\partial V}{\partial T}\right)_P dT + \left(\frac{\partial V}{\partial P}\right)_T dP.$$

(1.25)

The two derivatives on the right-hand side (RHS) of Eq.(1.25) are generally expressed in terms of α, the isobaric thermal expansivity, and β, the isothermal compressibility, defined as

$$\alpha \equiv \frac{1}{V_{298}^{\circ}} \left(\frac{\partial V}{\partial T} \right)_P, \quad \text{and} \tag{1.26}$$

$$\beta \equiv -\frac{1}{V_{298}^{\circ}} \left(\frac{\partial V}{\partial P} \right)_T. \tag{1.27}$$

Because volume decreases with increasing pressure, a negative sign is introduced on the RHS of (1.27) to facilitate tabulation of β as a positive quantity. Note also that the subscript 298 appearing in these equations is an abbreviation for the standard temperature of 298.15 K; this abbreviation *will be used throughout this book*.

From the above discussions it is evident that measurement and tabulation of H, S, C_P, and V are of fundamental importance to us. In Chapter 2 we shall see how they are measured, numerically manipulated, and tabulated for mineral equilibria calculations.

1.2.2 Composition Dependence of Gibbs Energy

1.2.2.1 Chemical Potential and the Gibbs-Duhem Equation

So far, we have handled a given amount of a pure homogeneous substance. Geologically more relevant, however, are substances of variable compositions, be they minerals or fluids. Consequently, we now examine the properties of a given amount of a homogeneous substance of variable composition, a solution. Thermodynamics of solutions will be developed in two steps. The first step, given in this chapter, though general in nature, is applicable to the molecular mixing characteristic of fluids. It also applies to isostructural crystalline solutions *to the extent* that they involve site mixing of atoms or ions on just *one* crystal structure site, like we have in (Na,K)Cl. The second step, detailed in Chapter 8, extends this treatment to crystalline solutions in general.

The Gibbs energy of a solution is a function of temperature, pressure, and composition. Its composition is usually expressed in terms of the numbers of moles of the components n_1, n_2, n_3, and so on. Thus, n, the total number of moles of components, is

$$n = \sum_i n_i. \tag{1.28}$$

The state function G, an extensive quantity, is then expressed as

$$G = f(T,P,n_1,n_2, ...). \tag{1.29}$$

Introducing the intensive variable G (a molar quantity) defined by

$$G \equiv \frac{G}{n},$$ (1.30)

the Eq.(1.29) may be recast to

$$nG = f(T,P,n_1,n_2,...).$$ (1.29a)

The total differential of nG is

$$d(nG) = \left[\frac{\partial(nG)}{\partial T}\right]_{P,n_1,n_2...} dT + \left[\frac{\partial(nG)}{\partial P}\right]_{T,n_1,n_2...} dP$$

$$+ \left[\frac{\partial(nG)}{\partial n_1}\right]_{T,P,n_2...} dn_1 + \cdots .$$ (1.31)

The partial derivative with respect to n_i in the above equation is the chemical potential[3], μ_i (identical to the partial molar Gibbs energy, g_i) of the i-th component. Thus,

$$\mu_i \equiv \left[\frac{\partial(nG)}{\partial n_i}\right]_{T,P,n_{j(j\neq i)}} \equiv g_i,$$ (1.32)

where n_j indicates the mole numbers of all components other than i, $(n_j \neq n_i)$. Note that μ_1 corresponds to the total response of the system to the addition of an infinitesimal amount of i at constant T, P, and n_j. Substituting Eqs.(1.8), (1.9), and (1.32) into Eq.(1.31), the well known Gibbs fundamental equation is recovered,

$$d(nG) = - (nS)dT + (nV)dP + \sum_i \mu_i dn_i.$$ (1.33)

For the rest of this section, we shall treat *isothermal and isobaric processes* (dT = 0 and dP = 0), restricting our treatment, for simplicity, to a *binary system* comprising components 1 and 2, and bearing in mind that all equations derived hereafter will be amenable to extension to multicomponent systems. Under isothermal-isobaric conditions, the total differential of the

---

[3] For the sake of completeness, it is pointed out that μ_1, the chemical potential of the i-th component, may be alternatively defined as
$$\mu_1 \equiv [\partial(nU)/\partial n_i]_{S,V,nj},$$ (1.32i)
$$\mu_i \equiv [\partial(nH)/\partial n_i]_{S,P,nj}, \text{ or}$$ (1.32ii)
$$\mu_1 \equiv [\partial(nA)/\partial n_1]_{V,T,nj}.$$ (1.32iii)
Because T, P, and n_1 are the variables of interest to geology, we shall always make use of the definition of μ_i given in Eq.(1.32).

Gibbs energy reduces to

$$d(nG) = \mu_1 dn_1 + \mu_2 dn_2. \tag{1.33a}$$

Now, like all other state properties of a solution, its total Gibbs energy, nG, is a homogeneous function of the first degree in the mole numbers of the components at any specified T and P. This implies that μ_i is dependent only upon the relative amounts of the individual components, not on their absolute amounts. For such a case, Euler's theorem (for its derivation and application to the state functions, see Klotz and Rosenberg 1972, pp.14-15) for a homogeneous function of the first degree may be shown to apply, whence it follows

$$nG = n_1 \left[\frac{\partial(nG)}{\partial n_1}\right]_{n_2} + n_2 \left[\frac{\partial(nG)}{\partial n_2}\right]_{n_1}. \tag{1.34}$$

Substitution of Eq.(1.32) into (1.34) yields

$$nG = n_1\mu_1 + n_2\mu_2. \tag{1.35}$$

Differentiating Eq.(1.35),

$$d(nG) = n_1 d\mu_1 + \mu_1 dn_1 + n_2 d\mu_2 + \mu_2 dn_2. \tag{1.36}$$

Equating (1.33a) to (1.36), the well known Gibbs-Duhem equation is recovered,

$$n_1 d\mu_1 + n_2 d\mu_2 = 0. \tag{1.37}$$

From now onward, the composition variables will be handled in terms of the mole fraction, X_i, of the i-th component, rather than its mole number, n_i. The former is defined as

$$X_i \equiv \frac{n_i}{n}. \tag{1.38}$$

Introduction of X_i leads to a normalization of this variable to

$$\sum_i X_i = 1, \tag{1.39}$$

such that, in a system of k components, there are k-1 independent composition variables. Introduction of X_i for n_i also requires that we express the state functions as intensive quantities. Dividing (1.29a) by n,

$$G = f(T,P,X_1,X_2 ...); \tag{1.29b}$$

the intensive quantity G now replacing the extensive quantity **G**. Therefore, Eq.(1.37) needs to be rewritten as

$$X_1 d\mu_1 + X_2 d\mu_2 = 0. \tag{1.40}$$

Note that the Gibbs-Duhem equation, derived above, applies to all state functions, Y ($\equiv$ U, S, V, H, A, G). While rewriting it in terms of y_i, the appropriate partial molar quantity, we may use X_2 as the master composition variable, such that the general form of the Gibbs-Duhem equation for a binary system becomes

$$(1\text{-}X_2)dy_1 + X_2 dy_2 = 0, \tag{1.41}$$

or alternatively, dividing both sides by dX_2,[4)]

$$(1 - X_2)\frac{dy_1}{dX_2} + X_2\frac{dy_2}{dX_2} = 0. \tag{1.42}$$

The Gibbs-Duhem equation is applied in whichever of these forms is appropriate for our purpose (see Sect. 8.5.3). For example, when μ_2 has been measured as a function of X_2 at any fixed T and P, μ_1 is obtained by integrating an equation analogous to Eq.(1.42) between the limits $X_2=0$ to $X_2=X_2$. Transposing it, and integrating,

$$\int_{X_2=0}^{X_2=X_2} \frac{d\mu_1}{dX_2} dX_2 = - \int_{X_2=0}^{X_2=X_2} \frac{X_2}{(1 - X_2)} \frac{d\mu_2}{dX_2} dX_2. \tag{1.43a}$$

Noting that μ_1 (at $X_2=0$) $\equiv \mu^*_1$ (for the details, see Sect. 1.2.2.3), the standard chemical potential of the component 1, the integration of (1.43a) may be accomplished by applying the rule of substitution, yielding,

$$\mu_1(\text{at } X_2) = \mu_1^*(\text{at } X_2 = 0) - \int_{\mu_2(\text{at } X_2=0)}^{\mu_2(\text{at } X_2=X_2)} \frac{X_2}{(1 - X_2)} d\mu_2. \tag{1.43}$$

[4)] The derivation of Eq.(1.42) from (1.41) involves the division of total differentials (like dy_1 and dy_2) by dX_2. To check its validity, we start with (1.41), explicitly writing out the total differentials (the two terms in square brackets of the next line):
$(1\text{-}X_2)[(\partial y_1/\partial X_1)dX_1+(\partial y_1/\partial X_2)dX_2]+X_2[(\partial y_2/\partial X_1)dX_1+(\partial y_2/\partial X_2)dX_2]=0.$
Substituting $-dX_2$ for dX_1, since $X_1 = (1\text{-}X_2)$, it follows that
$(1\text{-}X_2)[(\partial y_1/\partial X_1)(-dX_2)+(\partial y_1/\partial X_2)dX_2]+ X_2[(\partial y_2/\partial X_1)(-dX_2)+(\partial y_2/\partial X_2)dX_2] = 0$, or
$(1\text{-}X_2)[\{(-\partial y_1/\partial X_1)+(\partial y_1/\partial X_2)\}dX_2]+X_2[\{(-\partial y_2/\partial X_1)+(\partial y_2/\partial X_2)\}dX_2]=0.$
Noting that the two square bracketed terms in the last line are dy_1 and dy_2, respectively, and that each of them comprise dX_2, we divide both sides of it by dX_2 (thereby eliminating dX_2) to write the final equation
$(1\text{-}X_2) dy_1/dX_2 + X_2 dy_2/dX_2 = 0.$ \hfill (1.42)

The standard chemical potential of the pure substance i, μ^*_i, which is identical to its molar Gibbs energy at T and P, G^*_i, has been identified here by the superscript (*) to emphasize that it applies to *any* P *other than* P°. It is calculated from an equation analogous to Eq.(1.10),

$$\mu^*_i(P, T) = G^\circ_{f,298,i} - \int_{298}^{T} S^\circ_{f,i}(T)\,dT + \int_{P^\circ}^{P} V_{T,i}(P)\,dP$$

$$= G^\circ_{f,T,i} + \int_{P^\circ}^{P} V_{T,i}(P)\,dP \equiv G^*_i(P, T). \tag{1.44}$$

Note that the temperature integral is taken here over the entropy *of formation*, $S^\circ_{f,i}$, so that the sum of the first two terms on the RHS yields the standard molar Gibbs energy *of formation* at desired T, $G^\circ_{f,T,i}$. The pressure integral, however, is *not* taken over the volume of formation, $V_{f,i}$, but the molar volume of i, V_i. Thus, the sum of all the terms on the RHS is *no longer* a Gibbs energy *of formation*; it is the molar Gibbs energy, G^*_i. It applies to P and T, the upper limit of the integral. When P→P°, G^*_i simply reduces to $G^\circ_{f,T,i}$ and μ^*_i to μ°_i. That is, μ°_i is synonymous to $G^\circ_{f,T,i}$.

The molar Gibbs energy of the pure substance i at P and T, or its standard chemical potential, is a cornerstone in all practical calculations of mineral equilibria. Therefore, we shall dwell upon it some more to see how it is obtained. Thermodynamic data tables (e.g. Robie et al. 1979) do not provide the functional dependence of $S^\circ_{f,i}$ (or S°_i) on T, enabling a straightforward use of (1.44). Rather, they tabulate $H^\circ_{f,298}$, S°_{298}, and $C^\circ_P(T)$ for the minerals and their constituent elements. To use them, Eq. (1.44) is recast to

$$G^*_i = H^\circ_{f,T,i} - TS^\circ_{f,T,i} + \int_{P^\circ}^{P} V_{T,i}(P)\,dP. \tag{1.45a}$$

Recalling (1.14) and (1.15), the above equation is then brought to the final form

$$G^*_i = H^\circ_{f,298,i} + \int_{298}^{T} C^\circ_{P,f,i}(T)\,dT - T\left[S^\circ_{f,298,i} + \int_{298}^{T} \left[\frac{C^\circ_{P,f,i}(T)}{T}\right]dT\right]$$

$$+ \int_{P^\circ}^{P} V_{T,i}(P)\,dP. \tag{1.45}$$

Again, the enthalpy, entropy, and heat capacity terms are those *of formation* of i from the constituent elements; thus, the first three terms on the RHS of (1.45) sum to $G^\circ_{f,T,i}$ [see Eq.(1.44)]. In other words, to compute $G^*_i(T,P)$, we need to manipulate S°_{298} and $C^\circ_P(T)$ data for both phase i and its constituent elements.

It has been argued that the computational procedure indicated for Eq.(1.45) can be simplified if we use $\Delta_a G^{P,T}{}_i$, the apparent molar Gibbs energy of formation of i at P and T from the elements, which is a quantity analogous to $G^*{}_i$. The concept of $\Delta_a G^{P,T}{}_i$ was applied to mineralogical thermodynamics for the first time by Helgeson et al. (1978), and later adopted in a modified form by Berman et al. (1986). Its use permits *dropping* the $S°_{298}$ and $C°_P(T)$ terms of the *elements* from the above calculations. To appreciate this point, we examine the Berman et al. (1986) definition of $\Delta_a G^{P,T}{}_i$,

$$\Delta_a G_i^{P,T} \equiv \Delta_a H_i^{P,T} - TS_i^*(P,T), \tag{1.46}$$

$\Delta_a H^{P,T}{}_i$ denoting the apparent molar enthalpy of formation of i at P and T. The term $\Delta_a H^{P,T}{}_i$ is derived by integrating (1.23) between $P°$ (= 1 bar) and $T°$ (= 298.15 K) to P and T, using $H°_{f,298}$ as the constant of integration. Note that this integration uses $C°_{P,i}(T)$, rather than $C°_{P,f,i}(T)$. Thus,

$$\Delta_a H_i^{P,T} \equiv H_{f,298,i}^\circ + \int_{298}^T C_{P,i}^\circ(T)\,dT + \int_{P\circ}^P \left[V - T\left(\frac{\partial V}{\partial T}\right)_P \right]_{T,i} (P)\,dP. \tag{1.47}$$

The molar entropy of i at P and T, $S^*{}_i(P,T)$, of (1.46) is obtained by integrating Eq.(1.22) between $P°$ (= 1 bar) and $T°$ (= 298.15 K) to P and T, $S°_{298,i}$ being the constant of integration,

$$S_i^*(P,T) = S_{298,i}^\circ + \int_{298}^T \left[\frac{C_{P,i}^\circ(T)}{T}\right] dT - \int_{P\circ}^P \left[\left(\frac{\partial V}{\partial T}\right)_P\right]_{T,i} (P)\,dP. \tag{1.48}$$

Inserting Eqs.(1.47) and (1.48) into Eq.(1.46), the latter reduces to

$$\Delta_a G_i^{P,T} = H_{f,298,i}^\circ + \int_{298}^T C_{P,i}^\circ(T)\,dT$$

$$- T\left[S_{298,i}^\circ + \int_{298}^T \left[\frac{C_{P,i}^\circ(T)}{T}\right] dT \right] + \int_{P\circ}^P V_{T,i}(P)\,dP. \tag{1.49}$$

Comparison of (1.45) and (1.49) drives the point home that $\Delta_a G^{P,T}{}_i$ and $G^*{}_i$ are analogous quantities. Nevertheless, for any phase at a specified set of P and T, the values of $\Delta_a G^{P,T}{}_i$ and $G^*{}_i$ are not identical, because the latter is computed including the $S°_{298}$ and $C°_P(T)$ of the elements, the former is not. Despite that, both can be used to specify the chemical potential of a substance.

Though advantageous, we refrain from utilizing $\Delta_a G^{P,T}{}_i$ in this book for two reasons:
1. The use of $G^*{}_i$ (or $\mu^*{}_i$) is deeply entrenched in the literature, and beginners should be familiar with its manipulation.

2. More importantly, however, in computing phase equilibria, which is the central thrust in this book, the quantity of interest is ΔG, the Gibbs energy difference of the reaction. ΔG *remains the same* irrespective of whether it is calculated over G^*_i or $\Delta_a G^{P,T}_i$, because the *elements* of the reactants and the products *cancel out*. For this reason, computation of ΔG uses a format akin to (1.49); the elements are not considered at all [cf. Eq. (4.4)].

The bottom line is that using G^*_i or $\Delta_a G^{P,T}_i$ is not a question of predilection for one or the other; the latter has a winning edge if we are required to compute the chemical potentials, rather than their difference.

Following this digression, let us return to the consideration of the Gibbs-Duhem equation. Its utility should be amply clear by now (see also Chap. 8 for further details). A minimum amount of measurement of μ_2 [$= f(X_2)$], for instance, permits obtaining data on μ_1[$= f(X_2)$]. Knowing the partial molar quantities, μ_1 and μ_2, we may derive the integral (or total) molar Gibbs energy, G, of the solution. The equation required to calculate G is recovered by dividing (1.35) by n, and substituting $(1-X_2)$ for X_1,

$$G = (1-X_2)\mu_1 + X_2\mu_2. \tag{1.50}$$

The analogous identity

$$Y = (1-X_2)y_1 + X_2y_2 \tag{1.50a}$$

may be used to convert any partial molar state quantity into the corresponding integral molar quantity. For the inverse process of converting an integral molar into partial molar quantities, we may proceed as follows. Differentiating (1.50a) with respect to X_2,

$$\left(\frac{\partial Y}{\partial X_2}\right) = -y_1 + y_2. \tag{1.51}$$

Simultaneous solution of Eqs.(1.50a) and (1.51) yields[5]

$$y_1 = Y - X_2\left(\frac{\partial Y}{\partial X_2}\right), \quad \text{and} \tag{1.52}$$

$$y_2 = Y + (1 - X_2)\left(\frac{\partial Y}{\partial X_2}\right). \tag{1.53}$$

---

[5] Substituting y_1 [from Eq.(1.51)] into (1.50a)
$Y = (1-X_2)[y_2-(\partial Y/\partial X_2)] + X_2y_2$
   $= (1-X_2)y_2 + X_2y_2 - (1-X_2)(\partial Y/\partial X_2)$
   $= y_2 - (1-X_2)(\partial Y/\partial X_2)$, whence
$y_2 = Y + (1-X_2)(\partial Y/\partial X_2)$.   (1.53)
Likewise, substituting y_2 [from Eq.(1.51)] into the LHS of (1.53),
$y_1 + (\partial Y/\partial X_2) = Y + (1-X_2)(\partial Y/\partial X_2)$
$y_1 + (\partial Y/\partial X_2) = Y + (\partial Y/\partial X_2) - X_2(\partial Y/\partial X_2)$. Therefore,
$y_1 = Y - X_2(\partial Y/\partial X_2)$.   (1.52)

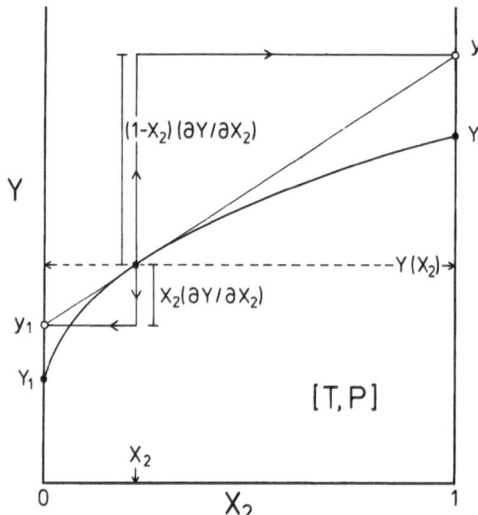

Fig. 1.1 The tangent intercept method for deriving the partial molar property, y_i, from the integral molar property Y at constant temperature and pressure. The composition dependence of Y is shown by the *heavy curve*; the tangent at $Y(X_2)$ by the light *straight line*

In other words, the partial molar properties are obtained from the intercepts of the tangent drawn to the Y vs X_2 curve at any given X_2; for this reason, this construction has come to be known as the method of tangent intercepts. Consequently, each category of state quantities can be immediately translated to the other; only one of the two has to be known. Figure 1.1 depicts how an integral molar and the two partial molar quantities are interrelated.

As is apparent from Fig. 1.1, by drawing a tangent to the Y vs X_2 curve at any desired X_2, y_i can be evaluated as a function of X_2. As X_2 increases from 0 to 1, y_1 increases and y_2 decreases continuously, and vice versa. Two special cases arise at each end of the concentration range. As $X_2 \rightarrow 0$ (that is, the component 2 is present at infinite dilution), $y_2 \rightarrow y^{\infty}_2$, y^{∞}_2 being the partial molar property at infinite dilution of component 2. For $X_2 \rightarrow 0$ (implying that the solution is made up essentially of component 1), we also have $y_1 \rightarrow Y_1$, Y_1 denoting the molar property of the pure substance 1 at T and P. Analogous relations hold for the opposite end of the concentration range; as $X_2 \rightarrow 1$, (i.e., $X_1 \rightarrow 0$), $y_2 \rightarrow Y_2$ and $y_1 \rightarrow y^{\infty}_1$.

1.2.2.2 Property Changes on Mixing

In this section, we shall consider the process of mixing of two pure substances (components 1 and 2) to generate a homogeneous solution at constant T and P. For this purpose, it is essential to distinguish between two categories of mixing processes: mechanical and chemical. The process of mechanical mixing preserves both the physical and the chemical identities of the pure substances being mixed. Consequently, the properties of mechanical mixtures are the linear sums of those of the pure components. By contrast, mixing of particles on a molecular (or atomic) scale, a process known as chemical mixing, wipes out the physical and chemical identities of both the pure substances, resulting in a homogeneous solution. The properties of a homogeneous solution often deviate from those of the linear sum of its components.

Keeping the background informations in mind, let us consider the properties of a homogeneous solution. As we already know, any molar property, Y, of a solution may be obtained from the identity

$$Y = (1-X_2)y_1 + X_2y_2. \tag{1.50a}$$

Alternatively, introducing the new quantity Y_m, the integral molar property of mixing, indicated in Fig. 1.2, Y may be expressed in terms of the molar property of the pure i-th component, Y_i,

$$Y = (1-X_2)Y_1 + X_2Y_2 + Y_m. \tag{1.54}$$

Note that the first two terms on the RHS of (1.54) correspond to the property change due to the mechanical mixing, while Y_m is that due to the chemical mixing. Note also that Y_m can be defined only with reference to the molar properties of the pure substances at system T and P,

$$Y_m \equiv Y - [(1-X_2)Y_1 + X_2Y_2]. \tag{1.54a}$$

Evidently, Y_m must vanish at both ends of the concentration range, as shown in Fig. 1.2. Equating (1.50a) and (1.54), and solving for Y_m,

$$Y_m = (1-X_2)(y_1-Y_1) + X_2(y_2-Y_2). \tag{1.55}$$

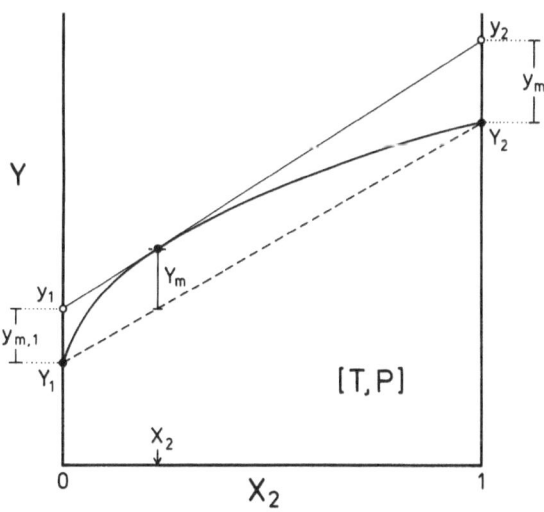

Fig. 1.2 A Y vs X_2 plot at constant T and P, showing the mutual relationship between Y_m, the integral molar quantity, and $y_{m,i}$, the partial molar quantity. Note that Y_m vanishes at either end of the concentration range

In analogy to Eq.(1.50a), we may also write

$$Y_m = (1-X_2)y_{m,1} + X_2 y_{m,2}, \tag{1.56}$$

$y_{m,i}$ being a partial molar quantity of mixing of i. The definition of $y_{m,i}$ emerges from a comparison of Eqs.(1.55) and (1.56):

$$y_{m,i} \equiv y_i\text{-}Y_i. \tag{1.57}$$

Application of the general relation (1.57) to Gibbs energy helps recover the defining equation for a_i (elaborated in the next section), the activity of the i-th component, in a solution,

$$g_{m,i} = \mu_i - \mu_i^* \equiv RT\ln a_i, \tag{1.57a}$$

where $g_{m,i}$ is the partial molar Gibbs energy of mixing of i, and R the universal gas constant. The concept of activity being central to the thermodynamics of solutions, we shall now focus on it.

1.2.2.3 Activity, Activity Coefficient, and Standard State

From Eq.(1.57a), it is apparent that data on the variation of μ_i in a solution in response to changing X_i at a given T and P lead to the a_i-X_i relation. A limiting behavior encountered in certain solutions suggests $a_i = X_i$ for *all* X_i. Such solutions are denoted as ideal solutions, and are defined by

$$\left(\frac{\partial \mu_i^{id}}{\partial \ln X_i}\right)_{P,T} \equiv RT, \tag{1.58}$$

the superscript "id" denoting the "ideal" behavior. Integrating it between the limits $X_i=1$ and $X_i=X_i$,

$$\int_{\mu_i(X_1=1)}^{\mu_i(X_1=X_1)} d\mu_i^{id} = RT \int_{\ln X_1(X_1=1)}^{\ln X_i(X_1=X_1)} d\ln X_i = RT\ln\frac{X_i}{1}. \tag{1.59a}$$

The utility of (1.59a) depends on the choice of the lower limit of integration, the standard state. The most convenient choice is the component i in its pure state ($X_i=1$) at system T and P.[6] For that choice, the lower limit of the integral

---

[6] Note that this new standard state is merely an extension of the 1 bar standard state chosen earlier (Sect. 1.2.1); it could be denoted as a "sliding P" standard state because of its validity at *all P other than* P°. As P→P° this standard state becomes identical to the 1 bar standard state.

on the left-hand side (LHS) reduces to μ^*_i, and the equation (1.59a) simplifies to

$$\mu^{id}_i - \mu^*_i = RT\ln X_i. \tag{1.59}$$

Here, μ^*_i denotes the standard chemical potential of i at T and P. As demonstrated in Section 1.2.2.1, μ^*_i at desired T and P can be evaluated from thermodynamic data of the pure substance i. That is true, however, providing pure i is available for measuring its properties, which is not always the case. A situation might arise in thermodynamics, where we must deal with a fictive - physically inaccessible - component. The standard chemical potential of such a component is indirectly determined. We shall give one such example in the following.

Real (or nonideal) solutions show perceptible deviation from the limiting ideal behavior over a more or less large range of X_i. For them, the deviation from ideality is expressed empirically by introducing a correction factor γ_i, such that

$$\mu_i - \mu^*_i = RT\ln X_i + RT\ln\gamma_i = RT\ln(X_i\gamma_i), \tag{1.60}$$

μ_i designating the chemical potential of the component i in a nonideal solution. Equating (1.57a) with (1.60), it is clear that a_i may also be expressed as

$$a_i \equiv X_i\gamma_i, \tag{1.61}$$

where γ_i, the correction factor, is the activity coefficient.

Evidently, introduction of a_i in lieu of X_i helps retain the simple algebraic form of the basic identity (1.59), also for a nonideal solution. From these considerations it is equally clear that an ideal solution is characterized by $\gamma_i=1$ or $RT\ln\gamma_i=0$ across the entire range of X_i. By contrast, in nonideal solutions, γ_i may be larger or smaller than 1, and is a function of X_i. The solution is said to exhibit a positive deviation from ideality if $\gamma_i>1$, while $\gamma_i<1$ means a negative deviation. Regardless of whether a solution shows a positive or a negative deviation, it invariably approaches the limiting ideal behavior both as $X_i\to1$ and $X_i\to0$.

Having elucidated the a_i-X_i relation for a solution, we focus on the identity

$$\mu_i - \mu^*_i = RT\ln X_i + RT\ln\gamma_i = RT\ln a_i. \tag{1.62}$$

In this relation, μ_i depends upon T, P, and X_i, as do γ_i and a_i. By contrast, μ^*_i is independent of X_i, being a function of T and P only. Note that for any measured value of μ_i, a numerical value cannot be ascribed to γ_i (and as such, also to a_i) as long as μ^*_i remains unknown. Thus, specification of μ^*_i is a prerequisite for a complete definition of γ_i.[7] In order to achieve this, we recall that both as $X_i \to 1$ and $X_i \to 0$, the limiting ideal solution behavior is approached; the former corresponds to Raoult's law, the latter, to Henry's law.

For a solution in which X_i may be varied continuously between 1 and 0 (the entire composition range is physically accessible), a specification of μ^*_i may follow Raoult's law. It demands that $\gamma_i \to 1$ (and $a_i \to 1$) as $X_i \to 1$. Thus, for $X_i = 1$, the RHS of (1.62) vanishes and μ_i becomes equal to μ^*_i. In other words, the pure substance i has unit activity at T and P. The activity and activity coefficients of i, defined on the basis of μ^*_i, are often called the Raoultian activity, $a_i(R)$ and the Raoultian activity coefficient, $\gamma_i(R)$. We shall use these extensively in the subsequent chapters. But how do we deal with the components of a solution, one of which is fictive or physically inaccessible? To appreciate the problem, consider a solution of $CsAlSi_3O_8$ in albite, $NaAlSi_3O_8$. A $CsAlSi_3O_8$ having an albite structure cannot be synthesized (physically inaccessible) as a pure phase. However, we may prepare a very dilute solution of $CsAlSi_3O_8$ in albite, and measure the chemical potentials of both the solvent and the solute components. To deal with the properties of the solvent ($NaAlSi_3O_8$), we may use the Raoultian a_i or γ_i. But for the solute species ($CsAlSi_3O_8$), the Henrian activity coefficient, $\gamma_i(H)$ must be employed, which demands that $\gamma_i(H) \to 1$ as $X_i \to 0$. Stated otherwise, the solution becomes ideal when i reaches the infinite dilution. The standard chemical potential for the *fictive pure* solute species i, $\mu^\square_i$, is derived by extrapolating back from the infinite dilution range along the slope $(\partial\mu^{id}_i/\partial\ln X_i = RT)$. Figure 1.3 depicts the nature of variation of the different interrelated quantities μ_i, $RT\ln X_i$, and $RT\ln\gamma_i$, as a function of X_i at constant T and P. From this, it is immediately clear that μ_i at any X_i can be unambiguously specified with reference to either μ^*_i or $\mu^\square_i$. We also grasp from this figure that μ^*_i and $\mu^\square_i$ are interrelated,

$$\mu^\square_i = \mu^*_i + RT\ln\gamma^\infty_i, \tag{1.63}$$

where γ^∞_i indicates the activity coefficient of the i-th component at infinite dilution. Because μ_i is unique, independent of whether $\gamma_i(R)$ or $\gamma_i(H)$ is being used, we may write

$$\mu^*_i + RT\ln X_i + RT\ln \gamma_i(R) = \mu^\square_i + RT\ln X_i + RT\ln\gamma_i(H). \tag{1.64}$$

As such, it is possible to translate one activity coefficient into the other. Specifically, from Eq.(1.64) we have

$$\frac{\gamma_i(R)}{\gamma_i(H)} = \exp\left[\frac{\mu^\square_i - \mu^*_i}{RT}\right]. \tag{1.65}$$

---

[7] What has been stated here applies in an analogous manner also to a system at 1 bar pressure. As demonstrated in Section 1.2.2.1, if $P \to P^\circ$, $\mu^*_i \to \mu^\circ_i$. Consequently, Equation (1.62) reduces to

$$\mu_i - \mu^\circ_i = RT\ln X_i + RT\ln\gamma_i = RT\ln a_i. \tag{1.62a}$$

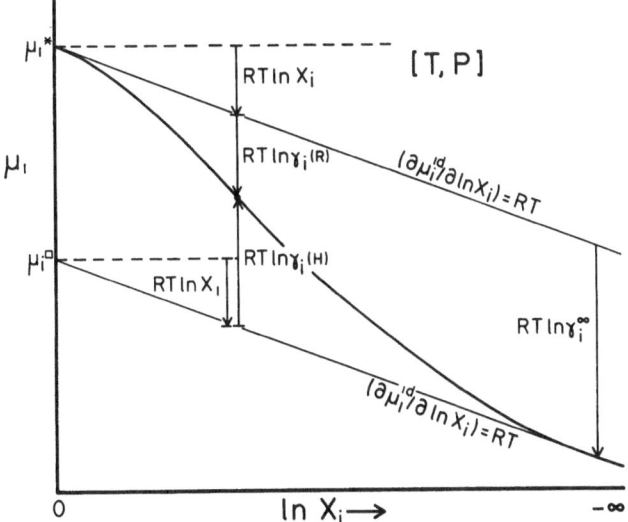

Fig. 1.3 A schematic plot of the functional dependence between μ_i and $\ln X_i$ at constant T and P (adapted from Froese 1981). The *heavy curve* depicts the behavior of a nonideal solution. Note how the chemical potential of the component i in this solution can be alternatively treated in terms of Raoultian or Henrian activity coefficients, $\gamma_i(R)$ and $\gamma_i(H)$, respectively, and the corresponding standard chemical potentials, μ_i^* and $\mu_i^{\square}$

To recast one activity to the other, (1.64) is first rearranged to

$$RT\ln a_i(H) = \mu_i^* - \mu_i^{\square} + RT\ln a_i(R). \tag{1.64a}$$

Substituting (1.63) into (1.64a)

$$RT\ln a_i(H) = - RT\ln\gamma_i^{\infty} + RT\ln a_i(R). \tag{1.66a}$$

Solving for $a_i(H)$,

$$a_i(H) = \frac{a_i(R)}{\gamma_i^{\infty}}. \tag{1.66}$$

As emphasized earlier, in the subsequent chapters, we shall extensively use the Raoultian activity. The corresponding a_i-X_i and γ_i-X_i relations are delineated in Fig. 1.4. As $X_i \rightarrow 1$ and $\gamma_i \rightarrow 1$, so does a_i. There is usually a short range of X_i in which $\gamma_i = 1$ (cf. Fig. 1.3), and therefore, $a_i = X_i$; this is the range of validity of Raoult's law. With increasing dilution of the i-th component, that is, with decreasing X_i, the deviation from ideality becomes conspicuous. The left side of the Fig. 1.4 depicts a positive deviation ($\gamma_i > 1$) from ideality, whereas the right side indicates a negative deviation ($\gamma_i < 1$). The ideal mixing

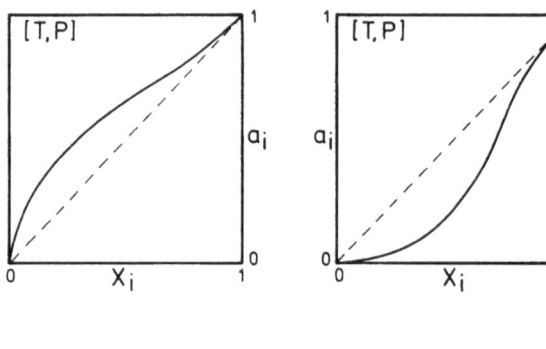

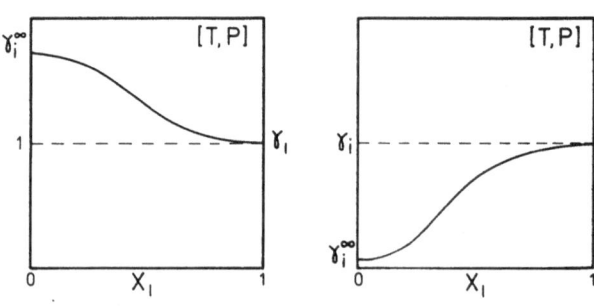

Fig. 1.4 A schematic plot of the isobaric-isothermal a_i-X_i and γ_i-X_i relations of nonideal solutions. The diagrams to the *left* illustrate a positive deviation from ideality, those to the *right* a negative deviation

behavior is represented in the a_i-X_i diagrams by the dashed diagonals; in the γ_i-X_i diagrams, it corresponds to the line $\gamma_i = 1$. As the infinite dilution range is approached, $RT\ln\gamma_i$ becomes a constant, equaling $RT\ln\gamma^\infty_i$ (see Fig. 1.3). Consequently, a_i becomes proportional to X_i (that is, $a_i = X_i\cdot$constant). In the a_i-X_i diagrams, this is recognized by a constant slope of the a_i-X_i curve for $X_i \rightarrow 0$; the intercept of this constant-slope straight line on the a_i-axis yields γ^∞_i. On the γ_i-X_i diagrams, the limiting behavior is implied by a tiny horizontal segment (Fig. 1.4).

1.2.2.4 Ideal, Excess, and Total Molar Properties

The partial molar Gibbs energy of mixing of the component i in a nonideal solution,

$$g_{m,i} = \mu_i - \mu_i^* = RT\ln a_i = RT\ln X_i + RT\ln\gamma_i, \tag{1.57a}$$

may be split up into two parts. The first part

$$g_{m,i}^{id} = \mu_i^{id} - \mu_i^* = RT\ln X_i, \tag{1.59b}$$

indicates the energy arising due to the ideal chemical mixing, the second part, that due to deviation from ideality, is known as the excess partial molar Gibbs energy of mixing of the i-th component; it is obtained by subtracting (1.59b) from (1.57a),

$$g_{m,i}^{ex} = \mu_i - \mu_i^{id} = RT\ln\gamma_i, \tag{1.67}$$

the superscript "ex" denoting an excess quantity. Consequently, we may also define an excess chemical potential, such that

$$\mu_i^{ex} \equiv RT\ln\gamma_i = g_{m,i}^{ex}. \tag{1.68}$$

Employing Eqs.(1.56), (1.59b), and (1.68), the corresponding parts of the integral molar Gibbs energy of mixing (abbreviated below as molar Gibbs energy of mixing) may be obtained:

$$G_m^{id} = (1-X_2)g_{m,1}^{id} + X_2 g_{m,2}^{id} = RT[(1-X_2)\ln(1-X_2) + X_2\ln X_2], \text{ and} \tag{1.69}$$

$$G_m^{ex} = (1-X_2)g_{m,1}^{ex} + X_2 g_{m,2}^{ex} = RT[(1-X_2)\ln\gamma_1 + X_2\ln\gamma_2]. \tag{1.70}$$

The molar Gibbs energy of mixing of a nonideal solution is then the sum of (1.69) and (1.70),

$$\begin{aligned}
G_m &= G_m^{id} + G_m^{ex} \\
&= RT[(1-X_2)\ln(1-X_2) + X_2\ln X_2)] + RT[(1-X_2)\ln\gamma_1 + X_2\ln\gamma_2] \\
&= RT[(1-X_2)\ln a_1 + X_2\ln a_2]. \tag{1.71}
\end{aligned}$$

Finally, the molar Gibbs energy of a binary solution is obtained in analogy with Eq.(1.54):

$$G = (1-X_2)\mu_1^* + X_2\mu_2^* + G_m. \tag{1.72}$$

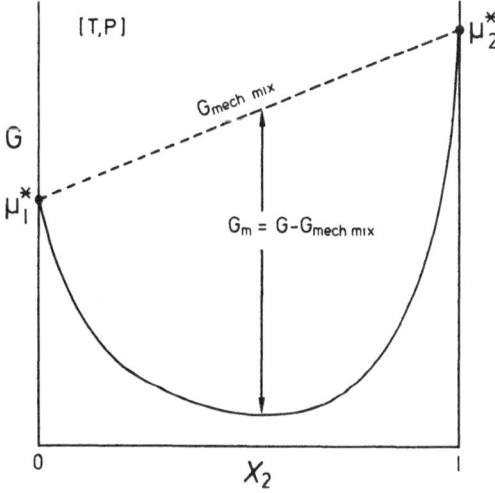

Fig. 1.5 A schematic plot of G vs X_2 of a solution at constant T and P

In this expression, the sum of the first two terms, depicted by the dashed straight line in Fig. 1.5, is the Gibbs energy due to mechanical mixing, $G_{mech.mix}$, of the pure substances. Throughout the composition range, G is lower than $G_{mech.mix}$, indicative of the solution being more stable than the mechanical mixture of the two pure components 1 and 2. The (composition-dependent) magnitude of G_m is given by the energy difference $G-G_{mech.mix}$. Note that at both ends of the composition range, G_m vanishes.

For a multicomponent solution, the molar Gibbs energy is obtained by inserting Eq.(1.71) into (1.72), and generalizing it to

$$G = \sum_i X_i\mu_i^* + RT \sum_i X_i\ln X_i + RT \sum_i X_i\ln\gamma_i. \tag{1.73}$$

Again, the first term on the RHS represents $G_{mech.mix}$, the second one the energy change due to ideal (chemical) mixing, and the last that due to the excess (chemical) mixing. We shall return to this identity in some of the subsequent chapters.

1.2.2.5 Some Properties of Ideal Solutions

Following Eq.(1.73), the molar Gibbs energy of an ideal binary solution may be written as

$$G^{id} = G_{mech.mix} + G_m^{id} = (1-X_2)\mu_1^* + X_2\mu_2^* + RT[(1-X_2)\ln(1-X_2) + X_2\ln X_2]. \tag{1.74}$$

Because $0<X_i<1$, G^{id}_m is always a negative quantity (i.e. a G^{id}_m vs X_2 curve is concave upward), symmetrical with respect to $X_i = 0.5$. With increasing temperature, G^{id}_m becomes more and more negative. Moreover, as $X_2 \rightarrow 0$, $(\partial G^{id}_m/\partial X_2) \rightarrow -\infty$, while as $X_2 \rightarrow 1$, $(\partial G^{id}_m/\partial X_2) \rightarrow +\infty$. This general behavior of G^{id}_m vs X_2 is apparent in Fig. 1.6. In other words, G of an ideal solution is always lower than that of a mechanical mixture of the pure components; that is, ideal chemical mixing enhances the relative stability.

Starting with Eq.(1.74), all important properties of ideal binary solutions may be derived as follows.

$$S^{id} = -\left(\frac{\partial G^{id}}{\partial T}\right)_{P,X_i}$$

$$= -(1 - X_2)\left(\frac{\partial\mu_1^*}{\partial T}\right)_P - X_2\left(\frac{\partial\mu_2^*}{\partial T}\right)_P - R[(1 - X_2)\ln(1 - X_2) + X_2\ln X_2]$$

$$= (1 - X_2)S_1^* + X_2S_2^* - R[(1 - X_2)\ln(1 - X_2) + X_2\ln X_2], \tag{1.75}$$

with S_i^* denoting the molar entropy of the pure i-th component at T and P of interest. The physical significance inherent to the last line on the RHS of

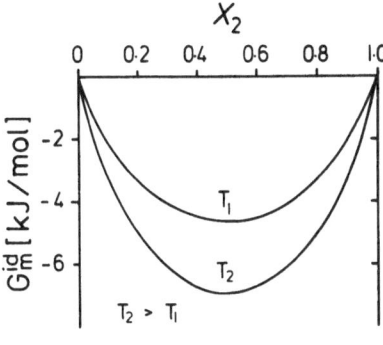

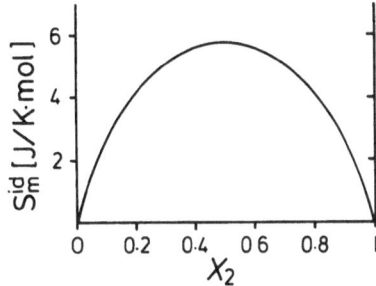

Fig. 1.6 Plots of G_m^{id} vs X_2 of an ideal solution at $T_1 = 800$ K and $T_2 = 1200$ K (*top*), and S_m^{id} vs X_2 (*bottom*)

(1.75) may be carefully noted. It implies that the entropy change due to a mechanical mixing, $S_{mech.mix}$, is given by the linear sum of S^*_i. The chemical mixing increases the entropy relative to the linear sum of S^*_i. From (1.75), it is also apparent that the ideal molar entropy of mixing, S^{id}_m, is given by

$$S_m^{id} = -R[(1-X_2)\ln(1-X_2) + X_2\ln X_2].\qquad(1.76)$$

Since $0<X_i<1$, S^{id}_m is always a positive quantity (i.e. $S^{id}m$ vs X_2 curve is concave downward), symmetrical with respect to $X_i = 0.5$. Furthermore, $(\partial S^{id}_m/\partial X_2)\rightarrow-\infty$ when $X_2\rightarrow1$, whereas $(\partial S^{id}_m/\partial X_2)\rightarrow+\infty$ for $X_2\rightarrow0$ (Fig. 1.6). Contrary to G^{id}_m, S^{id}_m is, however, independent of T. Note that the treatment of G^{id}_m and S^{id}_m given here, though sufficient in the context of molecular solutions, will require an extension to deal with the thermodynamic properties of crystalline solution; this will be achieved in Chap. 8.

To derive the molar enthalpy of an ideal binary solution, we start with an identity analogous to Eq.(1.3)

$$H^{id} = G^{id} + TS^{id}.\qquad(1.3b)$$

Substituting G^{id} (1.74) and S^{id} (1.75) into (1.3b), and collecting terms,

$$\begin{aligned}H^{id} &= (1-X_2)(\mu^*_1 + TS^*_1) + X_2(\mu^*_2 + TS^*_2) + RT[(1-X_2)\ln(1-X_2) + X_2\ln X_2] \\ &+ T\{-R[(1-X_2)\ln(1-X_2) +X_2\ln X_2]\} = (1-X_2)H^*_1 + X_2H^*_2,\end{aligned}\qquad(1.77)$$

where H^*_i $(= \mu^*_i + TS^*_i)$ indicates molar enthalpy of pure i at T and P. Note that the ideal mixing terms cancel out in Eq.(1.77), such that

$$H_m^{id} = 0. \tag{1.78}$$

And finally, the molar volume of an ideal binary solution is obtained by differentiating Eq.(1.74) with respect to P,

$$V^{id} = \left(\frac{\partial G^{id}}{\partial P}\right)_{T,X_i} = (1 - X_2)\left(\frac{\partial \mu^*_1}{\partial P}\right)_T + X_2\left(\frac{\partial \mu^*_2}{\partial P}\right)_T$$

$$= (1 - X_2)V_1 + X_2V_2, \tag{1.79}$$

where V_i denotes the molar volume of pure i at T and P. Thus, the molar volume of any ideal solution, like its molar enthalpy, is identical to that due to mechanical mixing. And like H^{id}_m,

$$V_m^{id} = 0. \tag{1.80}$$

Consequently, any solution with nonzero H_m or V_m must be a nonideal solution.

1.2.2.6 Excess Molar Properties of Mixing

An excess molar property of mixing is the difference between the actual property and that of the ideal solution. Alternatively, it may be defined as Y_m - Y^{id}_m. Therefore, in analogy to Eq.(1.71), we may write

$$Y_m^{ex} \equiv Y - Y^{id} = Y_m - Y_m^{id}, \tag{1.71a}$$

Y denoting any molar property of a real solution. Thus, Y^{ex}_m is a measure of the deviation of a real solution from its hypothetical ideal behavior. Applying this general relation to Gibbs energy,

$$G_m^{ex} = G - G^{id} = G_m - G_m^{id} = (1-X_2)g_{m,1}^{ex} + X_2g_{m,2}^{ex}$$
$$= RT[(1-X_2)\ln\gamma_1 + X_2\ln\gamma_2]. \tag{1.70}$$

Figure 1.7 delineates, in a general manner, the composition dependence of G_m and G^{id}_m (above) and G^{ex}_m (below) for a solution at constant T and P. In this example, a positive deviation from ideality is encountered $(G^{ex}_m>0)$; that is, the real solution is metastable relative to the ideal one at any composition, $G_m>G^{id}_m$. The opposite would be true for $G^{ex}_m<0$ (a negative deviation from ideality). Using relations analogous to those in Section 1.2.2.5, other excess mixing quantities may be derived.

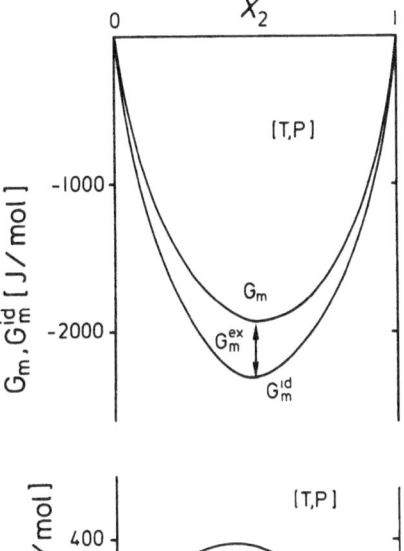

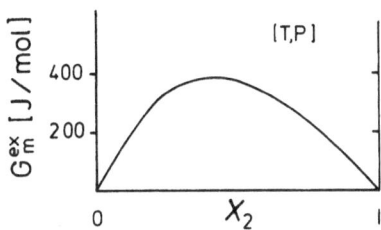

Fig. 1.7 Plots of G_m and G_m^{id} vs X_2 (*top*), G_m^{ex} vs X_2 (*bottom*) of a solution at constant T and P. The scale illustrating G_m^{ex} has been doubled for the sake of clarity

From the general expression obtained earlier for molar Gibbs energy of a solution [cf. Eq.(1.73)], it is evident that besides data on μ_i^* at T and P, knowledge of the activity coefficient, γ_i, at that T and P as a function of X_i is of fundamental importance. Knowledge of those two quantities sets the stage for calculation of any heterogeneous phase relation. For this reason, we shall now address the problem of how best to express the dependence of γ_i on X_i. Unfortunately, classical thermodynamics does *not* provide adequate insight into the γ_i vs X_i relation (Denbigh 1971, p.215). We can fortunately measure it at some fixed T and P, and express it empirically by the so called solution models. The majority of the solution models (Wisniak and Tamir 1978; Prausnitz et al. 1986) proposed thus far were originally devised for expressing the γ_i vs X_i relation of molecular solutions. Increasing numbers of these are now being applied to crystalline solutions (see Fei et al. 1986). In this treatise, we shall utilize only two of those models, which have proved to be quite successful in dealing with nonelectrolyte solutions.

Restricting ourselves, for simplicity, to a binary solution, in which the chemical potentials are known as a function of the master variable X_2 at any fixed T and P, the initial step is to compute $g^{ex}_{m,i}$. These are then put together according to Eq.(1.70) to obtain G^{ex}_m vs X_2. The composition dependence of G^{ex}_m is then expressed empirically.[8] Recalling that $G^{ex}_m \to 0$ as $X_2 \to 1$ *and*

---

[8] The composition dependence of other excess mixing quantities (U^{ex}_m, H^{ex}_m, S^{ex}_m, V^{ex}_m etc.) is expressed in an analogous manner.

$X_2 \to 0$, a power series such as

$$G_m^{ex} = (1-X_2)X_2[a + bX_2 + c(X_2)^2 + ...]$$ (1.81)

would seem to be meaningful. Note that introduction of the factor $(1-X_2)X_2$ into the RHS of Eq.(1.81) serves to make G^{ex}_m vanish both at $X_2 = 0$ and $X_2 = 1$. The number of significant terms in the power series in Eq.(1.81) would depend upon how complex the composition dependence of G^{ex}_m is, given an adequate amount of measured data points of sufficient accuracy to fit the function. In rare cases, where quadratic or higher order terms are justified, the quality of data fitting could be improved by transposing the origin of the coordinate from $X_2 = 0$ to the center of the composition axis ($X_2 = 0.5$), such that the G^{ex}_m data points are uniformly distributed on both sides of the origin. This goal is accomplished by modifying Eq.(1.81) to

$$G_m^{ex} = (1-X_2)X_2[A + B(1-2X_2) + C(1-2X_2)^2 + ...],$$ (1.82)

A, B, C being adjustable parameters, constants at fixed T and P. Note that $(1-2X_2)$ equals zero when $X_2 = 0.5$, thereby satisfying the necessity of rendering the G^{ex}_m data symmetrically distributed with reference to 0.5 X_2. Equation (1.82) will be referred to as the Redlich-Kister equation (Redlich and Kister 1948). It has been employed for analytically rendering isobaric-isothermal G^{ex}_m vs X_i data of fluid mixtures (Prausnitz et al. 1986) and mineral crystalline solutions (Gasparik 1984; Cohen 1986b). In the majority of mineral crystalline solutions, the accuracy of measurements of γ_i precludes use of more than two (A and B) or even one (A) such parameter. Clearly, a total lack of nonzero parameters implies an ideal mixing behavior.

Given the three-parameter expression for G^{ex}_m (1.82), activity coefficients of the components 1 and 2 are obtained by the method of tangent intercepts (see Eqs.(1.52) and (1.53)),

$$RT\ln\gamma_1 = G_m^{ex} - X_2\left(\frac{\partial G_m^{ex}}{\partial X_2}\right)$$

$$= (X_2)^2[A + B(3 - 4X_2) + C(1 - 2X_2)(5 - 6X_2)], \quad \text{and} \quad (1.83)$$

$$RT\ln\gamma_2 = G_m^{ex} + (1 - X_2)\left(\frac{\partial G_m^{ex}}{\partial X_2}\right)$$

$$= (1 - X_2)^2[A + B(1 - 4X_2) + C(1 - 2X_2)(1 - 6X_2)].$$ (1.84)

The Redlich-Kister expression for G^{ex}_m, coupled with μ^*_i, provides μ_i as a function of X_2. To obtain μ_i for any P and T, a polybaric-polythermal Redlich-Kister equation will be necessary, demanding knowledge of the T- and P-dependences of the fit parameters A, B, and C. That goal is accomplished by splitting them in analogy to the defining equation of Gibbs energy (1.3),

$$A(T,P) = A_U - TA_S + PA_V,$$ (1.85a)
$$B(T,P) = B_U - TB_S + PB_V, \text{ and}$$ (1.85b)
$$C(T,P) = C_U - TC_S + PC_V.$$ (1.85c)

Note that A_S, B_S, C_S, A_V, B_V and C_V are, in reality, functions of T and P. However, the data on crystalline solutions do not permit resolution of those functional dependencies. Thus, the entropy and the volume terms in Eqs.(1.85a-c) are handled as if they were constants, independent of T and P.

The Redlich-Kister equation has been applied to minerals from time to time. By contrast, the Margules equation has been far more popular, ever since Thompson (1967) introduced it to mineralogy. As will be demonstrated shortly, the Margules equation, although formulated much earlier (Margules 1895), must be regarded as a special case of the more general Redlich-Kister equation. Nevertheless, we shall explicitly treat it, primarily because it has been widely applied to deal with mineral systems. The Margules equation follows immediately from Eq.(1.81) when the power series is truncated beginning with the quadratic term, and the parameters a and b are redefined as $W_{G,2}$ and $(W_{G,1} - W_{G,2})$, respectively,

$$\begin{aligned} G_m^{ex} &= (1-X_2)X_2[W_{G,2} + (W_{G,1} - W_{G,2})X_2] \\ &= (1-X_2)X_2[W_{G,1}X_2 + W_{G,2}(1-X_2)]. \end{aligned}$$ (1.86a)

The $W_{G,i}$ appearing in Eq. (1.86a) is the Margules parameter; it is identical to the excess partial molar Gibbs energy of mixing of the i-th component at infinite dilution,

$$W_{G,i} \equiv g_{m,i}^{ex,\infty} = RT\ln\gamma_i^{\infty}.$$ (1.87)

Figure 1.8 shows a plot of $G_m^{ex}/(1-X_2)X_2$ vs X_2 at constant T and P, using the same set of G_m^{ex} data as in Fig. 1.7. Note that $G_m^{ex}/(1-X_2)X_2$ is a linear function of X_2. From Eq. (1.86a), it is immediately apparent that the intercepts of the straight line at $X_2 = 0$ and $X_2 = 1$ correspond to $W_{G,2}$ and $W_{G,1}$, respectively.

The subscript notation for the Margules parameters [in Eq.(1.86a] remains unequivocal as long as we are dealing with one binary at a time. In other cases, it may become ambiguous. That happens, for instance, when we deal with two limiting binaries, 1-2 and 1-3, of a ternary solution of components 1, 2, and 3. That is because $W_{G,1}$ may now refer to the infinite dilution of component 1 in 1-2 or in 1-3. This demands that the component other than 1 (i.e. 2 or 3) be specified. Such an extended notation has caused some confusion in the literature (see Andersen and Lindsley 1981, p.847). In this book, the notation introduced by Wohl (1946) will be adopted, according to which $W_{G,1}$ and $W_{G,2}$ (of 1-2) are written as $W_{G,12}$ and $W_{G,21}$, respectively.

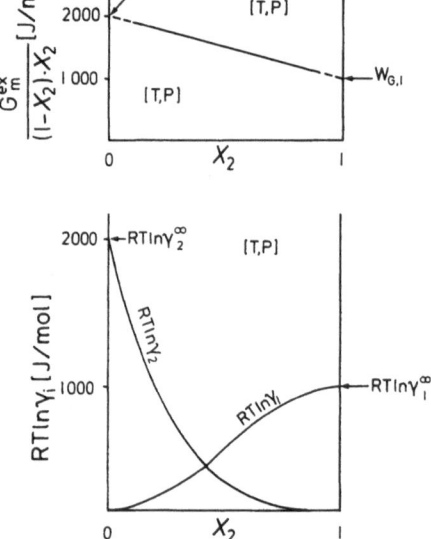

Fig. 1.8 Plots of the functions $G_m^{ex}/(1-X_2)X_2$ vs X_2 (*top*), and $RT\ln\gamma_i$ vs X_2 (*bottom*) of a solution at constant T and P. Note that $W_{G,i}$ equals $RT\ln\gamma_i^\infty$

Note that the component at infinite dilution appears first in succession. Thus, Eq.(1.86a) may be rewritten as

$$G_m^{ex} = (1-X_2)X_2[W_{G,21} + (W_{G,12} - W_{G,21})X_2]$$
$$= (1-X_2)X_2[W_{G,12}X_2 + W_{G,21}(1-X_2)]. \tag{1.86}$$

Because of its popularity in mineralogy, we shall mostly use the Margules equation in Chapters 8 and 9. It is emphasized though that the G^{ex}_m of a solution may be alternatively expressed in terms of the Margules or Redlich-Kister equation, since the former is a special case of the latter. The conversion of one into the other could be achieved simply by reformulating the parameters,

$$A = 0.5(W_{G,12} + W_{G,21}), \text{ and} \tag{1.88a}$$

$$B = 0.5(W_{G,21} - W_{G,12}). \tag{1.88b}$$

Given the composition dependence of G^{ex}_m (1.86) of a solution at constant T and P, the activity coefficients of its components are obtained by the tangent intercept method,

$$RT\ln\gamma_1 = [W_{G,12} + 2(W_{G,21} - W_{G,12})(1-X_2)](X_2)^2, \tag{1.89}$$

$$RT\ln\gamma_2 = [W_{G,21} + 2(W_{G,12} - W_{G,21})X_2](1-X_2)^2. \tag{1.90}$$

Figure 1.8 (bottom) shows the composition dependence of the $RT\ln\gamma_i$ terms for the same solution represented in the upper part of that diagram. Several points must be made here. First, as $X_i \to 1$, $RT\ln\gamma_i \to 0$ (or $\gamma_i \to 1$, i.e. the solution becomes ideal), agreeing with Raoult's law. Conversely, as $X_i \to 0$, $RT\ln\gamma_i \to RT\ln\gamma^{\infty}_i$, in agreement with Henry's law (cf. Fig. 1.3). Secondly, the slopes of $RT\ln\gamma_i$ curves in this plot are opposite in sign. That is consonant with the Gibbs-Duhem equation [see Eq. (1.42]. All these features are due to the fact that the functional form chosen to represent the G^{ex}_m vs X_2 data takes care of its boundary constraints discussed earlier. And finally, as anticipated, the values of $RT\ln\gamma^{\infty}_i$ (lower figure) agree with the corresponding $W_{G,i}$ terms (upper figure).

Within the limits of uncertainties of measurements, $W_{G,12}$ and $W_{G,21}$ are sometimes indistinguishable. In such cases, *one* Margules parameter, W_G, emerges,

$$W_G \equiv W_{G,12} = W_{G,21}. \tag{1.91}$$

In such cases, the Margules equation (1.86) simplifies to

$$G^{ex}_m = (1-X_2)X_2 W_G, \tag{1.92}$$

and the corresponding expressions for activity coefficients reduce to

$$RT\ln\gamma_1 = W_G(X_2)^2, \text{ and} \tag{1.93}$$

$$RT\ln\gamma_2 = W_G(1-X_2)^2. \tag{1.94}$$

The manipulation of Margules parameters is analogous to that of state functions. For instance, the temperature and pressure dependences of $W_{G,i}$ are given by equations like (1.8) and (1.9),

$$\left(\frac{\partial W_{G,i}}{\partial T}\right)_P = -W_{S,i}, \quad \text{and} \tag{1.95}$$

$$\left(\frac{\partial W_{G,i}}{\partial P}\right)_T = W_{V,i}, \tag{1.96}$$

$W_{S,i}$ and $W_{V,i}$ being the Margules entropy and volume parameters, respectively. For crystalline solutions, the T- and P-dependences of $W_{S,i}$, $W_{V,i}$ (and $W_{H,i}$) are seldom resolved experimentally; as a result, they are treated as constants. Using the defining equation $W_{H,i} \equiv W_{U,i} + PW_{V,i}$ [cf. Eq.(1.1)], $W_{G,i}$ is written as

$$W_{G,i}(T,P) = W_{H,i} - TW_{S,i} = W_{U,i} - TW_{S,i} + PW_{V,i}. \tag{1.97}$$

Solutions are sometimes classified according to the functional forms of their excess properties; those obeying the two-parameter Margules equation (1.86) are known as asymmetric solutions, while symmetric solutions follow the one-parameter equation (1.92). This classification derives its justification from the fact that in the former case, a Y^{ex}_m vs composition plot is asymmetric with respect to 0.5 X_2, whereas in the latter, it is symmetric. The symmetric form is the simplest type of nonideality in solutions; therefore, they have also been called simple solutions. Regular solutions are simple solutions with $S^{ex}_m = 0$ and $V^{ex}_m = 0$; their G^{ex}_m is, thus, independent of T and P. And finally, a solution with effectively zero H^{ex}_m is an athermal solution. By implication, no solution is ideal, unless its H^{ex}_m, S^{ex}_m, and V^{ex}_m, vanish at the same time.

From the above expositions, it is apparent that the determination of the thermodynamic mixing properties of a solution is tantamount to obtaining data on

$$G^{ex}_m = f(T,P,X_i), \tag{1.98}$$

expressing the functional dependences between the state function, G, and its characteristic variables, T, P, and X_i. Thus, Eq.(1.98) is an equation of state. We shall see later on (Chap. 8) that the thermodynamics of solutions pivots around obtaining equations of state of solutions.

1.3 Mineral Equilibria, Equilibrium Constant

Under isothermal-isobaric conditions, an equilibrium in a nonreacting system of variable composition, and capable of doing only P-V work, is given by

$$\sum_i \mu_i dn_i = 0. \tag{1.99}$$

If a chemical reaction takes place in such a system, the condition of equilibrium becomes (cf. Klotz and Rosenberg 1972, p.228)

$$\sum_i v_i \mu_i = 0, \tag{1.100}$$

where v_i, the stoichiometric coefficient of the i-th component, is counted negative for the reactants, and positive for the products. The energy balance inherent to Eq.(1.100) can also be expressed as

$$\Delta\mu = 0. \tag{1.100a}$$

The thermodynamic treatment of heterogeneous equilibria, for example, equilibria between minerals and fluids, must commence by writing a mass balance equation, a stoichiometric equation between the components of the phases, formalized by

$$\sum_i v_i M_i = 0, \tag{1.101}$$

with M_i denoting the molecular mass of the i-th component. Again, v_i is counted negative for the reacting components, and positive for the products. It is evident from the above considerations that every equilibrium is characterized by a mass balance *and* an energy balance relation. In some mineral systems, an interplay of several simultaneous equilibria may be operative; for them, an appropriate number of mass and energy balance relations must be handled simultaneously. Such an example will be given in Chap. 9.

To examine the nature of this energy balance relation, let us insert the expression for chemical potential from (1.57a) into (1.100), and recast that in terms of $\Delta\mu$,

$$\Delta\mu = \sum_i v_i(\mu_i^* + RT\ln a_i) = 0. \tag{1.102}$$

Giving due consideration to the sign conventions for the reactants and products, and collecting terms, (1.102) reduces to

$$\Delta\mu = \Delta\mu^* + RT\ln \prod_i (a_i)^{v_i} = 0. \tag{1.103}$$

Note that the first term on the RHS of (1.103) is identical to the Gibbs energy difference of the reaction (1.101) at P and T, ΔG^*. The second term comprises K, the equilibrium constant,

$$K \equiv \prod_i (a_i)^{v_i}, \tag{1.104}$$

in which the a_i of the products appear in the numerator, those of the reactants in the denominator. Inserting (1.104) into (1.103), we recover the familiar relation

$$\Delta\mu = \Delta G^* + RT\ln K = 0. \tag{1.105}$$

Note that ΔG^* in Eq. (1.105) refers to the pure substance standard state at system P and T. To relate it to the 1 bar standard state chosen earlier, the ΔG^* term needs to be split up,

$$\Delta\mu = \Delta G_T^\circ + \int_1^P \Delta V_T(P)dP + RT\ln K = 0. \tag{1.106}$$

In this equation, ΔG°_T is the standard Gibbs energy difference of the reaction at 1 bar and T (K). We shall invariably employ it for *all* mineral equilibria calculations beginning Chap. 4.

Chapter 2

Measurement, Evaluation, and Tabulation of Thermodynamic Properties

2.1 Introduction

Adequate knowledge of the nature of the thermodynamic data reproduced in tables is a necessary first step for calculating mineralogical phase equilibria. Thermodynamic data tables, in vogue today, fall into two categories. The first of these seeks to deal with *each phase individually*, drawing heavily on calorimetric (and a few electrochemical cell) measurements. The compilation of thermodynamic data of minerals and related substances by Robie et al. (1979) belongs to this category. The second type of database *strives to relate the thermodynamic properties[1] of the minerals to each other*, to come up with what is known as an internally consistent set of thermodynamic data. Two recent examples of such databases are Holland and Powell (1985) and Berman (1988). In the Chapters 4 to 6, we shall employ the Robie et al. (1979) table to calculate a variety of phase diagrams. Though calorimetry is, and will remain, the fundamental and an indispensable source of thermodynamic data, the exercises in those three chapters will reveal to us the perils of depending exclusively on such data, and urge us to develop the internally consistent thermodynamic database, which is achieved by a simultaneous treatment of the calorimetric *and* phase equilibria measurements. One of the methods of deriving internally consistent data will be discussed in detail in Chap. 7. The availability of such a database is a prerequisite for generating geologically meaningful phase diagrams. But more on that later on.

The objective of this chapter is twofold: to get acquainted with the Robie et al. (1979) tables, and to learn about the source of the data compiled and the numerical manipulations involved in tabulating them.

The Robie et al. (1979) tables are divided into two sections. The first section summarizes critically appraised values of S°_{298}, V°_{298}, $H^\circ_{f,298}$, $G^\circ_{f,298}$, and log $K^\circ_{f,298}$, the logarithm of the equilibrium constant of formation at 1 bar and 298.15 K, for various phases of interest to earth science. In the second section they list both the thermophysical parameters, $(H^\circ_T - H^\circ_{298})/T$, S°_T, $-(G^\circ_T - H^\circ_{298})/T$, $C^\circ_p(T)$, and the thermochemical parameters, $H^\circ_{f,T}$, $G^\circ_{f,T}$, and log $K^\circ_{f,T}$ at 100 K intervals to temperatures up to 1800 K. In order to derive these quantities from the data set forth in the preceding section, the only

---

[1] The term thermodynamic property will be used in this book as a collective designation for the thermophysical, thermochemical, and volumetric properties of a phase.

additional information required is that of the standard state heat capacity power function, $C°_P(T)$, for T above 298.15 K. It is important to note that all these data apply to a pressure of 1 bar only. The pressure dependence of those quantities must be calculated using data from other sources; they will be indicated whenever the necessity arises.

2.2 Outlines of Some Calorimetric Methods

As indicated above, the vast majority of the data tabulated by Robie et al. (1979) stems from calorimetry. Thus, knowledge of the bare essentials of calorimetry appears to be desirable. Those willing to go into details may refer to a standard monograph (e.g. Hemminger and Höhne 1979). The brief outline given below follows Kleppa's (1982) treatment, subdividing the calorimetric techniques into two broad categories: the calorimetry of non-reacting and reacting systems.

2.2.1 Calorimetry of Non-Reacting Systems

The most important experimental techniques include adiabatic, differential scanning, and drop calorimetry. They provide us with the thermophysical data.

Barring a few exceptions (for an example, see Sect. 2.4), adiabatic calorimetry is used for measuring $C°_P(T)$ in the "low-temperature" range, which is below 350 K. To accomplish this, an adiabatically shielded (that is, guaranteeing a negligible heat exchange with the surroundings) calorimeter is precooled to some desired initial temperature by an appropriate liquefied gas. The heat capacity of the sample is then measured by monitoring at short intervals the temperature increment of the sample in the calorimeter in response to a known amount of heat input. The precision of low-temperature $C°_P(T)$ measurements is usually better than 0.1%. The relevance of these measurements is appreciated from the fact that the standard entropy, $S°_{298}$, of a substance (without a phase transition between 0 and 298.15 K) is evaluated from the measured $C°_P(T)$ via

$$S°_{298} = S°_0 + \int_0^{298} \left[\frac{C°_P(T)}{T} \right] dT . \tag{2.1}$$

Note that the upper limit of integration in (2.1) is 298.15 rather than 298 K,[2] and the lower limit is 0 K, emphasizing the necessity of $C°_P(T)$ measurements to very low temperatures.

---

[2] As indicated in Chap. 1, 298.15 K is abbreviated as 298 K in this book.

In practice - at least for geologically relevant substances - $C°_P(T)$ used to be measured down to approximately 50 K prior to the 1970s (e.g. King and Weller 1961), and then extrapolated to 0 K. Assuming that the entire $C°_P$ of the phase arises from the lattice vibrations (other contributions being totally negligible), such extrapolations were made in analogy to Debye's law, according to which, in the vicinity of 0 K, the $C°_P$ of a mono-atomic solid is proportional to T^3. Therefore, $C°_P/T$ vs T^2 plots were used to extrapolate the low-temperature $C°_P$ data to 0 K. Unfortunately, the $C°_P$ arising from "other" sources cannot always be neglected. To give one example, near 0 K, the electronic contribution to $C°_P$ may become important in certain phases. This term is proportional to T and is superposed on the vibrational $C°_P$ (Hultgren et al. 1973), thereby introducing errors in extrapolations. Uncertainties might arise in such extrapolations due to another reason. In the absence of $C°_P$ measurements below 50 K, any phase transition in that range will go undetected. For example, Robie et al. (1984, Fig.1) detected a distinct λ-type phase transition in Ni_2SiO_4-olivine at 29.15 K, which they attributed to an antiferromagnetic ordering of the magnetic moments of the Ni^{+2} ions with rising temperature. To avoid those pitfalls, the "low-temperature" $C°_P$ measurements are now routinely executed down to around 5 K.

If a phase transition intervenes between 0 and 298.15 K, the calculation of $S°_{298}$ must take the entropy difference into account due to the phase transition, $\Delta S°_{tr}$. To include that into Eq.(2.1), it is expanded prior to evaluation to the general form

$$S°_{298} = S°_0 + \int_0^{298} \left[\frac{C°_P(T)}{T}\right] dT + \Delta S°_{tr}. \qquad (2.2a)$$

For now, we shall not worry about the technique needed to evaluate $\Delta S°_{tr}$; fortunately for us, all standard thermodynamic data tables invariably give $S°_{298}$, inclusive of $\Delta S°_{tr}$, if any. However, in the context of manipulation of the "high-temperature" $C°_P(T)$ data, we must cope with this problem (see Sect. 2.4).

To the extent that the substance under investigation behaves as an ideal crystal, the third law of thermodynamics,

$$\lim_{T \to 0} S = 0, \qquad (2.3)$$

applies. For that category of substances, $S°_0 = 0$, and as such, knowledge of $C°_P(T)$ suffices to derive their $S°_{298}$. Because $C°_P$ is a positive quantity tending to zero as T goes to zero, it is also apparent that $S°_T$ is a positive quantity measurable on an absolute scale.

Unfortunately, many minerals are not as well behaved as ideal crystals. They involve yet another entropy term, stemming from the site mixing of ions in the individual sublattices. This is the so-called configurational entropy,

$S°_{cfg}$. Because the order-disorder (site mixing) processes in minerals are often very sluggish, high-temperature disordered states tend to persist even at exceedingly low temperatures. This frozen-in $S°_{cfg}$ cannot be measured by an adiabatic calorimeter. It can, however, be computed from the site-occupancy data on each sublattice. In the event of a *random mixing* of ions, the *ideal molar configurational entropy*, $S°,^{id}_{cfg}$, may be derived starting from the Boltzmann equation (for derivation, see Cohen 1986a, p.176),

$$S^{°,id}_{cfg} = - R \sum_j q_j \sum_i X_{ij} \ln X_{ij}, \tag{2.4a}$$

with q_j denoting the number of sites in the sublattice j, X_{ij} the mole fraction of the ion i on the j-th sublattice, the summation extending over all ions i on the j-th sublattice, and over all the sublattices j. Note that in the case of mixing of ions on only *one* sublattice, Eq.(2.4a) simplifies to

$$S^{°,id}_{cfg} = - qR \sum_i X_i \ln X_i. \tag{2.4}$$

Whenever a configurational entropy term arises, $S°_0$ becomes nonzero. Therefore, the $S°,^{id}_{cfg}$ term must be substituted for $S°_0$ in (2.2a). To facilitate those calculations, Ulbrich and Waldbaum (1976, Tables 1-10) have compiled data on chemical site mixing in minerals. Robie et al. (1979, p.8) have utilized a part of their data. A case in point is the standard entropy of a K-feldspar. It has four tetrahedral(T)-sites per formula unit; these are occupied by 1 Al and 3 Si. The microcline structure is characterized by an ordered Al/Si distribution, the high sanidine by a random mixing of these ions over all four T-sites. In other words, the mole fractions of Al and Si in each of those four T-sites in the high sanidines are 0.25 and 0.75, respectively. From (2.4), it is seen immediately that $S°,^{id}_{cfg}$ of a microcline is zero, while that of a high sanidine is

$$S^{°,id}_{cfg} = -4R(0.25\ln 0.25 + 0.75\ln 0.75) = 18.70 \text{ J K}^{-1} \text{ mol}^{-1}. \tag{2.5}$$

Consequently, the $S°_{298}$ of high sanidine and microcline, quoted by Robie et al. (1979) as 232.9 and 214.2 J K^{-1} mol^{-1}, respectively, differ by 18.7 J K^{-1} mol^{-1}. In summary then, the final expression for $S°_{298}$ should be revised from Eq.(2.2a) to

$$S^{°}_{298} = S^{°,id}_{cfg} + \int_0^{298} \left[\frac{C^{°}_P(T)}{T} \right] dT + \Delta S^{°}_{tr} . \tag{2.2}$$

Another problem encountered in deriving $S°_{298}$ data of mineral end-members deserves to be mentioned. A typical low-temperature $C°_P$ measurement by adiabatic calorimetry requires nearly 50 g of material. Because it is rarely possible to synthesize such a large quantity of any pure end-member phase, many calorimetric runs are executed on samples of natural materials. Although the samples are selected such that they have near end-member compositions, small deviations therefrom have to be tolerated. Before deriving $S°_{298}$ of the end-member from (2.2), the measured $C°_p$ data are adjusted for compositional deviations (for details, see Robie and Hemingway 1984b), which may raise the uncertainty of the data. Fortunately, it has become possible to scale down the material requirement to about 10 g (Haselton and Westrum 1980), setting the stage for low-temperature $C°_p$ measurements more often on synthetic phases.

As pointed out earlier, the $C°_p(T)$ function at $T > 298.15$ K is required to extend 1 bar, 298.15 K data to higher temperatures. Such data are seldom obtained by an adiabatic calorimeter, for its range of operation rarely extends beyond 400 K. In that T range, the differential scanning calorimetry and drop calorimetry are the methods of choice.

Differential scanning calorimetry (DSC) has become quite a popular technique during the last two decades for measuring "high-temperature" $C°_p(T)$ functions of minerals. The principle on which it is based is simple. If a sample is subjected to a time-linear rise of temperature, usually on the order of 10 K min^{-1}, the heat flow rate into the sample is proportional to its instantaneous specific heat. Thus, the heat flow rate into the sample, placed in one of the cells of this twin-cell device, compared with that into a standard (commonly a disc of synthetic sapphire) loaded into the other cell, provides the specific heat of the sample as a function of temperature (see O'Neill 1966 for details). These measurements require no more than 10-30 mg of material, permitting $C°_p(T)$ data to be collected on pure synthetic materials on a routine basis. Although the accuracy of the $C°_p$ data obtained by this technique is seldom better than 1%, it is now widely used, in particular, to perform measurements in the 350-1000 K range. More recently, Bennington et al. (1987) reported having extended that range to the vicinity of 1400 K.

Drop calorimetry is the other important source of information on $C°_p(T)$ in the "high-temperature" range. In this technique, the sample to be measured is thermally equilibrated in a furnace at some high temperature prior to dropping it into a calorimeter bath operated at or near room temperature. The quantity measured by this process is *not* $C°_p$ *but* the *heat content*, $H°_T-H°_{298}$; $C°_p(T)$ is derived from it by differentiation [see Eq.(1-15)]. Typically, drop calorimetry is executed in the 350-1800 K range. It uses a sample size on the order of 10 g, and achieves a precision of about 0.1% (cf. Pankratz and Kelley 1963). The measured $H°_T-H°_{298}$ data are often fitted to a polynomial, suggested originally by Maier and Kelley (1932),

$$H°_T-H°_{298} = Q + AT + BT^2 + CT^{-1}. \tag{2.6}$$

In many cases, the Maier-Kelley polynomial does not fit the data well enough. This is true, in particular, for highly accurate measurements. For instance, the heat content of periclase reported by Pankratz and Kelley (1963) has a precision of 0.1%; the Maier-Kelley polynomial reproduces them with a deviation of 0.4% only. To do justice to high quality heat content data, Haas and Fisher (1976) extended the Maier-Kelley polynomial by two more terms,

$$H_T^\circ - H_{298}^\circ = Q + AT + BT^2 + CT^{-1} + DT^{0.5} + ET^3. \tag{2.7}$$

Differentiating (2.7) with respect to T, the corresponding $C_P^\circ(T)$ function is obtained,

$$C_P^\circ(T) = A + 2BT - CT^{-2} + 0.5DT^{-0.5} + 3ET^2. \tag{2.8a}$$

Rewriting the coefficients of T of Eq.(2.8a) as a, b, c, d, and e, respectively, the heat capacity power function is rewritten as

$$C_P^\circ(T) = a + bT + cT^{-2} + dT^{-0.5} + eT^2. \tag{2.8}$$

This five-parameter representation of $C_P^\circ(T)$ has been adopted by Robie et al. (1979) in their thermodynamic data table. Although it reproduces the observed C_P° values faithfully, serious problems arise if such data need to be extrapolated beyond the experimental range of $C_P^\circ(T)$ measurements. To appreciate this fact, recall that at sufficiently high temperature, C_P° approaches the limit (Berman and Brown 1985)

$$C_P^\circ = 3R + \frac{\alpha^2 V}{\beta} \cdot T. \tag{2.9}$$

Depending upon the sign and magnitude of the coefficient "e" in Eq. (2.8), $C_P^\circ(T)$ might show a more or less pronounced maximum, or an inflection, beyond the T-range of experimental C_P° data, both of which run counter to the theory. In order to caution against such extrapolations, Robie et al. (1979) always specify the "range of validity" of their $C_P^\circ(T)$ functions. In practical phase equilibria calculations, however, we often require C_P° data at temperatures beyond the range of validity indicated by Robie et al. (1979). In such cases the $C_P^\circ(T)$ polynomial recommended by Berman and Brown (1985), and demonstrated to be compatible with the theory, will be advantageous,

$$C_P^\circ(T) = a + cT^{-2} + dT^{-0.5} + fT^{-3}, \tag{2.10}$$

with c and d $\leq$ 0. Given these alternative possibilities, use will be made of the *combined expressions* for the $C_P^\circ(T)$ function throughout this book,

$$C_P^\circ(T) = a + bT + cT^{-2} + dT^{-0.5} + eT^2 + fT^{-3}, \tag{2.11}$$

in which up to three (Maier-Kelley), four (Berman-Brown), or five (Haas-Fisher) of the appropriate coefficients will be nonzero for any given phase. In some exceptional cases, yet other coefficients may have to be used; $C°_P(T)$ of hematite and magnetites in Robie et al. (1979) are good examples.

Given $C°_P(T)$ for the low-temperature range, and $H°_T$-$H°_{298}$ for the high-temperature range, data fitting is achieved by satisfying the following criteria (Robie et al. 1979). The heat content data are fitted to the polynomial of Eq.(2.7) and constrained such that at 298.15 K, $H°_T$-$H°_{298}$ equals zero and, moreover, the first derivative of the $H°_T$-$H°_{298}$ function at 298.15 K equals $C°_{P,298}$ obtained from low-temperature adiabatic calorimetry. Owing to this data fitting procedure, $C°_P(T)$ obtained from the heat content measurement joins smoothly with the highly accurate $C°_P(T)$ function measured by adiabatic cryogenic calorimetry.

Once the $C°_P(T)$ function is available, other thermophysical quantities like the standard entropy, $S°_T$, the enthalpy function, $(H°_T$-$H°_{298})/T$, and the Gibbs energy function, $(G°_T$-$H°_{298})/T$, may be derived easily (see Sect. 2.4).

2.2.2 Calorimetry of Reacting Systems

This category of calorimetric techniques includes combustion and solution calorimetry. They are utilized for obtaining data on the enthalpy of formation, phase transition, order-disorder processes, and excess enthalpy of mixing of mineral crystalline solutions.

Combustion calorimetry is the classical technique of reaction calorimetry which, under favorable conditions, provides data with a precision of 0.01 to 0.1%. Such precision is generally attained if the reaction products are gases. An example of this has been given by Wise et al. (1963, Table V), who determined the $H°_{f,298}$ of quartz by combustion calorimetry, using fluorine gas as an oxidation agent. $H°_{f,298}$ of quartz was obtained from a combination of the measured $\Delta H°_{298}$ of the two reactions:

$$Si + 2F_2 = SiF_4; \Delta H°_{298} = -1614940 \pm 795 \text{ J} \tag{2.12a}$$

$$SiO_2 + 2F_2 = SiF_4 + O_2; \Delta H°_{298} = -704000 \pm 1172 \text{ J} \tag{2.12b}$$

Because enthalpy is a state function, it is path-independent. In other words, $H°_{f,298}$ of quartz may be obtained by linear addition (subtracting the second reaction from the first one):

$$Si + O_2 = SiO_2; \Delta H°_{298} \equiv H°_{f,298} = (-1614940 + 704000)$$
$$= -910940 \pm 1416 \text{ J mol}^{-1}. \tag{2.12}$$

The standard deviation of $H°_{f,298}$, $\sigma_{H°f,298}$, follows from Eq. (4.30),

$$\sigma_{H°_{f,298}} = [\sum_i v_i^2 \sigma_i^2 (\Delta H°_{298})]^{0.5}, \tag{2.13}$$

v_i denoting the stoichiometric coefficient for the i-th reaction. Note that by convention, all sigmas given in thermodynamic data tables (e.g. Robie et al. 1979) correspond to two standard deviations. Unfortunately, combustion calorimetry cannot be applied to measure the $H°_{f,298}$ of all substances, including the vast majority of the minerals. Solution calorimetry is the alternative technique utilized in such cases.

Solution calorimetry is performed at different temperatures depending upon the nature of the phase to be dissolved and type of solvent used. Dissolution of certain minerals is already achieved between 25-80°C in an aqueous acidic solution of HF or HCl. This type of calorimetry has been used with success since Torgeson and Sahama's (1948) pioneering study. A highly sophisticated version of such a calorimeter has been described by Robie and Hemingway (1972). This type of device is now routinely utilized to procure $H°_{f,298}$ (e.g. Hemingway and Robie 1977) and H^{ex}_m data (e.g. Hovis and Waldbaum 1977). Unfortunately, quite a few refractory oxides and silicates do not dissolve completely under those circumstances within a reasonable amount of time. To deal with them, oxide-melt high-temperature (700-900°C) solution calorimetry is applied (see review by Navrotsky 1977), which commonly uses melts of $2PbO·B_2O_3$ or $3Na_2O·4MoO_3$ compositions as solvents.

In solution calorimetric experiments, the basic requirement is that the final solution approach the condition of infinite dilution, such that the solute-solute interactions are negligible. This condition is fulfilled by dissolving 20-30 mg of a mineral in 30 g or more of the solvent. The enthalpy of formation of the compound AB from A and B is then calculated from the measured enthalpies of solution, $H°_{soln}$, as follows (Kleppa 1982):

$$A + \text{solvent} = \text{solution}; \quad H°_{soln}(A) \tag{2.14a}$$
$$B + \text{solvent} = \text{solution}; \quad H°_{soln}(B) \tag{2.14b}$$
$$AB + \text{solvent} = \text{solution}; \quad H°_{soln}(AB) \tag{2.14c}$$

$$A + B = AB; \quad \Delta H° = H°_{soln}(A) + H°_{soln}(B) - H°_{soln}(AB). \tag{2.14}$$

Thus, $\Delta H°$ of the reaction A + B = AB is obtained from a linear sum of three $H°_{soln}$ measurements. Note that $\Delta H°$ is defined here as reactants minus products. To appreciate why the sign convention is just the reverse of what we use elsewhere (products minus reactants), recall that in solution calorimetry, the process under study is one of dissolution, just the opposite of putting together the building blocks to form the compound of interest.

To give a simple example of the application of Eq.(2.14), let us consider the enthalpy of formation of forsterite (Mg_2SiO_4) from periclase (MgO) and quartz (SiO_2), according to the reaction,

$$2\,MgO + SiO_2 = Mg_2SiO_4. \tag{2.15}$$

The enthalpies of solution for synthetic periclase, quartz, and forsterite were measured by Kiseleva et al. (1979), using oxide-melt ($2PbO \cdot B_2O_3$) solution calorimetry at 1170 K. The results are:

Phases	$H^{\circ}_{soln,1170}$ (J mol^{-1})	No. of runs
Periclase	8661 ± 1715	10
Quartz	3766 ± 335	5
Forsterite	80500 ± 795	8

Following the reaction scheme outlined in Eq.(2.14), ΔH°_{1170} of the reaction (2.15) is

$$\Delta H^{\circ}_{1170} = 2\,(8661) + 3776 - 80500 = -59402 \pm 3537 \text{ J}. \qquad (2.16)$$

It is essential to note that Eq.(2.16) is the standard enthalpy of formation of forsterite *from the oxides* at 1 bar and 1170 K. It should be carefully distinguished from the standard enthalpy of formation of forsterite *from the elements*, $H^{\circ}_{f,T}$, as explained in Chap. 1. The latter refers to ΔH°_T of the reaction

$$2\,Mg + Si + 2\,O_2 = Mg_2SiO_4 \qquad (2.17)$$

at 1 bar pressure and T K, the constituent elements being in their respective stable standard states. The worked example in Sect. 2.5 will demonstrate how the data obtained by solution calorimetry are combined with combustion calorimetric data on $H^{\circ}_{f,298}$ for the two reactants, periclase and quartz, to obtain $H^{\circ}_{f,298}$ of forsterite.

Attention is drawn to the fact that for oxygen-base minerals, knowledge of $H^{\circ}_{f,T}$ (and $G^{\circ}_{f,T}$) from the oxides suffices. In order to emphasize that Robie et al. (1979) even tabulate duplicate sets of thermochemical data for those minerals, one referring to the elements, the other to the oxides. Nevertheless, the convention of referring $H^{\circ}_{f,T}$ and $G^{\circ}_{f,T}$ to the elements is more versatile; it is appropriate for all substances, regardless of their compositions. *In our subsequent exercises, we shall exclusively use the tables of thermochemical data referring to the constituent elements.*

An attractive feature of the oxide-melt solution calorimetry is that only 20-30 mg of material is required per run (compared to 1 g for each HF solution calorimetry experiment), so that routine measurement on synthetic materials becomes feasible. Although the precision of the H°_{soln} data is only 1% (compared to 0.1% in HF solution calorimetry), the cumulative standard deviation of $H^{\circ}_{f,298}$ is comparable in both cases (Navrotsky 1977). This is due to two reasons. First, the heat effects measured in oxide-melt solution calorimetry are much smaller than in the HF solution experiments. And secondly, the reaction schemes in the former are far simpler; consequently, the uncertainties of fewer reactions combine to make up that of the $H°_{f,T}$ data. Therefore, oxide-melt solution calorimetry is very important for studying mineral equilibria. Other important recent developments in this field include the study of order-disorder phenomena in minerals (Carpenter et al. 1983) and measurement of $H°_{f,298}$ of hydroxyl-bearing silicates (Kiseleva and Ogorodova 1984) and sulfides (Cemic and Kleppa 1986).

2.3 Outlines of Electrochemical Cell Measurements

As noted earlier, the electrochemical cell measurements have provided some of the thermodynamic data tabulated by Robie et al. (1979). In this section, only the basic principles underlying this technique will be touched. Readers wishing to delve deeper are referred to Subbarao (1980) or Sato (1971); the latter emphasizes many mineralogical applications.

The electrochemical cells are utilized to measure $\Delta G°_T$ of an appropriate reaction or a_i-X_i relationship of suitable crystalline solutions. In this chapter we shall be concerned with the former, Chapter 8 providing an example of the latter. The prerequisite for measuring the $\Delta G°_T$ is the availability of an appropriate galvanic cell, in which the desired reaction takes place electrochemically. For measurements at low temperatures, aqueous electrolyte galvanic cells are used, while, at high temperatures, the solid electrolyte sensors (CaO- or Y_2O_3-doped ZrO_2) are employed to construct those cells. If one is designed such that no side reactions occur, then the reversible electromotive force (emf), E, measured across that cell, can be translated into $\Delta G°_T$ of the virtual chemical reaction by the identity

$$\Delta G°_T = - zFE(T),\qquad(2.18)$$

z, denoting the number of electrons involved in cell transfer, and F, the Faraday constant (96487 J Volt^{-1} mol^{-1}).

To give an example, Holmes et al. (1986) used a CaO-doped ZrO_2 solid electrolyte galvanic cell to obtain $\Delta G°_T$ of reaction

$$Ni + \frac{1}{2}O_2 = NiO\qquad(2.19)$$

in the range 923 to 1411 K. The temperature dependence of $\Delta G°$ for Eq.(2.19) was expressed as

$$\Delta G°_T(2.19) = -239885 + 122.35\ T - 4.584\ T\ \ln T,\qquad(2.20)$$

with T in K, and $\Delta G°_T$ in J mol^{-1}. Since $\Delta G°_T$ of the reaction (2.19) is identical to $G°_{f,T}$ of NiO (bunsenite), we have at 1200 K,

$$G°_{f,1200,bunsenite} = -132066\ J\ mol^{-1}.\qquad(2.21)$$

This compares excellently with the calorimetrically obtained value of -132122 J mol^{-1} (Robie et al. 1979) for $G°_{f,1200}$ of bunsenite.

2.4 Evaluation and Tabulation of Thermodynamic Data

Having acquired a rudimentary knowledge of the most important calorimetric and electrochemical methods, let us return to the manipulation and tabulation of standard state thermodynamic data. We start by taking a look at the basic equations used to calculate the thermophysical properties as a function of temperature.

The first of these temperature-dependent quantities listed in Robie et al. (1979) is the enthalpy function, $(H°_T-H°_{298})/T$. An intermediate stage in the calculation of the enthalpy function is the standard state heat content, $H°_T-H°_{298}$, derived by integrating Eq.(1.15):

$$H°_T - H°_{298} = \int_{298}^{T} C°_P(T)dT .$$

(2.22)

Inserting (2.11) into (2.22), and denoting 298.15 K as T_o,

$$H°_T - H°_{298} = \int_{T_o}^{T} (a + bT + cT^{-2} + dT^{-0.5} + eT^2 + fT^{-3})dT$$

$$= a[T - T_o] + \frac{b}{2}[T^2 - T_o^2] + c\left[\frac{1}{T_o} - \frac{1}{T}\right]$$

$$+ 2d[T^{0.5} - T_o^{0.5}] + \frac{e}{3}[T^3 - T_o^3] + \frac{f}{2}\left[\frac{1}{T_o^2} - \frac{1}{T^2}\right].$$

(2.23)

Given the $C°_P(T)$ polynomials, $H°_T-H°_{298}$ and $(H°_T-H°_{298})/T$ are thus easily obtained. Computing those quantities becomes somewhat more complicated, however, when a phase transition intervenes between 298.15 K and T, be it a first-order or a "higher-order" one. Prior to dealing with that, let us take a quick look at the nature of a phase transition.

A phase transition is said to be of the *first order*, when the *first derivative* of Gibbs energy with respect to T (or P) shows a finite difference at the transition temperature, T_{tr}, or pressure, P_{tr}. The Gibbs energy itself remains continuous at that point, but no longer continuously differentiable, resulting in the appearance of discontinuities in the first derivatives of G with respect to T and P. In order to compute ΔS_{tr} [$= -\partial G/\partial T)_P$] and ΔV_{tr} [$= \partial G/\partial P)_T$], the partial derivatives are taken infinitesimally above and below the transition point. Moreover, because G ($\equiv$ H - TS) is continuous at that point, the discontinuity in T·S must be compensated by one in H (and hence, in $C°_P$). These general

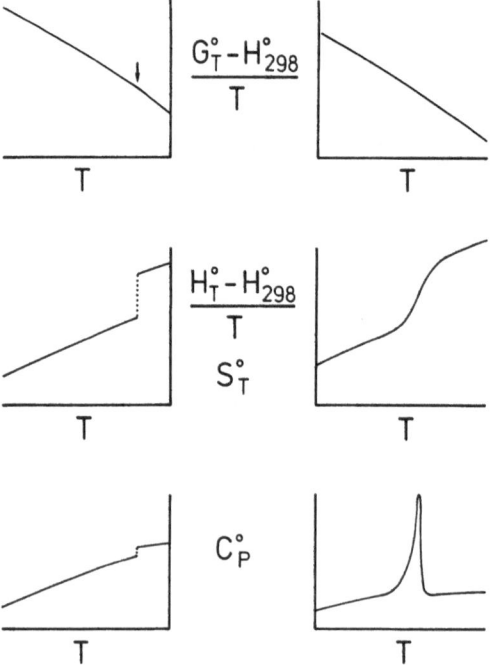

Fig. 2.1 A schematic depiction of the temperature dependence of the thermophysical properties of a first-order (*left column*) and a lambda-type (*right column*) phase transition. The *arrow* highlights the abrupt change in the slope of $(G_T^\circ - H_{298}^\circ)/T$ vs T curve at T_{tr}

features of a first order phase transition are shown in Fig. 2.1 (left column) by plotting the relevant thermophysical properties schematically. Melting and vaporization are generally cited as examples of first order phase transitions. However, some reconstructive phase transitions might also belong to this category, although this is very difficult to demonstrate experimentally (see Thompson and Perkins 1981). In a *higher-order* phase transition, G is not only continuous, but also continuously differentiable at the transition point; consequently, discontinuities can appear only in the *higher derivative* of G. One phenomenon, encountered in some minerals undergoing a displacive (like low-quartz/high-quartz) or an order-disorder transformation, is a "lambda"-shaped $C_P^\circ(T)$ curve (Fig. 2.1, right column). For these, the Gibbs energy is differentiable at the transition point; thus, λ-transition is also a higher-order phase transition. Since λ-transition is of considerable importance to mineralogy, we shall only distinguish between a first order and a λ-transition in this book.

To calculate $H_T^\circ - H_{298}^\circ$ across a first order phase transition, Eq. (2.22) is extended to

$$H_T^\circ - H_{298}^\circ = \int_{298}^{T_{tr}} C_{P,1}^\circ(T)dT + \Delta H_{tr}^\circ + \int_{T_{tr}}^{T} C_{P,2}^\circ(T)dT, \qquad (2.24)$$

ΔH_{tr}° being its standard enthalpy of transition, $C_{P,1}^\circ$ and $C_{P,2}^\circ$ the $C_P^\circ(T)$ functions for those phases at 298 K $<T <T_{tr}$ and at $T > T_{tr}$, respectively.

A similar integration across a λ-transition may be performed in two fundamentally different fashions. In the first procedure, advocated by Berman and Brown (1985), the lattice vibrations part of the $C^{\circ}_p(T)$ is fit by Eq.(2.10), which serves as a base line. Then, the C°_p (T) due to the λ-transition, $C^{\circ}_p(\lambda)$, is superposed on it. For this purpose, Berman and Brown (1985)[3] fit $C^{\circ}_p(\lambda)$ in the range $T_{ref} < T < T_{tr}$, as

$$C^{\circ}_p(\lambda) = T(l_1 + l_2T)^2, \qquad (2.25)$$

T_{ref} being a suitably chosen reference temperature, commonly, but not necessarily, 298.15 K (cf. Berman and Brown 1986, Table 1). At $T > T_{tr}$, $C^{\circ}_p(T)$ is again given by the base function (2.10). Thus, integrating (2.10) and (2.25) between the appropriate limits, the enthalpy function is obtained across a λ-transition. As emphasized by Berman (1988), the simple form of (2.25) does not permit a very good reproduction of the thermophysical properties over a range of ± 30 K on either side of the "lambda point", T_{tr}.

Robie et al. (1979) tackled this problem in a different manner, obtaining similar results. Regardless of whether a first order or a λ-transition is observed, they handled them alike, as if they are both first order phenomena taking place at a unique T. But how can a λ-transition be handled like a first order phase transition? To appreciate this, refer to Fig. 2.2. The upper diagram shows the $C^{\circ}_p(T)$ data for quartz, based mainly on direct $C^{\circ}_p(T)$ measurements by high-temperature adiabatic calorimetry (Moser 1936). A lambda-like C°_p curve is observed, indicative of the low-quartz to high-quartz transformation. When low-quartz is subjected to increasing temperature, its SiO_4-tetrahedra rotate and tilt with respect to each other. A clear sign of thermal response to such a structural adjustment is the appearance of the inflection point on the $C^{\circ}_p(T)$ curve at 670 K (Moser 1936). With a further rise in temperature, C°_p increases rapidly until at 844 K the maximum is reached. After that, C°_p plummets to a much lower value, characteristic of high-quartz. The $H^{\circ}_T-H^{\circ}_{298}$ curve displayed in Fig. 2.2, was obtained by numerical integration of this C°_p data by the Kronrod technique (Piessens et al. 1983). Note that $H^{\circ}_T-H^{\circ}_{298}$ shows an inflection at 844 K, the T_{tr}, typical of a higher-order phase transition.

Given such a phase transition, Robie et al. (1979) obtained two $C^{\circ}_p(T)$ functions, one each for low- and high-quartz. These $C^{\circ}_p(T)$ data are shown by the dashed lines in the upper diagram; note the sharp discontinuity (dotted line) between the two at T_{tr}. The area between the true $C^{\circ}_p(T)$ curve (shown as a solid line) and the base lines (the dashed lines) yields ΔH_{tr}, which is

---

[3] Ghiorso et al. (1979, Appendix) suggested an analogous procedure, in which the T-dependence of $C^{\circ}_p(\lambda)$ is expressed as:

$$C^{\circ}_p(\lambda) = \sum_{i=1}^{n} l_i[T/(T-T_{tr})^i],$$

l_i being the i-th coefficient. It is not introduced in this book, because the approach is basically similar to that of Berman and Brown (1985).

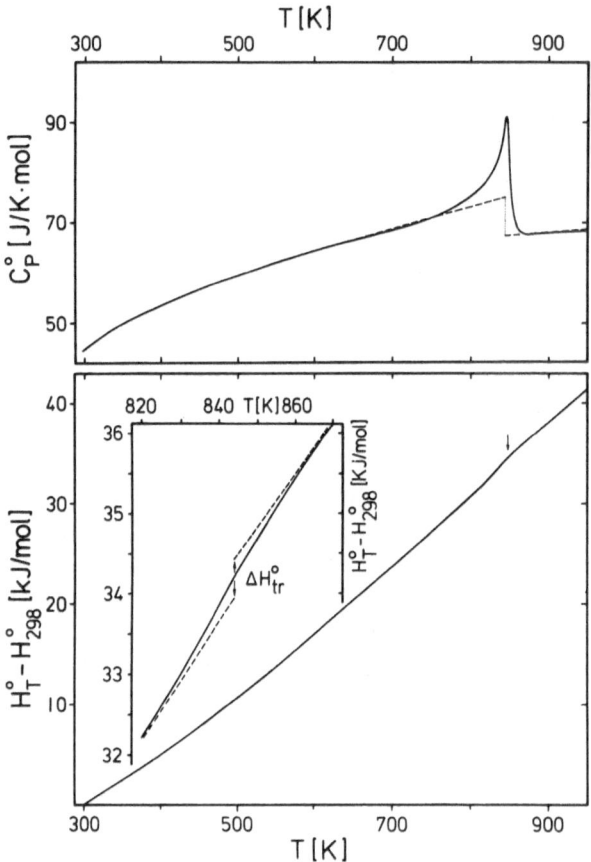

Fig. 2.2 The $C_P^{\circ}(T)$ (*top*) and H_T°-H_{298}° (*bottom*) functions for low- and high-quartz based primarily on high-temperature adiabatic calorimetry by Moser (1936). The inflection point on the H_T°-H_{298}° curve has been highlighted by an *inverted arrow*. It corresponds to the $C_P^{\circ}(T)$ maximum at 844 K. The *solid line* in the *inset* shows the results of numerical integration of $_{298}\!\int^{T}C_P^{\circ}(T)dT$ across T_{tr}; for comparison, the *dashed lines* display the results obtained from the $C_P^{\circ}(T)$ data of low- and high-quartz, and ΔH_{tr}°, tabulated by Robie et al. (1979)

tabulated by Robie et al. (1979). Given the $C_{P}^{\circ}(T)$ data for the ranges 298.15 K-T_{tr} and T_{tr}-T, in addition to ΔH°_{tr}, the evaluation of H°_{T}-H°_{298} makes use of (2.24). The result thus obtained compares excellently with those derived by numerical integration, except for the range 820-870 K, that is, ± 25 K on either side of T_{tr} (see inset of Fig. 2.2). The maximum difference is at T_{tr}, which tapers off on both sides. This is due to the fact that Robie et al. (1979) treated the transition as though it takes place at T_{tr} sharp.

We next address the remaining thermophysical quantities listed by Robie et al. (1979). The property given in the second column of their tables is the standard entropy, $S°_T$. At $T > 298.15$ K, it is obtained from

$$S°_T = S°_{298} + \int_{298}^{T} \left[\frac{C°_P(T)}{T} \right] dT .$$ (2.26)

Substituting (2.11) into (2.26), and with 298.15 K $\equiv T_o$, we have the following general relation:

$$S°_T = S°_{298} + \int_{T_o}^{T} \left[\frac{(a + bT + cT^{-2} + dT^{-0.5} + eT^2 + fT^{-3})}{T} \right] dT$$

$$= S°_{298} + a\ln\frac{T}{T_o} + b[T - T_o] + \frac{c}{2}\left[\frac{1}{T_o^2} - \frac{1}{T^2} \right]$$

$$+ 2d\left[\frac{1}{T_o^{0.5}} - \frac{1}{T^{0.5}} \right] + \frac{e}{2}[T^2 - T_o^2] + \frac{f}{3}\left[\frac{1}{T_o^3} - \frac{1}{T^3} \right].$$ (2.27)

Given $S°_{298}$ and $C°_P(T)$, $S°_T$ is thus easily evaluated for a desired temperature. If a first order phase transition intervenes between 298.15 and T, Eq. (2.26) is rewritten as

$$S°_T = S°_{298} + \int_{298}^{T_{tr}} \left[\frac{C°_{P,1}(T)}{T} \right] dT + \Delta S°_{tr} + \int_{T_{tr}}^{T} \left[\frac{C°_{P,2}(T)}{T} \right] dT ,$$ (2.28)

with $\Delta S°_{tr}$ representing the entropy of transition. In the event of a λ-transition, procedures analogous to those discussed above may be employed. Figure 2.3 displays the result obtained by simulating the λ-transition in quartz as a first order phase transition. The standard entropy, $S°_T$, computed from the numerical integration of $C°_P(T)/T$ reveals an inflection point at 844 K, commensurate to the maximum on the $C°_P(T)/T$ curve of the upper diagram. The details of simulating the λ-transition of quartz as a first order transition are shown in the inset.

The Gibbs energy function, $(G°_T-H°_{298})/T$, tabulated next by Robie et al. (1979), is derived by combining the enthalpy function and the standard entropy as follows,

$$\frac{G°_T - H°_{298}}{T} = \left[\frac{(H°_T - TS°_T) - H°_{298}}{T} \right] = \frac{H°_T - H°_{298}}{T} - S°_T .$$ (2.29)

Contrary to both $(H°_T-H°_{298})/T$ and $S°_T$, the Gibbs energy function is always a negative quantity. To facilitate its tabulation, Robie et al. (1979) listed the negative of $(G°_T-H°_{298})/T$.

In the thermodynamic tables (cf. Robie et al. 1979) a dashed horizontal line highlights the T_{tr} for every phase transition taking place above 298.15 K.

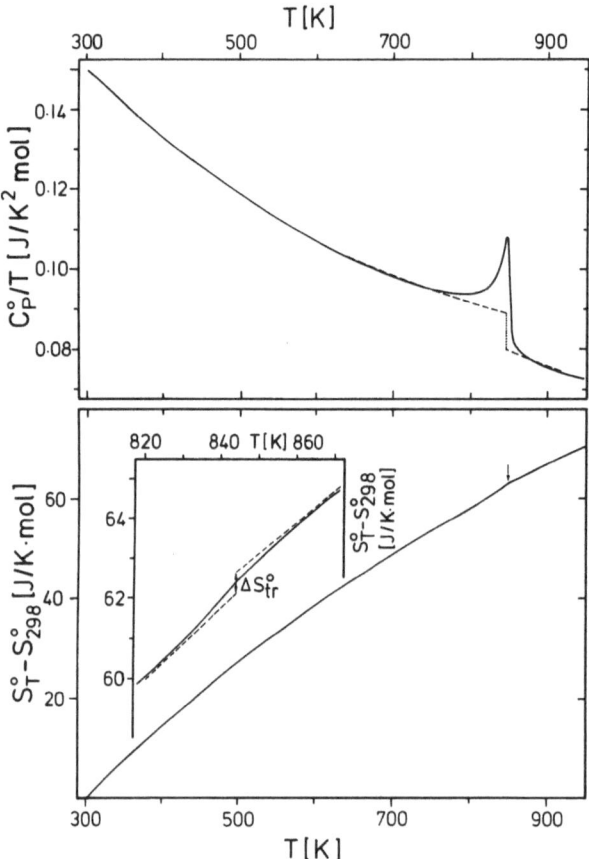

Fig. 2.3 The function $C_P^\circ(T)/T$ (*top*) and $S_T^\circ - S_{298}^\circ$ (*bottom*) for low- and high-quartz based primarily on high-temperature adiabatic calorimetry (Moser 1936). The inflection point on the $S_T^\circ - S_{298}^\circ$ curve is indicated by an *arrow*. The *inset* contrasts the $S_T^\circ - S_{298}^\circ$ data derived by a numerical integration of Moser's C_P° data (*solid line*) with those obtained from the tables of Robie et al. (1979) (*dashed lines*)

Consider, for example, the case of quartz. Its low-high phase transition occurs at 844 K. Therefore, $(H_T^\circ - H_{298}^\circ)/T$ appears twice at that temperature; for low-quartz it is 40.059, for high-quartz, 40.623 J K^{-1} mol^{-1}. The stepwise change of $(H_T^\circ - H_{298}^\circ)/T$ at T_{tr} equals a ΔH_{tr}° of 476 J mol^{-1}. A similar discontinuous change also appears in the S_T° column, from which we may compute a ΔS_{tr}° of 0.55 J K^{-1} mol^{-1}. Because both the polymorphs are at equilibrium at T_{tr}, $(H_T^\circ - H_{298}^\circ)/T$ and S_T° offset each other and, therefore, no discontinuity appears in $(G_T^\circ - H_{298}^\circ)/T$. Note, however, that the temperature derivative of $(G_T^\circ - H_{298}^\circ)/T$ at T_{tr} is clearly discontinuous, reflecting ΔS_{tr}°.

Apart from the thermophysical properties, Robie et al. (1979) also tabulated the thermochemical parameters $H_{f,T}^\circ$, $G_{f,T}^\circ$, and log $K_{f,T}^\circ$ at 100 K intervals. The first two are obtained by combining $H°_{f,298}$ with the enthalpy and the Gibbs energy functions:

$$H°_{f,T} = H°_{f,298} + T\Delta_f \left[\frac{H°_T - H°_{298}}{T} \right], \quad \text{and} \tag{2.30}$$

$$G°_{f,T} = H°_{f,298} + T\Delta_f \left[\frac{G°_T - H°_{298}}{T} \right], \tag{2.31}$$

where Δ_f refers to the reaction of formation of the phase from the constituent stable elements. Note that $H°_{f,T}$ shows a discontinuity whenever a phase change is involved, whether for the phase itself *or* in one of its constituent elements; by contrast, $G°_{f,T}$ invariably remains single-valued. Nevertheless, the first derivative of $G°_{f,T}$ shows an abrupt change at T_{tr}, the temperature at which the phase itself undergoes a transititon, reflecting the $\Delta S°_{tr}$. In the tables of Robie et al. (1979), a phase transition is highlighted by dashed lines drawn across the columns of both thermophysical and thermochemical quantities. Transitions of the constituent elements are indicated by lines across the columns of only the thermochemical properties; only these are affected by a transition in the elements.

And finally, Robie et al. (1979) also listed the logarithm of the equilibrium constant of formation (from the stable elements) of compounds at 1 bar pressure, $\log K°_{f,T}$, at 100 K intervals. It is computed from the equation

$$\log K°_{f,T} = -\frac{G°_{f,T}}{2.3026 \ RT} . \tag{2.32}$$

The nature of the various thermophysical and thermochemical quantities and their temperature dependences, discussed above, are of great importance for thermodynamic calculations. In particular, property changes in response to a phase transition or a transition in any of the constituent elements must be taken into account if data are interpolated between the 100 K intervals of tabulations.

2.5 Worked Examples of Evaluation and Manipulation of Data

Three examples of evaluation of thermodynamic properties are given below to familiarize the reader with the nature of the data reproduced in standard tables.

Example 1. Derivation of $H°_{f,298}$ of forsterite from solution calorimetry.

In Section 2.2.2, it was demonstrated how Kiseleva et al. (1979) utilized high-temperature solution calorimetry to measure $\Delta H°_{1170}$ of the reaction

$$2 \ MgO + SiO_2 = Mg_2SiO_4. \tag{2.15}$$

Table 2.1. Heat capacity power functions, $C_P^\circ(T)$ (in J K^{-1} mol^{-1}), for periclase, quartz, and forsterite

Phase (Validity)	a	b	c	d	e
Periclase[b] (298-1800 K)	65.211	-1.2699E-3[a]	-4.6185E5	-387.24	
Low quartz[b] (298-844 K)	44.603	3.7754E-2	-1.0018E6		
ΔH_{844}° (low quartz = high quartz) = 476 J					
High quartz[b] (844-1800 K)	58.928	1.0031E-2			
Forsterite[c] (298-1800 K)	87.36	0.08717	-3.699E6	843.6	-2.237E-5

[a] "E-3" is an abbreviation for "x 10^{-3}, "E5" for "x 10^5", etc.
[b] Robie et al. (1979).
[c] Robie et al. (1982).

Starting with this, we shall now derive $H^\circ_{f,298}$ of forsterite. The first step toward that goal is to convert ΔH°_{1170} to ΔH°_{298}. Using Eq.(2.22), we have

$$\Delta H_{298}^\circ = \Delta H_{1170}^\circ - \int_{298}^{1170} \Delta C_P^\circ(T)dT = \Delta H_{1170}^\circ - \Delta(H_{1170}^\circ - H_{298}^\circ), \quad (2.33)$$

where Δ denotes the difference due to the reaction (2.15). The $C_P^\circ(T)$ functions necessary to evaluate $\Delta(H_T^\circ - H_{298}^\circ)$ in Eq. (2.33) are listed in Table 2.1. Using these data, along with Eqs. (2.22), (2.23), and (2.24), we obtain

$$H_{1170}^\circ - H_{298}^\circ(\text{periclase}) = 65.211[1170-298.15]$$
$$+ [(-1.2699\text{E-}3)/2][(1170)^2-(298.15)^2]$$
$$+ (-4.6185\text{E}5)[(1/298.15)-(1/1170)]$$
$$+ 2(-387.24)[(1170)^{0.5}-(298.15)^{0.5}]\text{ J}$$
$$= 41768.8\text{ J}.$$

$$H_{1170}^\circ - H_{298}^\circ(\text{quartz}) = 44.603[844-298.15]$$
$$+ [(3.7754\text{E-}2)/2][(844)^2-(298.15)^2]$$
$$+ (-1.0018\text{E}6)[(1/298.15)-(1/844)]$$
$$+ 476 + 58.928[1170-844]$$
$$+ [(1.0031\text{E-}2)/2][(1170)^2-(844)^2]\text{ J}$$
$$= 56921.7\text{ J}.$$

$H_{1170}^{\circ}-H_{298}^{\circ}(\text{forsterite}) = 87.36[1170-298.15]$
$\qquad + [(0.08717)/2][(1170)^2-(298.15)^2]$
$\qquad + (-3.699E6)[(1/298.15)-(1/1170)]$
$\qquad + 2(843.6)[(1170)^{0.5}-(298.15)^{0.5}]$
$\qquad + [(-2.237E-5)/3][(1170)^3-(298.15)^3]$ J
$\qquad = 139542.1$ J.

Substituting these into Eq.(2.33), and ignoring the uncertainties of the $C^{\circ}_p(T)$ data,

$\Delta H_{298}^{\circ}(2.15) = -59402 - [139542.1-\{2(41768.8)+56921.7\}]$ J
$\qquad\qquad\quad = -58485 \pm 3537$ J.[4]   (2.34)

The enthalpy of formation of forsterite from the constituent elements, $H^{\circ}_{f,298}$, is now recovered by linearly adding (2.34) with known $H^{\circ}_{f,298}$ for periclase and quartz. The appropriate figure for periclase is taken from Robie et al. (1979), while the $H^{\circ}_{f,298}$ of quartz, derived above in Eq.(2.12) from combustion calorimetry, is used:

$2(Mg + 0.5O_2) = 2MgO$;	$2H_{f,298}^{\circ} = 2(-601490 \pm 290)$ J	(2.35)
$Si + O_2 \qquad = SiO_2$;	$H_{f,298}^{\circ} = -910940 \pm 1416$ J	(2.12)
$2MgO + SiO_2 = Mg_2SiO_4$;	$\Delta H_{298}^{\circ} = -58485 \pm 3537$ J	(2.34)

$2Mg+Si+2O_2 = Mg_2SiO_4$;	$H_{f,298}^{\circ} = -2172405 \pm 3854$ J.	(2.17)

This result may be compared to -2170400 ± 1400 J, quoted by Robie et al. (1982) for $H^{\circ}_{f,298}$ of forsterite. The agreement between the two is reassuring especially because the Robie et al. (1982) datum is based on the alternative technique of HF solution calorimetry by King et al. (1967). It is pointed out in this context that the general practice is *not* to round off the tabulated $H^{\circ}_{f,298}$ values on the basis of their estimated uncertainties to utilize the full accuracy of the information on $C^{\circ}_p(T)$ (Robie et al. 1979).

Example 2. Computation of the thermophysical and thermochemical properties of Mg, Si, O_2, and forsterite.

From Section 2.1, it is known that the basic data required to compute the thermophysical and thermochemical properties of a phase are its $H^{\circ}_{f,298}$, S°_{298}, and $C^{\circ}_p(T)$ functions. Employing the $H^{\circ}_{f,298}$ of forsterite obtained above, and the S°_{298} and $C^{\circ}_p(T)$ for forsterite and its constituent elements given in Table 2.2, let us compute $H^{\circ}_{f,T}$ and $G^{\circ}_{f,T}$ of forsterite for a number of temperatures including 298.15, 500, 700, 900, 922, 1000, and 1200 K. Regarding $H^{\circ}_{f,T}$ and $G^{\circ}_{f,T}$ of the reference elements, all we need to do is to

---

[4] This is the standard enthalpy of formation of forsterite "from the oxides" at 298.15 K, comparable to -56690 ± 610 J mol^{-1} quoted by Robie et al. (1979).

Table 2.2. Heat capacity power functions and some related data for Mg, Si, O_2, and forsterite

	Mg(solid)[a]	Mg(melt)[a]	Phases Si(solid)[a]	O_2(gas)[a]	Forsterite[b]
$C_p^\circ(T) = a + bT + cT^{-2} + dT^{-0.5} + eT^2$ ($J\,K^{-1}\,mol^{-1}$)					-
a	16.095	32.64	31.778	48.318	87.36
b	1.3795E-2		5.3878E-4	-6.9132E-4	0.08717
c	-1.5759E5		-1.4654E5	4.9923E5	-3.699E6
d	1.1053E2		-1.7864E2	-4.2066E2	843.6
e					-2.237E-5
Validity (in K)	298-922	922-1361	298-1685	298-1800	298-1800
S_{298}° ($J\,K^{-1}\,mol^{-1}$)	32.68		18.81	205.15	94.11
ΔH_{922}° ($J\,mol^{-1}$)	8954				
ΔS_{922}° ($J\,K^{-1}\,mol^{-1}$)	9.71				

[a] Robie et al. (1979).
[b] Robie et al. (1982).

bear in mind that they are zero by definition at all temperatures (see Sect. 1.2.1).

Computation of $(H^\circ_T - H^\circ_{298})/T$ and $(G^\circ_T - H^\circ_{298})/T$ for forsterite and its constituent elements is a prerequisite for applying Eqs. (2.30) and (2.31) to obtain $H^\circ_{f,T}$ and $G^\circ_{f,T}$ of forsterite. Table 2.3 reproduces the results of calculation of those properties at the desired temperatures. The equations used to compute them have been discussed earlier, and require no reiteration. The functions S°_T, $(H^\circ_T - H^\circ_{298})/T$, and $-(G^\circ_T - H^\circ_{298})/T$ for Si, O_2, and forsterite are continuous between 298 and 1200 K. However, discontinuities in S°_T and $(H^\circ_T - H^\circ_{298})/T$ appear at 922 K for Mg, documenting its melting at that T. Note that $-(G^\circ_T - H^\circ_{298})/T$ remains single-valued even at that temperature. The phase transition in Mg has been highlighted in Table 2.3 by the horizontal dashed line. These data, along with $H^\circ_{f,298}$ of forsterite, were used next to calculate $H^\circ_{f,T}$ and $G^\circ_{f,T}$ for forsterite. The dashed line at 922 K, crossing the columns of $H^\circ_{f,T}$ and $G^\circ_{f,T}$ of forsterite, is not to imply a phase transition in it, but in one of its constituent elements, Mg. Because $H^\circ_{f,T}$ is obtained via (2.30), any discontinuity in $(H^\circ_T - H^\circ_{298})/T$ of any of the constituent elements necessarily shows up in $H^\circ_{f,T}$ of that compound. By contrast, $G^\circ_{f,T}$, obtained via (2.31), does not show a stepwise change, because $(G^\circ_T - H^\circ_{298})/T$ of Mg remains single-valued despite phase transition. Note that the tables of Robie et

Table 2.3. Calculated thermodynamic functions for Mg, Si, O_2, and forsterite

T (K)	$(H_T^\circ-H_{298}^\circ)/T$ (J K^{-1} mol^{-1})	S_T° (J K^{-1} mol^{-1})	$-(G_T^\circ-H_{298}^\circ)/T$ (J K^{-1} mol^{-1})	$H_{f,T}^\circ$ (J mol^{-1})	$G_{f,T}^\circ$ (J mol^{-1})
		Magnesium: 298-922 K (solid), 922-1361 K (liquid)			
298.15	0.00	32.68	32.68	0.00	0.00
500	10.55	46.13	35.59	0.00	0.00
700	15.66	55.68	40.02	0.00	0.00
900	19.02	63.41	44.38	0.00	0.00
922	19.34	64.18	44.84	0.00	0.00
922	29.05	73.89	44.84	0.00	0.00
1000	29.33	76.54	47.21	0.00	0.00
1200	29.88	82.49	52.61	0.00	0.00
		Silicon: 298-1685 K (solid)			
298.15	0.00	18.81	18.81	0.00	0.00
500	8.88	30.10	21.22	0.00	0.00
700	13.30	38.29	24.98	0.00	0.00
900	16.05	44.73	28.68	0.00	0.00
922	16.29	45.36	29.07	0.00	0.00
1000	17.07	47.50	30.43	0.00	0.00
1200	18.71	52.40	33.69	0.00	0.00
		Oxygen: 298-1800 K (diatomic gas)			
298.15	0.00	205.15	205.15	0.00	0.00
500	12.18	220.70	208.53	0.00	0.00
700	17.87	231.48	213.62	0.00	0.00
900	21.38	239.94	218.56	0.00	0.00
922	21.69	240.76	219.08	0.00	0.00
1000	22.69	243.58	220.88	0.00	0.00
1200	24.79	250.01	225.21	0.00	0.00
		Forsterite: 298-1800 K (solid)			
298.15	0.00	94.11	94.11	-2172415	-2053048
500	55.01	163.92	108.91	-2172080	-1972140
700	83.73	216.12	132.39	-2170056	-1892506
900	102.17	257.97	155.80	-2167627	-1813531
922	103.83	262.12	158.29	-2167362	-1804887
922	103.83	262.12	158.29	-2185268	-1804887
1000	109.28	276.22	166.94	-2184245	-1772745
1200	120.81	308.74	187.93	-2181103	-1690735

al. (1979) do indicate all transitions in the constituent elements, but they do not tabulate the corresponding values of $H°_{f,T}$ and $G°_{f,T}$ *at* the transition temperatures. These features of the tables have to be borne in mind when interpolating any of those quantities.

Example 3. Derivation of the $H°_{f,298}$ for bunsenite from electrochemical cell measurements.

In Section 2.3, we have seen how Holmes et al. (1986) employed high-temperature electrochemical cell measurements to obtain $G°_{f,T}$ for NiO, bunsenite, in the range 923-1411 K. Utilizing their data, we have

$$G°_{f,1200,bunsenite} = -132066 \text{ J mol}^{-1}. \tag{2.21}$$

Let us derive $H°_{f,298}$ of bunsenite, starting from its $G°_{f,1200}$. To do this, Eq.(2.31) is first rearranged to

$$H°_{f,298} = G°_{f,T} - T\Delta_f \left[\frac{G°_T - H°_{298}}{T} \right], \tag{2.31a}$$

where Δ_f refers to the reaction of formation of bunsenite from the constituent elements. Recalling its formation reaction,

$$Ni + \frac{1}{2}O_2 = NiO, \tag{2.19}$$

Eq.(2.31a) may be explicitly rewritten as

$$H°_{f,298,NiO} = G°_{f,T,NiO} - T\left[\left\{ \frac{G°_T - H°_{298}}{T} \right\}_{NiO} - \left\{ \frac{G°_T - H°_{298}}{T} \right\}_{Ni} \right. $$
$$\left. - 0.5\left\{ \frac{G°_T - H°_{298}}{T} \right\}_{O_2} \right]. \tag{2.36}$$

Inserting the appropriate $(G°_T-H°_{298})/T$ data at 1200 K from Robie et al. (1979) into Eq.(2.36),

$$H°_{f,298,NiO} = -132066 - 1200[-72.08 + 49.16 + 0.5(225.22)]$$
$$= -239694 \text{ J mol}^{-1}. \tag{2.37}$$

Because $(G°_T-H°_{298})/T$ is derived by coupling $C°_p(T)$ and $S°_{298}$ [cf. Eq.(2.29)], it is evident that independent data on $C°_p(T)$ and $S°_{298}$ are required to convert $G°_{f,T}$ to $H°_{f,298}$. That is in essence the method of "third law analysis", as applied by Holmes et al. (1986, p.2442 ff) to their emf data. Note that $H°_{f,298}$, computed by that method, agrees excellently with the calorimetric value of -239743 J mol^{-1} quoted by Robie et al. (1979). By implication, the $C°_p(T)$ and $S°_{298}$ data are also compatible with the emf measurements.

Chapter 3

Equations of State for Fluids and Fluid Mixtures

3.1 Introduction

Fluids[1] abound during the formation of most rocks and mineral deposits. In the majority of geological environments, they tend to be mixtures of a number of species like H_2O, CO_2, CH_4, CO, H_2, N_2, H_2S, etc. The presence of NaCl or other salts may add to their complexity. The computation of mineral equilibria involving a fluid species i requires that its chemical potential be known at a given set of T, P, and X_i [cf. Eq.(1.100)].

This chapter opens with a discussion of two alternative ways in which to deal with the chemical potential of a fluid species in mixed fluids, and by deriving the expressions for its fugacity and activity. Calculation of these quantities requires knowledge of the volume equations of state for pure fluids and fluid mixtures. The equations of state, popular in geochemical literature, will be reviewed, emphasizing the modified Redlich-Kwong (MRK) equation.

3.2 Some Fundamental Concepts: Standard States, Fugacity, Activity

The chemical potential of the i-th species, μ_i, in any nonideal solution (irrespective of its state of aggregation) has been defined earlier (Sect. 1.2.2.3) as

$$\mu_i \equiv \mu_i^* + RT\ln a_i = \mu_i^* + RT\ln X_i + RT\ln \gamma_i. \tag{1.62}$$

In this definition, the standard chemical potential, μ_i^*, applies to any P (and T), the pure species i ($X_i = 1$) having unit activity at that P and T.

Although the above definition is of a perfectly general nature, geochemists commonly use an *alternative* definition of the chemical potential of the i-th species in fluid mixtures. Rather than using the "sliding" P standard state implied by (1.62), the hypothetical ideal gas i at 1 bar ($P^\circ = 1$ bar) is chosen as the standard state. The chemical potential of the i-th fluid is then expressed in terms of the fugacity of the species i, f'_i, in the fluid mixture. Because μ_i is independent of the choice of the standard state, its two alternative definitions must be mutually compatible. We shall demonstrate this by deriving the

---

[1] Fluid will be used here as a collective term for subcritical and supercritical physical states.

alternative defining equation of μ_i (3.9) starting from (1.62). This exercise will have the benefit of clarifying our concept of fugacity of i in a fluid mixture, and its relation to pressure, P, and activity, a_i.

As an initial step in deriving Eq.(3.9), let us split the μ^*_i term of (1.62) and rewrite it as

$$\mu_i = G^\circ_{f,T,i} + \int_{P^\circ}^{P} V_{T,i}(P)dP + RT\ln a_i \,, \tag{3.1}$$

with V_i denoting the molar volume of pure i at T and P. Note that the change in energy level from $G^\circ_{f,T,i}$ to $\mu_i(T,P,X_i)$ is due to *two* changes of state; first, a change of pressure from $P^\circ(= 1\text{ bar})$ to P bar, and second, dilution of pure i ($X_i = 1$) to X_i.

Next, consider the chemical potential of the pure ideal fluid species i, $\mu^{*,id}_i$, at some T and P. From the ideal-gas equation of state (3.21), we have

$$V^{id} = \frac{RT}{P} \,. \tag{3.2}$$

Inserting (3.2) into (3.1), and noting that for pure i, the $RT\ln a_i$ term of (3.1) vanishes,

$$\mu^{*,id}_i = G^\circ_{f,T,i} + \int_{P^\circ}^{P} \frac{RT}{P}dP = G^\circ_{f,T,i} + RT\ln\frac{P}{P^\circ} \,. \tag{3.3}$$

Defining the ratio P/P° as P^*, (3.3) reduces to

$$\mu^{*,id}_i = G^\circ_{f,T,i} + RT\ln P^*. \tag{3.3a}$$

Since $P^\circ \equiv 1$ bar, P^* must be a dimensionless quantity, numerically equal to P (in bar). Although fluids do not behave ideally in most geological situations, (3.3a) is important because it represents a limiting behavior for them.

To handle the nonideal pure fluid i, a new quantity, called the fugacity of i, f_i, is introduced in lieu of P. Introduction of f_i helps retain the simple algebraic form of Eq.(3.3). The chemical potential of the nonideal pure fluid i at P and T, μ^*_i, is then defined as

$$\mu^*_i = G^\circ_{f,T,i} + RT\ln\frac{f_i}{f^\circ_i} = G^\circ_{f,T,i} + \int_{P^\circ}^{P} V_{T,i}(P)dP \,, \tag{3.4}$$

where f°_i, the standard state fugacity, equals 1 bar at all T. By implication, a nonideal gas having 1 bar fugacity at any T has an energy equivalent to $G^\circ_{f,T,i}$. Inserting f^*_i for f_i/f°_i, Eq.(3.4) also simplifies to

$$\mu^*_i = G^\circ_{f,T,i} + RT\ln f^*_i \tag{3.4a}$$

Thus, f^*_i ($\equiv f_i/f^\circ_i$) is also a dimensionless quantity, numerically equal to f_i (in bar).

The mutual relationship between f_i and P follows from Eqs.(3.3) and (3.4). Subtracting the former from the latter,

$$\mu^*_i - \mu^{*,id}_i = RT\ln\frac{f_i}{f^\circ_i} - RT\ln\frac{P}{P^\circ} = RT\ln\frac{f_i}{P} - RT\ln\frac{f^\circ_i}{P^\circ}. \tag{3.5}$$

The $RT\ln(f^\circ_i/P^\circ)$ term drops out since both f°_i and P° equal 1 bar, simplifying (3.5) to

$$\mu^*_i - \mu^{*,id}_i = RT\ln\frac{f_i}{P}. \tag{3.5a}$$

Evidently, (3.5a) measures the energy difference due to deviation of the pure nonideal gas i from its hypothetical ideal behavior at T and P. The fugacity coefficient for the pure fluid i, ϕ_i, is introduced into the RHS of (3.5a) to express this deviation term:

$$\mu^*_i - \mu^{*,id}_i = RT\ln\phi_i. \tag{3.6a}$$

In other words, the fugacity coefficient of the pure i-th species at T and P is defined as,

$$\phi_i \equiv \frac{f_i}{P}. \tag{3.6}$$

Because f_i and P have the units of bar, ϕ_i is also dimensionless, as it should be in order that $RT\ln\phi_i$ [in Eq. 3.6a] can have the dimension of energy. Moreover, like f_i, ϕ_i is also a function of T and P. Recalling that real gases behave ideally as P→0, we have

$$\lim_{P \to 0} \frac{f_i}{P} = \lim_{P \to 0} \phi_i = 1. \tag{3.7}$$

This implies that f_i approaches P as P→0. Equation (3.7) completes the definition of f_i as given in (3.4) by establishing the mutual relationship between f_i and P.

Next, we shall consider an ideal mixture of nonideal fluids. The chemical potential of the i-th species in such a mixture, at a given T, P, and X_i, is obtained from Eqs.(1.62) and (3.4),

$$\mu^{id\ mix}_i = G^\circ_{f,T,i} + RT\ln\frac{f_i}{f^\circ_i} + RT\ln X_i. \tag{3.8}$$

Under many geological conditions, the fluid mixtures are likely to deviate from the ideal mixing behavior. To account for a nonideal mixing, we have to include the $RTln\gamma_i$ term of Eq.(1.62). Thus, μ_i of a nonideal mixture of nonideal fluids is given by

$$\mu_i = G^\circ_{f,T,i} + RTln\frac{f_i}{f^\circ_i} + RTlnX_i + RTln\gamma_i$$

$$= G^\circ_{f,T,i} + RTln\frac{f_i}{f^\circ_i} + RTlna_i .$$

(3.9b)

Evidently, the change in energy level from $G^\circ_{f,T,i}$ to μ_i in (3.9b) is a *two* step process. The first step, $RTln(f_i/f^\circ_i)$, refers to the energy change due to a change in pressure from P° to P bar, while the second step, $RTlna_i$, takes care of the dilution of pure i from $(X_i = 1)$ to X_i. In the *alternative* definition of μ_i, the concept of fugacity of i in the fluid mixture, f'_i, is introduced in order to combine both steps of energy changes,

$$\mu_i = G^\circ_{f,T,i} + RTln\frac{f''_i}{f^\circ_i} .$$

(3.9)

Like the fugacity of pure i, the fugacity of the species i in the fluid mixture, f'_i, is given in bar. Introducing the dimensionless quantity f^∇_i $(\equiv f_i''/f^\circ_i)$, (3.9) simplifies to

$$\mu_i \equiv G^\circ_{f,T,i} + RTlnf^\nabla_i.$$

(3.9a)

Combining Eqs.(3.9), (3.9a), (3.9b) and (3.4a), we may deduce

$$f''_i = f_i\, a_i, \text{ and}$$

(3.10a)

$$f^\nabla_i = f^*_i a_i.$$

(3.10b)

The definition of activity of i in the fluid mixture at any P, T, and X_i, a_i, follows from (3.10a) and (3.10b),

$$a_i \equiv \frac{f''_i}{f_i} = \frac{f^\nabla_i}{f^*_i} .$$

(3.10)

Activity of i is the ratio of the fugacity of i in a fluid mixture to that of the pure fluid i at that P and T.

It is clear from the foregoing derivations that the chemical potential of the i-th fluid may be alternatively written as (1.62) *or* (3.9), depending upon the choice of the standard state. In the former case, the standard state chosen is one of unit activity of the pure fluid at P and T; in the latter, it is the hypothetical ideal gas at 1 bar pressure. It must be emphasized that the choice of a standard state is entirely arbitrary; the chemical potential, nevertheless, is uniquely determined regardless of which standard state is adopted. Finally,

(3.10a) helps derive an expression for the fugacity coefficient of i in the fluid mixture, $\phi''_i(T,P,X_i)$. Substituting Eqs.(3.6) and (1.61) into the RHS of Eq.(3.10a), we have

$$f''_i = P\phi_i X_i \gamma_i. \tag{3.11}$$

Note that for $X_i = 1$ (i.e. $\gamma_i = 1$), f''_i reduces to f_i as in (3.6). Rearrangement of Eq.(3.11) yields

$$\frac{f''_i}{PX_i} = \phi_i \gamma_i \equiv \phi''_i, \quad \text{or} \tag{3.12a}$$

$$\frac{f''_i}{p_i} \equiv \phi''_i, \tag{3.12b}$$

p_i being the partial pressure of i. Since all gas mixtures become ideal solutions as $P \to 0$, we have

$$\lim_{P \to 0} \frac{f''_i}{p_i} = \lim_{P \to 0} \phi''_i = 1. \tag{3.12}$$

This additional stipulation completes the definition of f''_i, given in Eq.(3.9).

Before passing on to the next section, we emphasize that the concept of fugacity of i, developed here in the context of a fluid mixture, also applies to solids. Thus, the fugacity of i in a pure solid, or that of the component i in any crystalline solution, is its fugacity in the fluid in equilibrium with the solid.

3.3 Utility of the Volume Equations of State

Having outlined the concepts of fugacity and activity, we may explore the techniques necessary to evaluate them. The computation of the fugacity of the pure fluid i, $f^*_i(=f_i/f^\circ_i)$, at any T and P, could be executed using (3.4). This demands that a volume equation of state for the pure fluid i be available. On the other hand, the calculation of the activity of the species i in any fluid mixture requires that a volume equation of state for the fluid mixture be available. To appreciate this, recall the equation

$$RT\ln\gamma_i = \mu_i - \mu_i^{id} = g^{ex}_{m,i}. \tag{1.67}$$

At any T and X_i, in analogy to Eq.(1.9), we have

$$\left(\frac{\partial g^{ex}_{m,i}}{\partial P}\right)_{T,X_i} = v^{ex}_{m,i}. \tag{3.13}$$

Integrating,

$$g^{ex}_{m,i} = \int_0^P v^{ex}_{m,i}|_{T,X_i}(P)dP = RT\ln\gamma_i. \tag{3.14}$$

Since $V^{id}_m = 0$ [Eq.(1.80)], $v^{ex}_{m,i}$ may be recast as $v_i - V_i$, where v_i is the partial molar volume of i in the nonideal fluid mixture at T, P, and X_i, and V_i is the molar volume of pure i at those T and P. Inserting this into (3.14), the equation employed to obtain $RTln\gamma_i$ is recovered,

$$RTln\gamma_i = \int_0^P (v_i - V_i)_{T,X_i}(P)dP. \tag{3.15}$$

The activity of i, a_i, then follows from (3.15) and (1.61). These considerations suffice to drive home the fundamental importance of the volume equations of state,

$$V = f(T,P,X_i), \tag{3.16}$$

for the thermodynamic treatment of fluids and fluid mixtures.

To be ideally suitable for geochemical applications, a volume equation of state should fulfill the following criteria:

1. The functional form of an equation of state should have a theoretical justification. That is best achieved by a molecular thermodynamic approach (see Prausnitz et al. 1986), a subject way beyond the scope of this book.

2. The equations of state for pure fluids should accurately reproduce experimental P-V-T data both in sub- and supercritical ranges of T and P, and permit extrapolations beyond the ranges of experimental data. Their validity to pressures of 100 kbar or more is sometimes required for geochemical calculations.

3. The pure fluid equations of state must be extendable to fluid mixtures. The latter should be capable of providing reliable a_i-X_i data for T and P of interest to geochemistry. It should also predict phase separation at low temperatures, as is common among geochemically relevant fluid mixtures.

For now the bottom line is that no equation of state proposed thus far can meet this tall order. Because the modified Redlich-Kwong (MRK) equation serves our purpose reasonably well, we shall emphasize it later on.

3.4 Volume Equations of State for Pure Fluids

3.4.1 General Comments and Literature Overview

For a pure fluid, volume is a function of T and P only,

$$V = f(T,P). \tag{3.16a}$$

In practice, the function (3.16a) is always derived by smoothing experimental P-V-T data. Geochemists have tried four different approaches to achieve this goal.

1. Smoothing of experimental P-V-T data by arbitrary polynomials.

The most notable effort belonging to this category is that by Burnham et al. (1969a), who employed arbitrary polynomials in T and P of the form

$$PV_{H_2O}(T,P) = \sum_{j=0}^{m} \sum_{i=0}^{m-j} c_{ij} T^i P^{r \cdot j - 1} , \tag{3.17}$$

to smooth their own (Burnham et al. 1969b) V_{H2O} data in the range 20 to 925°C at pressures up to 8500 bar, and to extrapolate them over the T-P range 0 to 1000°C and 1000 to 10000 bar. Different degrees of polynomials m and coefficients c_{ij} had to be employed in three subregions of the T-P space, along with a switch variable r alternatively assuming the values +1 and -1 in those subregions. These results were subsequently extended to 100 kbar pressure by Delany and Helgeson (1978) on the basis of V_{H2O}-data computed from shock wave experiments of Rice and Walsh (1957). Again, a similar polynomial was used for data smoothing in the 10-100 kbar range.

Besides being straightforward, the greatest advantage of data smoothing by arbitrary polynomials is that it permits an excellent control on the overall error within the P-T range of experimental P-V-T measurements. The drawbacks are: 1. Extrapolation beyond the range of experimental P-V-T data is not possible. 2. Thermodynamic treatment of a pure fluid can barely be extended to that of fluid mixtures.

The above equations of state by Burnham et al. (1969a) and by Delany and Helgeson (1978) were utilized by geochemists for a decade, until the modified Redlich-Kwong (MRK) equation, described later in considerable detail, more or less phased it out.

2. The virial equation of state.

In the virial equation of state, the compressibility factor, Z ($\equiv PV/RT$), of a fluid is rendered as a power series in pressure (or density),

$$Z = 1 + BP + CP^2 + DP^3 + ..., \tag{3.18}$$

where the coefficients B, C, etc are functions of temperature, but independent of pressure (or density). The coefficients for a pure fluid are obtained from the P-V-T data (Prausnitz et al. 1986). Truncated after the third term, this power series often provides a remarkably good representation of the P-V-T data for a variety of fluids to approximately one-half of their critical densities. Most mineral equilibria computations, however, require knowledge of the P-V-T properties of fluids at pressures of several kbar, where far higher fluid densities are encountered. Consequently, the virial equation is of limited use to geochemistry. The main advantages of using the virial equation are: 1. It has a very sound theoretical basis, and can be derived from the first principles.

2. It can be readily extended to deal with the fluid mixtures. An example of a geochemical application of the virial equation is that by Spycher and Reed (1988), who used it to model hydrothermal boiling. They observed that their virial equation is not meant for extrapolation beyond the range of the experimental P-V-T data employed to derive it, and that it fails to reproduce the P-V-T properties of fluids close to the critical point.

3. P-V-T correlation from the theorem of corresponding states.

A large body of experimental P-V-T data is a prerequisite to fit arbitrary polynomials or a virial equation of state. For a few fluids, the experimental P-V-T database is meager. In those cases, the theorem of corresponding states looks advantageous. It asserts that for *all* fluids there exists a relation of the form

$$V_r = f(T_r, P_r),$$ (3.16b)

where V_r ($\equiv V/V_c$), T_r ($\equiv T/T_c$), and P_r ($\equiv P/P_c$) denote the reduced variables, V_c, T_c, and P_c, the critical quantities of the relevant fluids. The implication is evident: at any T_r and P_r, *every* fluid exhibits an identical degree of nonideality. Were this strictly true, derivation of Eq. (3.16b) on the basis of the measured P-V-T data for any one fluid would have sufficed; P-V-T work on all other fluids would have become redundant, except for determining their critical conditions.

Several attempts have been made by geochemists to apply this theorem. Recently, Saxena and Fei (1987a) employed this concept to derive the following corresponding state equation, reminiscent of a virial equation (except for $A \neq 1$):

$$Z = A + BP_r + CP_r^2$$ (3.19)

Again, Z is the compressibility factor, and the coefficients A, B, and C are functions of T_r. They pointed out that (3.19) is *not* of universal validity; it succeeds in reproducing the P-V-T data for H_2, O_2, CO, CO_2, CH_4, and N_2, but fails to do so for H_2O. That is an immediate consequence of the fact that as $T \rightarrow T_c$ and $P \rightarrow P_c$, so does $Z \rightarrow Z_c$. Stated otherwise, for (3.19) to be strictly applicable, all fluids must have an identical Z_c. This, however, is not true. While Z_c of H_2, O_2, CO, CH_4, and N_2 is indeed tightly grouped near 0.295, that of CO_2 is 0.276. The Z_c of H_2O is yet further removed, being only 0.235. No wonder that (3.19), tailored to fit the other fluids, fails to reproduce the P-V-T properties of H_2O. Although Saxena and Fei (1987a) devised the above equation of state for the crustal T-P conditions, they observed that it does not apply below 400 K and 1 kbar. For $P \leq 1$ kbar, they advanced another equation:

$$Z = 1 + B'P_r + C'P_r^2$$ (3.19a)

In a second publication, Saxena and Fei (1987b) presented yet another equation of state specifically designed to be valid up to very high temperatures and pressures like 3000 K and 1 Mbar,

$$Z = A + BP_r + CP_r^2 + DP_r^3. \tag{3.20}$$

Two sets of temperature-dependent coefficients had to be derived; one applies to Ar, Xe, H_2, O_2, CO, CO_2, CH_4, and N_2, the other to H_2O. These equations of state clearly open up new perspectives for the computation of mineral equilibria in the earth's mantle.

4. Reproduction of P-V-T data by two-parameter equations of state

In recent years, a two-parameter equation of state, known as the modified Redlich-Kwong equation (MRK), has been used with some success for fitting P-V-T data of geochemically relevant fluids. Given a variety of bewilderingly complex equations of state (Reid et al. 1977, pp.36-55), none of which can meet all the criteria noted above (see Sect.3.3), even a partial success is encouraging.

Application of MRK to geochemistry was championed by Holloway (1977). He proved the feasibility of using it in the supercritical range of pure fluids and fluid mixtures. Following his cue, other workers including Bottinga and Richet (1981), Jacobs and Kerrick (1981a,b), and Kerrick and Jacobs (1981) proposed MRKs for CO_2, CH_4, and H_2O. All these efforts had one thing in common; they were predominantly geared to apply to temperatures above 350 to 400°C, i.e. at $T > T_c$ of H_2O. Halbach and Chatterjee (1982a) attempted to extend the MRK of H_2O to the subcritical region.

The advantages of MRK are: 1. It can faithfully reproduce the measured P-V-T data over a large range of T and P. 2. *If* a simple functional form is chosen to render the T-P dependences of its two parameters, it permits extrapolation of V to P and T outside the range of experimental P-V-T data. 3. It can be extended to handle fluid mixtures. Nevertheless, the MRK has been criticized because: (1) it lacks a rigorous theoretical foundation, and (2) due to the complex T-P-dependences of its empirical parameters, its numerical manipulation is rather cumbersome.

The ultimate aim is the development of equations of state based on molecular thermodynamics.[2] If successful, only they could be expected to fulfill the formidable list of criteria listed in Section 3.3. Although some progress has been made in recent years, much remains to be done until intermolecular forces operating in fluids can be quantitatively linked to their macroscopic thermodynamic properties (Prausnitz et al. 1986, p.48). For now, the MRK appears to hold the winning edge for geochemical applications. Therefore, we now turn to MRK and other related two-parameter equations.

---

[2] Molecular thermodynamics combines concepts of chemical physics with those of statistical and classical thermodynamics to deal with the properties of molecular mixtures. For a review of recent developments in that field, see Prausnitz et al. (1986).

3.4.2 Some Examples of Analytical Equations of State

The fundamental purpose of an analytical equation of state is to mathematically express the function $V = f(T,P)$. The ideal gas equation is the simplest of all analytical equations of state. To deal with nonideal fluids, the ideal gas equation is expanded by introducing two or more adjustable parameters. Examples of two-parameter equations of state are the van der Waals (vdW) and the Redlich-Kwong (RK) equations. In this section, we shall explore three analytical equations of state: the ideal gas, the vdW, and the RK equations. This will pave the path for understanding the advantages and limitations inherent to the modified Redlich-Kwong equation (MRK). The latter, popular among geochemists, will be treated in detail in the next section.

1. The ideal gas equation

The ideal gas equation is the simplest equation of state; it lacks adjustable parameters

$$P = \frac{RT}{V^{id}} . \tag{3.21}$$

Because *no* gas behaves ideally under geological conditions, it can not be applied to phase equilibria calculations. Nevertheless, it does represent a very useful limiting case, because all nonideal gases behave ideally as $P \rightarrow 0$. Consequently, all equations of state for nonideal gases must reduce to the ideal gas equation for $P \rightarrow 0$.

2. The van der Waals (vdW) equation

All two-parameter equations of state stipulate that P can be split into two parts, one due to the repulsive (P_R) and the other due to the attractive (P_A) interaction between the molecules,

$$P = P_R + P_A. \tag{3.22}$$

The vdW equation is the simplest of all two-parameter equations,

$$P = \frac{RT}{V - b} - \frac{a}{V^2} , \tag{3.23}$$

a and **b** denoting two specific constants, characterizing each fluid species. Comparing (3.22) and (3.23), it is clear that **a** and **b** are related, respectively, to the attractive and repulsive molecular interactions. The vdW equation only qualitatively accounts for the observed P-V-T properties of real fluids over a rather limited T-P range. Of greater relevance to us is the fact that vdW is a cubic equation in volume, as may be seen by recasting Eq.(3.23) as

$$V^3 - \left[b + \frac{RT}{P}\right]V^2 + \left[\frac{a}{P}\right]V - \frac{ab}{P} = 0.$$ (3.24)

Therefore, it will have one or three real roots depending upon the number of phases in the system. At $T \geq T_c$, only one supercritical fluid exists, and we obtain a unique solution for V. However, at $T < T_c$, three solutions for V are obtained, of which the largest and smallest refer to V_{gas} and V_{liquid}, respectively. Clearly, the vdW can handle both the gas and the liquid states. Furthermore, it can also be utilized to represent the isothermal P-V relations for the hypothetical continuous sequence of homogeneous states between the stable liquid and the stable gas for $T < T_c$ (van der Waals 1881). We shall have to come back to that later on.

An equation of state for a real fluid must also fulfill the two thermodynamic criteria of stability at the critical point,

$$\left(\frac{\partial P}{\partial V}\right)_{T_c} = 0 \quad \text{and}$$ (3.25)

$$\left(\frac{\partial^2 P}{\partial V^2}\right)_{T_c} = 0.$$ (3.26)

The consequences of applying (3.25) and (3.26) to (3.23), the vdW equation, are of considerable interest to us. Therefore, we take a careful look at them, following the treatment given by Abbott and van Ness (1972, p.155). Starting with Eq.(3.23), and taking the proper derivatives,

$$\left(\frac{\partial P}{\partial V}\right)_{T_c} = \frac{-RT_c}{(V_c - b_c)^2} + \frac{2a_c}{V_c^3} = 0 \quad \text{and}$$ (3.27)

$$\left(\frac{\partial^2 P}{\partial V^2}\right)_{T_c} = \frac{2RT_c}{(V_c - b_c)^3} - \frac{6a_c}{V_c^4} = 0,$$ (3.28)

a_c and b_c denoting a and b at the critical point. The solution of these identities for a_c and b_c yields,

$$a_c = \frac{9RT_c V_c}{8} \quad \text{and}$$ (3.29)

$$b_c = \frac{V_c}{3}.$$ (3.30)

Inserting (3.29) and (3.30) back into (3.23), we have

$$P_c = \frac{RT_c}{V_c - \dfrac{V_c}{3}} - \frac{\dfrac{9RT_cV_c}{8}}{V_c^2} = \frac{3RT_c}{8V_c},$$ (3.31)

which, in turn, may be rearranged to yield

$$Z_c = \frac{P_cV_c}{RT_c} = \frac{3}{8} = 0.375, \quad \text{and}$$ (3.32)

$$V_c = \frac{3RT_c}{8P_c}.$$ (3.33)

Substituting (3.33) into (3.29) and (3.30), the latter two may be rewritten as

$$a_c = \frac{27R^2T_c^2}{64P_c} \quad \text{and}$$ (3.34)

$$b_c = \frac{RT_c}{8P_c}.$$ (3.35)

Thus, merely by knowing T_c and P_c of a fluid, we can solve for a_c and b_c. For example, using its T_c and P_c (Angus et al. 1976), we derive the following a_c and b_c for CO_2,

$$a_c(CO_2) = 3.6558E6 \text{ [(cm}^3 \text{ mol}^{-1})^2 \text{ bar] and}$$ (3.36)

$$b_c(CO_2) = 42.826 \text{ (cm}^3 \text{ mol}^{-1}).$$ (3.37)

Since **a** and **b** are constants for each fluid in the vdW equation, we may then proceed to calculate its V at any desired set of T and P simply by inserting into Eq.(3.24) $a_c = a$ and $b_c = b$.

From the above, it is also evident that the prerequisites for the correct estimations of a_c and b_c are not only the validity of (3.25) and (3.26), but also that of (3.32), requiring that Z_c for the vdW equation be *unique*, equaling 0.375. The experimentally observed Z_c, however, differs from fluid to fluid (it is 0.295 for H_2, O_2, CO, CH_4, N_2; 0.276 for CO_2; 0.235 for H_2O), and moreover, it is far from the theoretical value of 0.375. The vdW equation is thus likely to perform poorly near the critical point. Indeed, all other *two-parameter* equations do the same, since none can satisfy these *three* conditions intrinsically valid at the critical point.

There is yet another factor exacerbating the poor performance of the vdW equation. That is easily appreciated by considering the behavior of the parameters **a** and **b** along the saturation curve[3] of any pure fluid. At each T

---

[3] The saturation curve is the univariant P-T curve for the liquid-gas equilibrium; it extends from the triple point (the three-phase solid-liquid-gas coexistence) to the critical point (where gaseous and liquid states are no longer distinguishable).

and P on the saturation curve, denoted here as T_{sat} and P_{sat}, two identities hold simultaneously,

$$P_{sat} = \frac{RT_{sat}}{V_{liq} - \mathbf{b}_{sat}} - \frac{\mathbf{a}_{sat}}{V_{liq}^2} \quad \text{and} \tag{3.38a}$$

$$P_{sat} = \frac{RT_{sat}}{V_{gas} - \mathbf{b}_{sat}} - \frac{\mathbf{a}_{sat}}{V_{gas}^2}, \tag{3.38b}$$

V_{liq} and V_{gas} being the molar volumes of the coexisting liquid and gas, respectively. These two equations can be solved for the two unknowns, $\mathbf{a}_{sat}$ and $\mathbf{b}_{sat}$ (a and b on the saturation curve), for any pair of T_{sat} and P_{sat}. This has been done for CO_2; the results are listed in Table 3.1. As we proceed along the saturation curve, V_{liq} and V_{gas} approach each other, until they become identical at the critical point. In other words, we have only *one* equation like (3.38) *at* the critical point, and therefore, a direct solution for both $\mathbf{a}_{sat}$ and $\mathbf{b}_{sat}$ is no longer possible. However, as $T_{sat} \rightarrow T_c$ and $P_{sat} \rightarrow P_c$, we also have $\mathbf{a}_{sat} \rightarrow \mathbf{a}_c$ and $\mathbf{b}_{sat} \rightarrow \mathbf{b}_c$; these limiting values of $\mathbf{a}_c$ and $\mathbf{b}_c$ are also displayed in Table 3.1.

Two facts emerge from these results. First, the parameters a and b are *not* constants after all; they are clearly functions of T and P, a(T,P) and b(T,P). No wonder that the reproduction of the measured P-V-T data employing the vdW equation remains very crude as long as a (= $\mathbf{a}_c$) and b (= $\mathbf{b}_c$) are treated

Table 3.1. The van der Waals $\mathbf{a}_{sat}$ and $\mathbf{b}_{sat}$ parameters for CO_2 as functions of T_{sat} and P_{sat}, computed from the P-V-T data of CO_2 in Angus et al. (1976). The last line of the table lists $\mathbf{a}_c$ and $\mathbf{b}_c$

T_{sat} (K)	P_{sat} (bar)	$\mathbf{a}_{sat}/10^6$ $((cm^3 mol^{-1})^2 \cdot bar)$	$\mathbf{b}_{sat}$ $(cm^3 mol^{-1})$
220.0	5.996	5.6464	33.131
230.0	8.935	4.9760	33.155
240.0	12.830	4.6869	33.480
250.0	17.856	4.4665	33.880
260.0	24.194	4.2643	34.312
270.0	32.034	4.0742	34.786
280.0	41.773	3.8631	35.233
290.0	53.152	3.6974	35.972
300.0	67.095	3.4729	36.883
302.0	70.220	3.4158	37.169
303.0	71.830	3.3826	37.361
304.0	73.475	3.3413	37.670
304.2 (T_c)	73.825 (P_c)	3.3303 ($\mathbf{a}_c/10^6$)	37.884 ($\mathbf{b}_c$)

as constants. If the vdW equation is to hold over a large P-T range, it must be modified to express **a** and **b** as functions of T and P. A good example of such a modified vdW equation is the one formulated for H_2O by Juza (1966). Secondly, the values of $\mathbf{a}_c$ and $\mathbf{b}_c$ obtained by the two different methods are contradictory; those given in Table 3.1 are quite different from the ones obtained earlier [Eqs.(3.36) and (3.37)] from the T_c and P_c data. Equally contradictory results are obtained when fluids other than CO_2 are used. This testifies that the two-parameter equations are fundamentally inadequate to handle the critical phenomenon.

3. The Redlich-Kwong (RK) equation

It was proven earlier that the performance of a two-parameter equation in the vicinity of the critical point highly depends upon its characteristic Z_c value. Therefore, to improve a two-parameter equation, its functional form must be chosen in such a manner that its characteristic Z_c approaches those observed in the majority of the fluids. Redlich and Kwong (1949) partially succeeded in doing this with their equation

$$P = \frac{RT}{V-b} - \frac{a}{V(V+b)T^{0.5}} . \tag{3.39}$$

Like the vdW equation, the RK equation has two species-specific parameters **a** and **b**. Although Redlich and Kwong (1949) emphasized that (3.39) is an empirical equation, the parameters **a** and **b** have the same physical significance as in the vdW equation. Indeed, much of what has been stated earlier regarding the vdW equation, holds also for the RK (and all other two-parameter) equations. Applying the criteria of stability at the critical point [Eqs. (3.25) and (3.26)] to (3.39), and proceeding in a manner analogous to the above treatment of the vdW equation, it may be demonstrated that for the RK equation,

$$\mathbf{a}_c = \frac{RT_c^{1.5}V_c}{3(2^{\frac{1}{3}}-1)} \quad \text{and} \tag{3.40}$$

$$\mathbf{b}_c = (2^{\frac{1}{3}}-1)V_c . \tag{3.41}$$

Substituting (3.40) and (3.41) into (3.39), it follows that

$$Z_c = \frac{P_cV_c}{RT_c} = \frac{1}{3} = 0.333, \quad \text{and} \tag{3.42}$$

$$V_c = \frac{RT_c}{3P_c} . \tag{3.43}$$

Inserting (3.43) into (3.40) and (3.41), we recover the equations

$$\mathbf{a_c} = \frac{R^2 T_c^{2.5}}{9(2^{\frac{1}{3}} - 1)P_c} \quad \text{and} \tag{3.44}$$

$$\mathbf{b_c} = \frac{(2^{\frac{1}{3}} - 1)RT_c}{3P_c} . \tag{3.45}$$

Once again, taking CO_2 as an example, and using its critical point data from Angus et al. (1976), we have

$$\mathbf{a_c}(CO_2) = 64.6097E6 \; [(cm^3 \; mol^{-1})^2 \; K^{0.5} \; bar] \; and \tag{3.46}$$

$$\mathbf{b_c}(CO_2) = 29.684 \; (cm^3 \; mol^{-1}). \tag{3.47}$$

Employing these values as the **a** and **b** parameters for the RK equation, V_{CO2} can be calculated at various T and P and compared with the experimental P-V-T data and with those calculated earlier from the vdW equation. Comparison of vdW and RK equations applied to various fluids clearly shows the improved performance of the RK equation (see Redlich and Kwong 1949, Fig. 1). This is expected because its characteristic Z_c (0.333) is in better agreement with the experimentally observed Z_c of various fluids species.[4]

Another aspect of the RK equation must be emphasized. Like the vdW equation, it is also a cubic equation in V:

$$[PT^{0.5}]V^3 - [RT^{1.5}]V^2 + [\mathbf{a} - RT^{1.5}\mathbf{b} - \mathbf{b}^2 T^{0.5}P]V - \mathbf{ab} = 0. \tag{3.48}$$

Consequently, it will yield three real solutions for V at $T < T_c$, and only one at $T \geq T_c$. By implication, it applies to the gaseous and the liquid states, in addition to the hypothetical sequence of homogeneous states intermediate between them (see Fig. 3.2).

Next, let us examine to what extent **a** and **b** are constants, as suggested by the original RK equation. For this, we again analyze the P-V-T data for the saturation curve of a fluid. To compare the results with those obtained above for the vdW equation, the P-V-T data for CO_2 (Angus et al. 1976) were used again. The results are listed in Table 3.2.

---

[4] The two-parameter equation proposed by Peng and Robinson (1976),

$$P = \frac{RT}{V-b} - \frac{a(T)}{V(V+b) + b(V-b)},$$

has a theoretical Z_c of 0.307, better approaching the experimental values. Evidently, it is a good candidate for use in geochemistry. Heyen (1980) modified the Peng-Robinson equation by introducing a third parameter **e**, to reproduce the experimental Z_c values. Such a three-parameter equation might prove quite promising. Having three parameters, such an equation could potentially meet all the three conditions [cf. Eqs.(3.25), (3.26), (3.32)] applying to a critical point.

Table 3.2. The Redlich-Kwong a_{sat} and b_{sat} parameters for CO_2 as functions of T_{sat} and P_{sat}, computed from the P-V-T data of CO_2 in Angus et al. (1976). The last line of the table lists a_c and b_c

T_{sat} (K)	P_{sat} (bar)	$a_{sat}/10^6$ $((cm^3\ mol^{-1})^2 \cdot K^{0.5} \cdot bar))$	b_{sat} $(cm^3\ mol^{-1})$
220.0	5.996	83.6387	29.526
230.0	8.935	75.3085	28.846
240.0	12.830	72.5713	28.612
250.0	17.856	70.7880	28.448
260.0	24.194	69.2028	28.278
270.0	32.034	67.7604	28.110
280.0	41.773	65.8895	27.835
290.0	53.152	64.9751	27.842
300.0	67.095	63.2931	27.868
302.0	70.220	62.8750	27.942
303.0	71.830	62.6432	28.012
304.0	73.475	62.3920	28.166
304.2 (T_c)	73.825 (P_c)	62.3865 ($a_c/10^6$)	28.298 (b_c)

    Two conclusions emerge from the data displayed in Table 3.2:
1. The **a** and **b** parameters of the RK equation are not constants; they are clearly functions of T and P. A good reproduction of the P-V-T data by the RK equation would require that $a(T,P)$ and $b(T,P)$ be employed. For this purpose, the T-P dependences of **a** and **b** have to be extracted first from experimental P-V-T data. That is what the MRK does (see Sect.3.4.3). Thus, by its very nature, an MRK is essentially an empirical equation of state.
2. To the extent that two-parameter equations are to be used for rendering P-V-T properties of pure fluids, the RK equation should be preferred to vdW. However, no amount of fine-tuning of the $a(T,P)$ and $b(T,P)$ functions can alleviate the problem of poor performance in the vicinity of the critical point.

3.4.3 A Modified Redlich-Kwong (MRK) Equation of State for H_2O

In Redlich and Kwong's (1949) original proposal, the equation

$$P = \frac{RT}{V-b} - \frac{a}{V(V+b)T^{0.5}} \tag{3.39}$$

was designed to be operative under supercritical conditions, **a** and **b** being constants for each specified fluid. It is amply clear from the above expositions why such an equation can apply to a limited range of T and P. To be of use to geochemistry, however, an equation of state must cover a large T-P range, including both the supercritical and the subcritical regions. This goal is achieved, more or less satisfactorily, by expressing **a** and **b** as functions of T and P. A Redlich-Kwong equation, in which **a** and **b** are no longer constants, is dubbed as a modified Redlich-Kwong equation, MRK. The objective of this section is to briefly elucidate three MRKs developed by geochemists for H_2O. The first two were designed only for its supercritical state, while the last attempted an extension to its subcritical range.

Holloway (1977) introduced the MRK for pure fluids and fluid mixtures to geochemistry. Relying on the molecular thermodynamic arguments of de Santis et al. (1974), he treated **a** as a function of T, and **b** as a constant. While processing the P-V-T data of H_2O to derive **a**(T), he observed that **a** ceases to be a smooth function of T in the vicinity of the critical point. This made him restrict his MRK to the supercritical region, recommending its application to T between 450-1800°C and P between 0.5-40 kbar. He also pointed out the advantage of keeping an MRK functionally simple; only they can be extrapolated beyond the range of the P-V-T measurements.

Kerrick and Jacobs (1981), in a study of the P-V-T properties of supercritical H_2O, CO_2, and H_2O-CO_2 mixtures, devised their MRK on the basis of Carnahan and Starling's (1969) "hard-sphere"[5] version of MRK. Carnahan and Starling's "hard-sphere" MRK is written as

$$P = \frac{RT(1 + y + y^2 - y^3)}{V(1 - y)^3} - \frac{a(T)}{V(V + b)T^{0.5}} , \qquad (3.49a)$$

where y = **b**/4V, **b** being a specific constant and **a** a function of T. Kerrick and Jacobs (1981) observed that the fit to the P-V-T data for H_2O and CO_2 can be improved beyond that achieved by Holloway (1977), if Eq.(3.49a) is still further modified by expressing **a** as a function of T and P (or V). Thus, Kerrick and Jacobs' MRK assumes the form

$$P = \frac{RT(1 + y + y^2 - y^3)}{V(1 - y)^3} - \frac{a(T, P)}{V(V + b)T^{0.5}} . \qquad (3.49)$$

Halbach and Chatterjee (1982a) attempted to extend the MRK to cover both the supercritical and the subcritical region of H_2O. A systematic analysis of all accurate P-V-T data for H_2O revealed that **a** is primarily a function of T (and to a lesser extent of P), while **b** is a monotonous function of P, with little T-dependence.[6] Eager to keep the MRK functionally simple, guaranteeing a reliable extrapolation outside the range of experimental P-V-T data, they chose to ignore the P-dependence of **a**, and formulated the MRK with **a**(T)

---

[5] Carnahan and Starling (1969) derived their hard sphere model by modifying the repulsive term of the original RK equation. In this model, the gas molecules are handled as if they behave like "hard" (rigid) spheres.
[6] In all previous versions of the MRK, **b** was a constant. Making **b** a function of P would imply that the molecules no longer behave as "rigid spheres" at sufficiently high pressures. The high-pressure P-V-T data corroborate this notion.

and $b(P)$:

$$P = \frac{RT}{V - b(P)} - \frac{a(T)}{V[V + b(P)]T^{0.5}} , \qquad (3.50)$$

where $a(T) = A_1 + A_2T + A_3/T$, and
$b(P) = (1 + B_1P + B_1P^2 + B_3P^3)/(B_4 + B_5P + B_6P^2)$,

the individual adjustable parameters being,

$A_1 = 1.616E8$ $B_1 = 3.4505E-4$ $B_4 = 6.3944E-2$
$A_2 = -4.989E4$ $B_2 = 3.8980E-9$ $B_5 = 2.3776E-5$
$A_3 = -7.358E9$ $B_3 = -2.7756E-15$ $B_6 = 4.5717E-10$.

The reproduction of the P-V-T data was deemed satisfactory for the temperature range 100-1000°C, all the way up to 200 kbar pressure, except for a P-T envelope along the saturation curve, and in its continuation beyond the critical point. Figure 3.1 illustrates two calculated isotherms with V_{H2O}

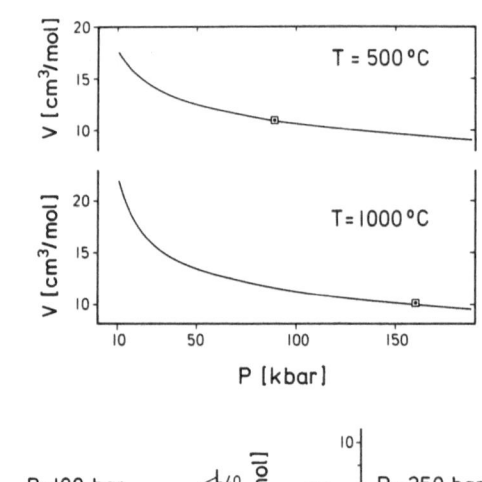

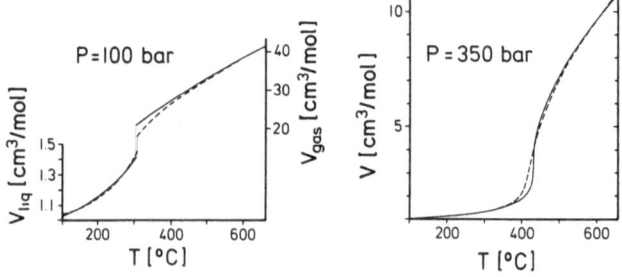

Fig. 3.1 The V_{H_2O} vs P curves at 500 and 1000°C (*above*), based on the MRK of Halbach and Chatterjee (1982a), compared with V_{H_2O} observed in shock experiments (*dots inside box symbols*). Note the excellent agreement between the experiment and the prediction. The figures below depict isobaric plots of V_{H_2O} vs T calculated by the MRK (*solid lines*), and compare them with the steam table (Schmidt 1979) data (*dashed lines*). The agreement between these two is less satisfactory, in particular, in the 300-500°C range

values superimposed from the shock experiments. An excellent agreement between the two vindicates the extrapolability of this MRK to pressures on the order of 200 kbar. A pair of isobaric V_{H2O} vs T°C plots, appearing in the lower half of Fig. 3.1, demonstrates that computed V_{H2O} is less satisfactory near the saturation curve, in particular, at T > 300°C.

In principle, a further improvement of this MRK is possible, especially for the low pressure region. This goal is achieved by rendering both **a** and **b** as functions of T and P. However, the only way to do this, without impairing the extrapolability of the MRK, is to derive an additional set of **a**(T,P) and **b**(T,P) parameters for the *residuals* of V_{H2O} near the saturation curve, and to impose it on the base function, **a**(T) and **b**(P). Every such attempt, however, will be subject to the law of diminishing returns.

3.4.4 Calculation of Phase Relations and Thermodynamic Properties of H₂O from MRK

The objective of this section is to provide insight into how an MRK for H_2O may be utilized to compute the phase relations and some thermodynamic properties of H_2O. While the calculation of the saturation curve, i.e. the liquid-gas phase relation, is described only briefly, the derivation of the fugacity will be presented in detail due to its fundamental importance in computing solid-fluid phase equilibria.

3.4.4.1 Calculation of the Saturation Curve of H₂O

The first step in calculating the saturation curve is to pick any one temperature in the subcritical range, and to solve for the corresponding saturation pressure. The saturation curve itself is then derived by repeating this process at several temperatures. An example may serve to illustrate the procedure.

Suppose we need to calculate the saturation pressure at 270°C. The initial step is to obtain the P-V isotherm at 270°C, combining Eqs.(3.50) and (3.48). Figure 3.2 displays this isotherm. The saturation pressure, P_{sat}, is then solved for by the identity

$$P_{sat}(V_{gas} - V_{liq}) = \int_{V_{liq}}^{V_{gas}} P_T(V)dV. \qquad (3.51)$$

The inset in Fig. 3.2 shows the solution to (3.51) schematically. P_{sat} is the pressure corresponding to the points F and B, the latter two indicating the molar volumes, V_{liq} and V_{gas}, respectively. The fulfillment of (3.51) is achieved when the areas F_1 and F_2 are mutually equal. This criterion is known as Maxwell's equal area rule (van der Waals 1881, p.92; Abbott and van Ness 1972, p.171).

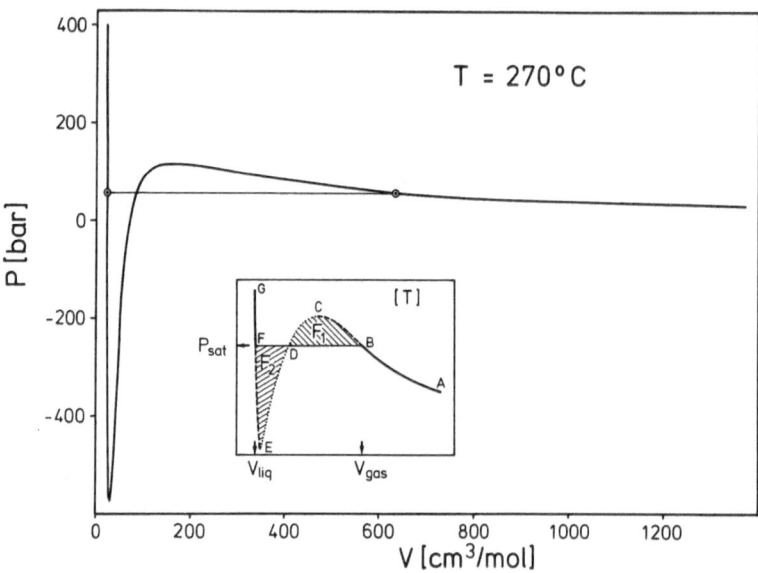

Fig. 3.2 The 270°C P-V_{H_2O} isotherm, calculated from the MRK for pure H_2O. The volumes of the liquid and gaseous H_2O coexisting on the saturation curve are indicated by *small circles*. The schematic diagram in the *inset* has been explained in the text

Before proceeding further, we clarify the significance of the dashed and dotted parts of the P-V curve in the inset of Fig. 3.2. The segment B-C-D-E-F depicts the hypothetical continuous sequence of homogeneous states intermediate between a stable gas (the solid curve A-B) and a stable liquid (the solid curve F-G). As indicated later (Fig. 3.4), the dashed and the dotted segments represent two different levels of metastabilities with respect to the stable gaseous and liquid states.

The practical procedure to solve for P_{sat} via Eq. (3.51) involves an execution of the ∫PdV integral, which is best done numerically. The integration in (3.51), however, requires that P be expressed as a function of V prior to computing the area under the P-V curve, a procedure that might add to the uncertainty of the computed value of P_{sat}. To avoid this, we shall avail ourselves of an alternative method, in which the variables are exchanged before the integration is performed. The exchange of variables for the P-V isotherm at 270°C is displayed in Fig. 3.3; referring to it, we may write

$$F_1 = -\int_{P_D}^{P_C} V_T(P)dP - \int_{P_C}^{P_B} V_T(P)dP, \quad \text{and} \tag{3.52a}$$

$$F_2 = \int_{P_F}^{P_E} V_T(P)dP + \int_{P_E}^{P_D} V_T(P)dP, \tag{3.52b}$$

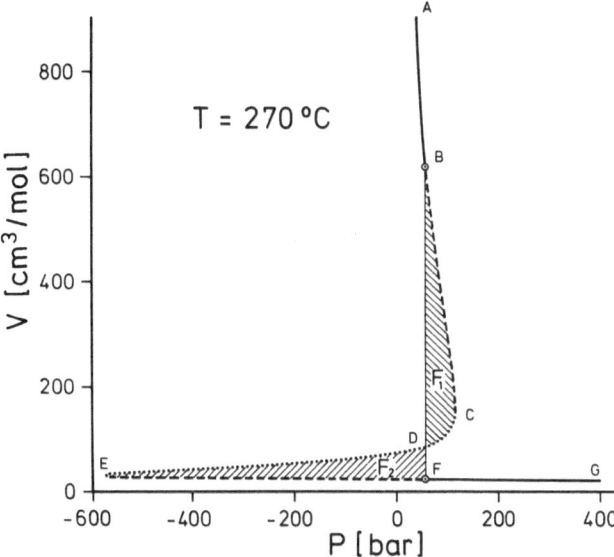

Fig. 3.3 The 270°C P-V$_{H_2O}$ data of Fig. 3.2 replotted with the variables exchanged. The *solid lines* A-B and *F-G* denote the stable gaseous and liquid states of H_2O, respectively, whereas the *dashed* and the *dotted lines* represent two levels of metastability

where P_B, P_C, P_D, etc designate the pressures corresponding to the points B, C, D, etc. Solution for P_{sat} requires that the areas F_1 and F_2 be compared to each other before every iteration cycle at a new value of $\hat{P}$. Performing the iteration yields a P_{sat} of 59 bar, compared to 55 bar indicated by Schmidt (1979) for 270°C. The agreement between the P_{sat} computed from the MRK, and those quoted by Schmidt (1979), deteriorates at higher temperatures as a result of the increasingly poor representation of the properties of H_2O by the MRK in the vicinity of its critical point.

3.4.4.2 Calculation of Fugacity and Molar Gibbs Energy of H_2O

Having delineated the phase relations of H_2O, we address the problem of obtaining f^*_{H2O} ($\equiv f_{H2O}/f^°_{H2O}$) and G^*_{H2O} ($= \mu^*_{H2O}$) at a specified set of T and P. Recalling the definition of fugacity of the fluid i, f_i, given earlier [Eq.(3.4)], we may write

$$G^*_i - G^°_{f,T,i} \equiv RT\ln\frac{f_i}{f^°_i} = \int_{P^°}^{P} V_{T,i}(P)dP , \qquad (3.53)$$

where $f^°_i$ is unity at any T. Splitting G^*_i on the LHS of (3.53) into an ideal (G^{*id}_i) and a deviation from the ideality term, $(G^*_i - G^{*id}_i)$,

$$RT\ln\frac{f_i}{f^°_i} = (G^{*id}_i - G^°_{f,T,i}) + (G^*_i - G^{*id}_i) . \qquad (3.54)$$

Choosing the hypothetical ideal gas i at 1 bar as the standard state (implying that such a gas at 1 bar pressure and at a given T has an energy equal to $G°_{f,T,i}$)[7], and substituting Eq.(3.3) into the first term on the RHS of (3.54), the latter is recast to

$$RT\ln\frac{f_i}{f_i°} = RT\ln\frac{P}{P°} + (G_i^* - G^{*id}).$$ (3.55)

Now, as stressed in Sect. 3.2, all real gases behave ideally as P→0. Consequently, at *zero* pressure, $G_i^* = G_i^{*id}$ (Rossini 1950, p.240). Accordingly, the second term on the RHS of (3.55) may be further split into

$$RT\ln\frac{f_i}{f_i°} = RT\ln\frac{P}{P°} + [G_i^*(\text{at } P = P) - G_i^*(\text{at } P = 0)]$$

$$- [G_i^{*id}(\text{at } P = P) - G_i^{*id}(\text{at } P = 0)].$$ (3.56)

Note that each of the two square bracketed terms on the RHS of (3.56) corresponds to an appropriate $\int_0^P VdP$ integral, such that

$$RT\ln\frac{f_i}{f_i°} = RT\ln\frac{P}{P°} + \int_0^P V_{T,i}(P)dP - \int_0^P V_{T,i}^{id}(P)dP.$$ (3.57)

Unfortunately, the two integrals on the RHS of (3.57) cannot be evaluated individually, because in both the cases V→∞ as P→0. However, the difference between the two, $(V_i - V^{id}_i)$, happens to be a finite, and commonly nonzero quantity; it may be integrated between 0 to P bar. Substituting (3.2) into $(V_i - V^{id}_i)$, we have

$$RT\ln\frac{f_i}{f_i°} - RT\ln\frac{P}{P°} = \int_0^P \left(V_{T,i}(P) - \frac{RT}{P}\right)dP.$$ (3.58)

Utilizing Eqs.(3.5), (3.5a), and (3.6), we rewrite (3.58) as

$$RT\ln\frac{f_i}{P} = RT\ln\phi_i = \int_0^P \left(V_{T,i}(P) - \frac{RT}{P}\right)dP.$$ (3.59)

Again, given an analytical equation of state for the fluid i, the integration on the RHS of (3.59) is best executed numerically. A fairly accurate graphical integration is also possible (Skippen 1977, Eq.5.10 and Fig.5.2), although that

---

[7] Note that Robie et al. (1979) listed $G°_{f,T,i}$ of (hypothetical) ideal gases [H_2O ("steam"), CO_2 etc] at 1 bar pressure, making our choice of a hypothetical ideal gas standard state compatible with their tables. Phase equilibria calculations between minerals and fluids in the subsequent chapters use this standard state together with $G°_{f,T,i}$ data from Robie et al. (1979).

is a tedious procedure. Once ϕ_i is known for a desired set of P and T, f_i is obtained from Eq.(3.6).

The next quantity to be computed is the molar Gibbs energy of H_2O, G^*_{H2O}, at T and P. Utilizing data for $G^\circ_{f,T,H2O}$ from Robie et al. (1979), and knowing f_{H2O} at that T and P, we obtain:

$$G^*_{H_2O} = G^\circ_{f,T,H_2O} + RT\ln\frac{f_{H_2O}}{f^\circ_{H_2O}} = G^\circ_{f,T,H_2O} + RT\ln f^*_{H_2O} . \qquad (3.60)$$

Figure 3.4 displays the results; G^*_{H2O} has been plotted against P for the 270°C isotherm. The individual segments of the G^*_{H2O} vs P curve of Fig. 3.4 are labeled in a manner similar to Fig. 3.2. Let us first focus on the segments A-B and F-G, shown by solid lines. Starting at 1 bar pressure (corresponding to the point A in Fig. 3.4, where G^*_{H2O} is identical to $G^\circ_{f,T,H2O}$), G^*_{H2O} increases very rapidly up to B. The solid line A-B represents the stable state (the lowest energy level) of gaseous H_2O. At B, P_{sat} is reached; H_2O gas is now in equilibrium with liquid H_2O. The solid line F-G depicts the G^*_{H2O} vs P relation for liquid H_2O. Notice the abrupt change in the slope of the G^*_{H2O} vs P curve at P_{sat}. Because $(\partial G/\partial P)_T = V$, this change in slope reveals a change in volume due to the gas→liquid phase transition in H_2O. The magnitude of

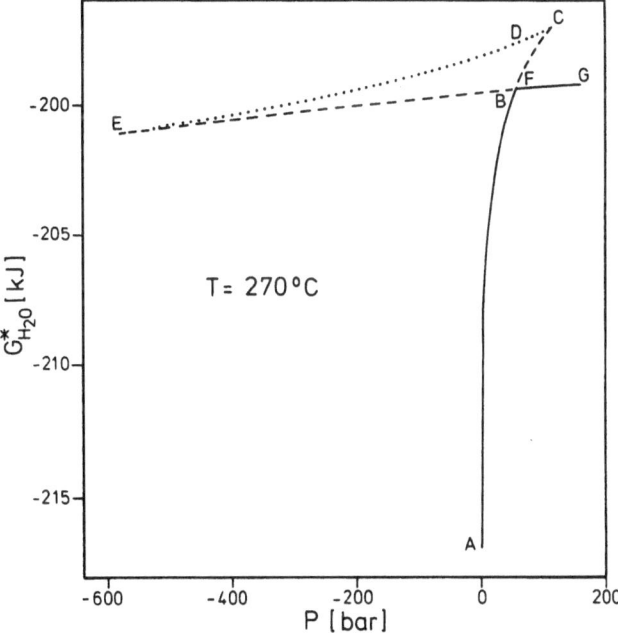

Fig. 3.4 $G^*_{H_2O}$ as a function of P at 270°C, calculated from the MRK. Once again, the *solid lines A-B* and *F-G* delineate the stable gaseous and liquid states of H_2O, respectively, whereas the *dashed* and the *dotted lines* represent two levels of metastability

this volume change is seen in Fig. 3.2. The segment B-C, shown as a dashed line in Fig. 3.4, denotes the first level of metastability for liquid H_2O; likewise, E-F represents the first level of metastability for H_2O gas. And finally, the dotted segment C-D-E illustrates the second level of metastability for the liquid (C-D) and the gaseous (D-E) form of H_2O.

3.5 Modified Redlich-Kwong Equation of State for a Fluid Mixture

3.5.1 Theoretical Background

The thermodynamic treatment of fluid mixtures requires that, in addition to the T-P-dependences, the composition dependences of **a** and **b** be known as well. Given these, the molar volume of the fluid mixtures can be computed at any set of T, P, and X_i, employing the Redlich-Kwong equation. The remaining thermodynamic properties are then derived from $V(T,P,X_i)$. In the following, we denote the **a** and **b** parameters of a fluid mixture as $\mathbf{a}_{mix}$ and $\mathbf{b}_{mix}$.

   A prerequisite for deriving an MRK for a fluid mixture is the availability of the MRK for the constituent pure components. Given these, the strategy adopted to obtain an MRK for the fluid mixture depends upon the quality of its P-V-T-X_i data.

Case I Good quality P-V-T-X_i data for the mixture are available

To the extent that sufficient and reliable P-V-T-X_i data for a fluid mixture are available, many authors (see Vidal 1978, p.787) derive $\mathbf{a}_{mix}$ and $\mathbf{b}_{mix}$, employing the following mixing rules:

$$\mathbf{a}_{mix} = \sum_{i=1}^{n} \sum_{j=1}^{n} \mathbf{a}_{ij} X_i X_j \text{ and} \tag{3.61a}$$

$$\mathbf{b}_{mix} = \sum_{i=1}^{n} \sum_{j=1}^{n} \mathbf{b}_{ij} X_i X_j. \tag{3.61b}$$

The $\mathbf{a}_{ii}$, $\mathbf{b}_{ii}$, $\mathbf{a}_{jj}$ and $\mathbf{b}_{jj}$ terms in Eqs.(3.61a and 3.61b) take care of the like-species interactions; they are identical to the **a** and **b** parameters of the pure species i and j at the desired T and P. The unlike-species interactions, $\mathbf{a}_{ij}$ and $\mathbf{b}_{ij}$, are calculated from $\mathbf{a}_{ii}$, $\mathbf{a}_{jj}$, $\mathbf{b}_{ii}$ and $\mathbf{b}_{jj}$ using the so-called combining rules,

$$\mathbf{a}_{ij} = (\mathbf{a}_{ii}\mathbf{a}_{jj})^{0.5}(1-\varepsilon_{ij}) \text{ and} \tag{3.62a}$$

$$\mathbf{b}_{ij} = 0.5(\mathbf{b}_{ii}+\mathbf{b}_{jj})(1-\tau_{ij}), \tag{3.62b}$$

ϵ_{ij} and τ_{ij} being adjustable parameters, dependent upon T and P. Their T-P-dependences are derived by least squares treatment of the P-V-T-X_i data of the fluid mixture. Given such an equation of state, $V(P,T,X_i)$ is very well reproduced *within* the P-T-X_i range of the experimental data. However, caution must be exercised when extrapolating the equation of state beyond the range of the experimental P-V-T-X_i data.

Case II P-V-T-X_i data for the fluid mixture are poor or scarce

It is evident from the above that deriving ϵ_{ij} and τ_{ij} becomes meaningless as the quality of the P-V-T-X_i data deteriorates. In such cases, ϵ_{ij} and τ_{ij} are set equal to zero, reducing (3.62a) to

$$a_{ij} = (a_{ii}a_{jj})^{0.5}, \tag{3.63}$$

a traditional combining rule. Likewise, $\tau_{ij} = 0$ reduces (3.62b) to

$$b_{ij} = 0.5(b_{ii}+b_{jj}). \tag{3.64}$$

Accordingly, for a fluid mixture with insufficient P-V-T-X_i data, the mixing and combining rules for a_{mix} simplify to

$$a_{mix} = \sum_{i=1}^{n} \sum_{j=1}^{n} a_{ij}X_iX_j, \text{ with } a_{ij} = (a_{ii}a_{jj})^{0.5}. \tag{3.65}$$

And substituting (3.64) into (3.61b), and bearing in mind that the mole fractions add up to one, Eq.(3.61b) reduces to

$$b_{mix} = \sum_{i=1}^{n} b_iX_i. \tag{3.66}$$

Equations (3.65) and (3.66) are the time-honored mixing rules, recommended by Redlich and Kwong (1949). While b_{mix} is obtained by the simple averaging of the component species, the derivation of a_{mix} requires that the pair interactions between the like- and the unlike-species be distinguished.

Having derived a volume equation of state, $V(P,T,X_i)$, for the fluid mixture, we may apply Eq.(3.15) to come up with $RT\ln\gamma_i$ for any P, T, and X_i. For this, we need to know V_i and v_i; while V_i follows from the MRK of pure i, v_i is obtained from the MRK of the fluid mixture by using the technique of tangent intercepts [(1.52) and (1.53)]. Once derived, $\gamma_i(P,T,X_i)$ can be readily converted via (1.61) into $a_i(P,T,X_i)$. On the other hand, the $RT\ln\gamma_i(P,T,X_i)$ data may also be used to formulate $G^{ex}(P,T,X_i)$, the equation of state for the fluid mixture (see Sect.1.2.2.6).

3.5.2 A Modified Redlich-Kwong Equation for the H_2O-CO_2 Mixture

Having outlined the theoretical basis for deriving an equation of state for a fluid mixture, we focus on the H_2O-CO_2 binary, for which Kerrick and Jacobs (1981) derived an MRK. Their MRK for the pure species, H_2O and CO_2, on which the MRK for the binary mixture is based, has been briefly discussed in Sect. 3.4.3.

In order to derive the MRK for the H_2O-CO_2 binary, Kerrick and Jacobs (1981) adopted the mixing rules (3.65) and (3.66), together with the combining rule of Eq. (3.63). This strategy is justified, because the P-V-T-X_i data on this binary are quite limited, and in part, contradictory. While the V(P,T,X_i) data of Greenwood (1969) are in agreement with those of Gehrig (1980), both seem to be only in partial agreement with the molar volumes reported previously by Franck and Tödheide (1959). A more recent set of V(P,T,X_i) data by Zhakirov (1984) even suggests a systematically higher value on the order of 5%. Given these controversies, it is sensible not to rely on the measured volumes, but to utilize those only for the sake of comparison with the volumes predicted by the MRK. Figure 3.5 shows a V-X_{CO2} plot at 500°C, 200 and 500 bar. The experimental data of Greenwood (1969) appear as circles, those of Gehrig (1980) as the inverted triangles, on this graph. The

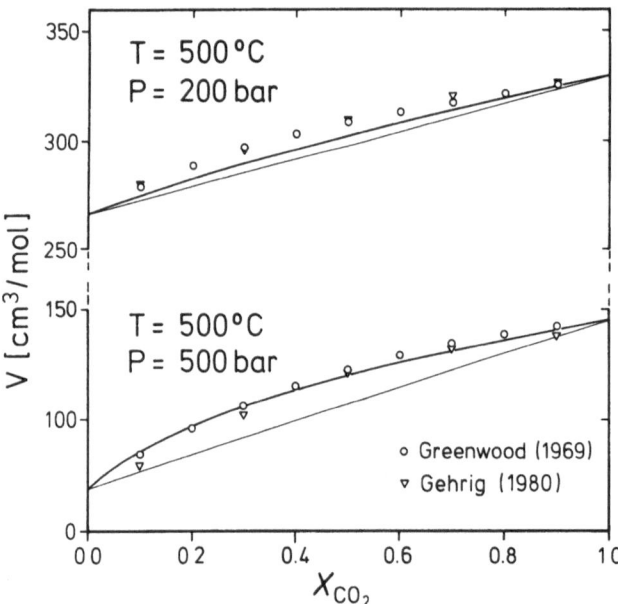

Fig. 3.5 Two isobaric-isothermal V-X_{CO_2} diagrams of mixtures of H_2O and CO_2. The *thin lines* show the ideal mixing behavior, while the *heavier lines* depict the V-X_{CO_2} function computed from the MRK (Kerrick and Jacobs 1981). The *circles* and the *inverted triangles* denote the volumes measured experimentally by Greenwood (1969) and Gehrig (1980), respectively

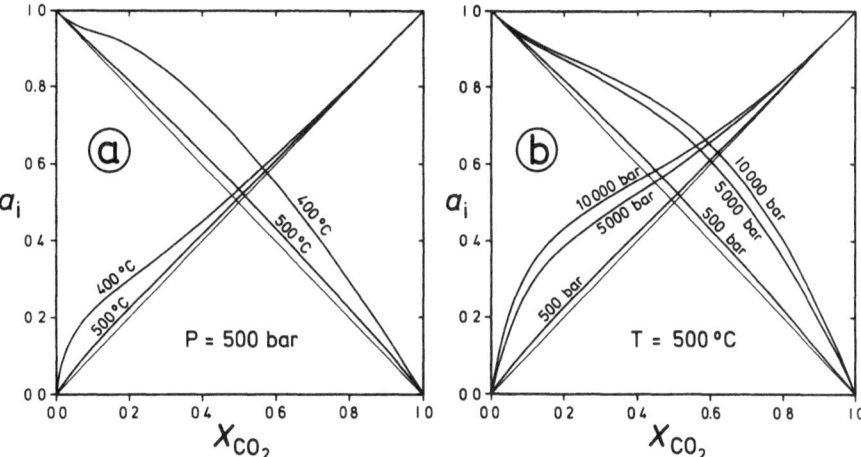

Fig. 3.6 The activity-composition relations for the system H_2O-CO_2, computed from the MRK of Kerrick and Jacobs (1981)

heavy solid lines show the volumes computed from the MRK, using a computer program by Jacobs and Kerrick (1981a). Note the significant positive deviation from ideality at both 200 and 500 bar. Although a remarkable agreement exists between the theory and the experiments at 500 bar pressure, the observational data suggest a higher degree of nonideality at 200 bar, compared to that predicted by the MRK. With increasing T, an agreement between the two will be expected, because solutions tend to behave ideally with rising T.

Next, we shall focus on the activity-composition relationships for the H_2O-CO_2 mixtures. Though the basic principles for deriving them are simple (see Sect. 3.5.1), the algorithms required to evaluate them are cumbersome (cf. Kerrick and Jacobs 1981; Jacobs and Kerrick 1981a for details). Figure 3.6a shows an isobaric (P = 500 bar)-polythermal (T = 400 and 500°C) a_i vs X_i plot, as derived from Jacobs and Kerrick's (1981a) computer program. A conspicuous positive deviation from ideality is encountered at 400°C, with the nonideality waning quickly as T rises to 500°C. Indeed, the a_i-X_i diagram is reminiscent of Greenwood's (1973) a_i-X_i data, and is indicative of the veracity of the MRK in the 450 to 800°C range explored by Greenwood (1969, 1973). It is primarily in this range that most of our mixed-volatile equilibria calculations will be done in Chapters 6 and 9. Jacobs and Kerrick (1981a, p. 133) recommended that their MRK be used from 300 to 1050°C, and 1 bar to 20 kbar pressure. The problem of judging the performance of the MRK at pressures of 10 or 20 kbar, or at temperatures near 300°C, looms large at present. Figure 3.6b displays an isothermal (T = 500°C)-polybaric (P = 500, 5000, and 10000 bar) a_i-X_i diagram for the H_2O-CO_2 mixture, based on the MRK. A dramatic rise in positive deviation from ideality is encountered with increasing pressure. In absence of experimental P-V-T-X_i data at these high pressures, we cannot corroborate these results, nor can we question them. By contrast, the performance of the MRK at temperatures down to 300°C can be verified based on some recent measurements of the H^{ex}_m for the H_2O-CO_2 gas mixtures by a flow mixing calorimeter (see Wormald et al. 1986). To achieve this, we compute the H^{ex}_m from MRK, and compare it with the experimental data of Wormald et al. (1986).

Let us start by computing the H^{ex}_m for the H_2O-CO_2 mixtures at just one set of P-T-X_i conditions, also explored calorimetrically by Wormald et al. (1986, Table 2). For example, we pick 598.15 K, 60 bar, and 0.5 X_{CO2}. The initial step is to calculate the G^{ex}_m at that P and X_{CO2} over a small range of T above and below 598.15 K, using a relation analogous to Eq. (1.70):

$$G^{ex}_m = RT \left[X_{H_2O} \ln \frac{a_{H_2O}}{X_{H_2O}} + X_{CO_2} \ln \frac{a_{CO_2}}{X_{CO_2}} \right].$$

(3.67)

Table 3.3 lists the G^{ex}_m data for T between 588.15 - 608.15 K, for a pressure of 60 bar and at X_{CO2} = 0.50. The a_{H2O} and a_{CO2} values, required for this purpose, were taken from the computer program by Jacobs and Kerrick (1981a). The next step is to convert G^{ex}_m into H^{ex}_m. Recalling the defining equation

$$G^{ex}_m = H^{ex}_m - TS^{ex}_m ,$$

(3.68)

it is immediately obvious that our problem boils down to obtaining the slope of the G^{ex}_m vs T curve, and insert that back into (3.68) to come up with the H^{ex}_m. The slope of the G^{ex}_m vs T curve is best determined from a least squares fit of the G^{ex}_m vs T data. Because our G^{ex}_m vs T data extend over a small range of T, over which both H^{ex}_m and S^{ex}_m are likely to remain constant, G^{ex}_m may turn out to be linear in T. Indeed, the least squares treatment vindicates the adequacy of the linear fit for this range of temperature. From the data in

Table 3.3. The G^{ex}_m of H_2O-CO_2 mixtures at 60 bar and 0.5 X_{CO_2} for the temperature range 588.15 to 608.15 K, computed from the MRK of Kerrick and Jacobs (1981)

P(bar)	X_{CO_2}	T(K)	a_{CO_2}	a_{H_2O}	G^{ex}_m (J mol⁻¹)
60	0.50	588.15	0.5028	0.5033	29.7381
60	0.50	590.15	0.5027	0.5032	28.8638
60	0.50	592.15	0.5026	0.5032	28.4718
60	0.50	594.15	0.5026	0.5031	28.0771
60	0.50	596.15	0.5025	0.5030	27.1858
60	0.50	598.15	0.5025	0.5029	26.7826
60	0.50	600.15	0.5024	0.5029	26.3756
60	0.50	602.15	0.5023	0.5028	25.4674
60	0.50	604.15	0.5023	0.5027	25.0524
60	0.50	606.15	0.5022	0.5026	24.1324
60	0.50	608.15	0.5022	0.5026	24.2120

Table 3.4. Comparison of the H_m^{ex} for H_2O-CO_2 mixtures calculated from the MRK of Kerrick and Jacobs (1981) with direct calorimetric measurement of H_m^{ex} by Wormald et al. (1986)

T(K)	X_{CO_2}	P(bar)	MRK H_m^{ex} (J mol^{-1})	Calorimetry H_m^{ex} (J mol^{-1})
698.15	0.50	50	56.8	241
698.15	0.50	108	213.8	676
698.15	0.50	134	313.5	930

Table 3.3, we obtain,

$$G_m^{ex} \ (J \ mol^{-1}) = 197.2 - 0.285 \ T. \tag{3.69}$$

Equating (3.68) with (3.69), it follows that

$$H_m^{ex} = 197.2 \ J \ mol^{-1} \ and \tag{3.70a}$$

$$S_m^{ex} = 0.285 \ J \ K^{-1} \ mol^{-1}. \tag{3.70b}$$

The H^{ex}_m calculated from the MRK is way below 584 J mol^{-1} measured at 598.15 K, 60 bar, and 0.50 X_{CO2}, by calorimetry. The same type of disparity occurs at 698.15 K (425°C), 0.5 X_{CO2}, 50 to 134 bar, as documented in Table 3.4. It is clear that H^{ex}_m returned by the MRK falls short of the calorimetric value by a substantial margin. Indications are that the MRK does not work well at these low T and P, presumably because it was not designed to be operative at these conditions.

3.6 Concluding Remarks

In concluding, we note that it is possible to formulate an MRK for supercritical H_2O such that it works well to 100 kbar or more. Its validity at those high pressures is corroborated by the V(P,T) data derived from shock wave experiments. However, its performance necessarily deteriorates in the subcritical region, and no amount of fine tuning can make it applicable to the critical point. What has been said here regarding H_2O is likely to apply to most other fluids.

Extending the MRK of pure fluids to that of a fluid mixture is a more difficult job, especially because of the limited P-V-T-X_i data available on fluid mixtures. The MRK for the H_2O-CO_2 mixture (Kerrick and Jacobs 1981) may serve as an example. Judged by the experimental V(P,T,X_i) data existing on this mixture, it does well at supercritical temperatures up to a pressure of a few kbar. Its performance at pressures in excess of 10 kbar cannot be tested at present due to a lack of experimental P-V-T-X_i data.

Chapter 4

Phase Relations among End-Member Solids

4.1 Introduction

The thermodynamic background acquired in the preceding three chapters enables us to compute phase diagrams among a set of end-member minerals, with or without a pure fluid or a fluid mixture. Yet another subject, which can be handled at the present stage, is the derivation of an internally consistent thermodynamic database, which is indispensable for obtaining a geologically reliable phase diagram. However, the computation of the phase relations involving crystalline solutions, a topic of fundamental interest to geology, must be deferred until the thermodynamics of crystalline solutions is introduced in Chapter 8.

This chapter will deal with the first topic in our agenda: phase diagram calculation between mineral end-members. The variables of relevance are P and T, the phase diagram being a depiction of the mutual stability relations on a P-T diagram. The objective will be to calculate the P-T curves of a few reactions, and to estimate their uncertainties. The worked examples, provided with supportive details for reference in later chapters, include polymorphic phase transitions as well as a reaction between mineral end-members.

4.2 Thermodynamic Formalism

Consider the reaction,

$$bB = cC + dD, \tag{4.1}$$

where b, c, and d represent the stoichiometric coefficients of the phases B, C, and D, respectively. To compute this equilibrium, we start with the equation derived in Chapter 1 as the starting point for computing every heterogeneous phase equilibrium,

$$\Delta\mu = 0 = \Delta G_T^\circ + \int_1^P \Delta V_T(P)dP + RT\ln K , \tag{1.106}$$

where ΔG_T° is the standard Gibbs energy difference of the reaction at T and 1 bar pressure. For reactions among the end-members, the $RT\ln K$ term drops out because the activities of the components are unity (see Sect. 1.2.2.3).

Thus, (1.106) reduces to

$$\Delta G^* = 0 = \Delta G^{\circ}_{T} + \int_{1}^{P} \Delta V_{T}(P)dP, \quad (4.2)$$

ΔG^* denoting the Gibbs energy difference of the reaction involving the end-members at system T and P.

Next, we successively consider the terms on the RHS of Eq. (4.2). Splitting ΔG°_{T} in analogy to Eq. (1.11),

$$\Delta G^{\circ}_{T} = \Delta H^{\circ}_{T} - T\Delta S^{\circ}_{T}, \quad (4.3)$$

and inserting Eqs.(2.22) and (2.26) into (4.3), we have

$$\Delta G^{\circ}_{T} = \Delta H^{\circ}_{298} + \int_{298}^{T} \Delta C^{\circ}_{P}(T)dT - T\Delta S^{\circ}_{298} - T\int_{298}^{T}\left[\frac{\Delta C^{\circ}_{P}(T)}{T}\right]dT. \quad (4.4)$$

Given $H^{\circ}_{f,298}$, S°_{298}, and $C^{\circ}_{P}(T)$, ΔG°_{T} is thus easily evaluated.

Let us now focus on the $_{1}\!\int^{P}\Delta V_{T}(P)dP$ term on the RHS of Eq.(4.2). It is a measure of the energy difference due to a change of state of the system from 1 bar to P bar at temperature T. Consequently, a rigorous execution of the integral requires that ΔV be known at T as a function of P. For an end-member phase, $V = f(T,P)$. Because V is a state function of the characteristic variables T and P,

$$dV = \left(\frac{\partial V}{\partial T}\right)_{P} dT + \left(\frac{\partial V}{\partial P}\right)_{T} dP. \quad (4.5)$$

The two derivatives on the RHS of (4.5) are generally expressed in terms of thermal expansivity, α, and compressibility, β, which may be written in analogy to Eqs.(1.26) and (1.27):

$$\alpha \equiv \frac{1}{V^{\circ}_{298}}\left(\frac{\partial V}{\partial T}\right)_{P}, \quad \text{and} \quad (4.6)$$

$$\beta = -\frac{1}{V^{\circ}_{298}}\left(\frac{\partial V}{\partial P}\right)_{T}, \quad (4.7)$$

V°_{298} indicating the standard molar volume. Substituting (4.6) and (4.7) into (4.5), the latter may be recast to

$$dV = V^{\circ}_{298}(\alpha dT - \beta dP). \quad (4.8)$$

The V(T,P) data available suggest that within the uncertainties of measurements, α and β of solids are adequately represented by

$$\alpha = \alpha(T) = \alpha_{o} + \alpha_{1}T, \quad \text{and} \quad (4.9)$$

$$\beta = \beta(P) = \beta_{o} + \beta_{1}P. \quad (4.10)$$

Inserting (4.9) and (4.10) into (4.8),

$$dV = V_{298}^{\circ}[\alpha(T)dT - \beta(P)dP]. \tag{4.11}$$

Stipulating that $(\partial\alpha/\partial P)_T = 0 = (\partial\beta/\partial T)_P$,[1]) integration of (4.11) yields

$$V_{T,P} - V_{298}^{\circ} = V_{298}^{\circ}\ [\alpha_o(T\text{-}298) + 0.5\cdot\alpha_1(T^2\text{-}298^2) \tag{4.12a}$$
$$- \beta_o(P\text{-}1) - 0.5\cdot\beta_1(P^2\text{-}1)],$$

which may be rewritten as

$$V_{T,P} = V_{298}^{\circ}\ [1 + \alpha_o(T\text{-}298) + 0.5\cdot\alpha_1(T^2\text{-}298^2) - \beta_o(P\text{-}1) - 0.5\cdot\beta_1(P^2\text{-}1)]. \tag{4.12}$$

Knowing the molar volume at system temperature, T, we may carry on with the integration.

$$\int_1^P V_T(P)dP = \int_1^P V_{298}^{\circ}\left[1 + \alpha_o(T - 298) + \frac{\alpha_1}{2}(T^2 - 298^2)\right]dP$$

$$- \int_1^P V_{298}^{\circ}\left[\beta_o(P - 1) + \frac{\beta_1}{2}(P^2 - 1)\right]dP$$

$$= V_{298}^{\circ}\left[1 + \alpha_o(T - 298) + \frac{\alpha_1}{2}(T^2 - 298^2)\right.$$

$$\left. - \left\{\frac{\beta_o}{2} + \frac{\beta_1}{6}(P + 2)\right\}(P - 1)\right](P - 1). \tag{4.13}$$

Observing the rule that the integral of a difference is identical to the difference of the integrals, we may then derive $\int_1^P\Delta V_T(P)dP$ from (4.13),

$$\int_1^P \Delta V_T(P)dP = \Delta\int_1^P V_T(P)dP$$

$$= \sum_i \nu_i\left\{V_{298,i}^{\circ}\left[1 + \alpha_{0,i}(T - 298) + \frac{\alpha_{1,i}}{2}(T^2 - 298^2)\right.\right.$$

$$\left.\left. - \left\{\frac{\beta_{0,i}}{2} + \frac{\beta_{1,i}}{6}(P + 2)\right\}(P - 1)\right](P - 1)\right\}$$

---

[1]) Note that $\alpha(T)$ and $\beta(P)$ are in truth also functions of P and T, respectively. These dependencies are ignored here because they are seldom resolved in volume measurements of the solids. Though $\alpha(T)$ and $\beta(P)$ are expressed as linear functions of T and P, their extrapolation beyond the T-P range of the experimental measurement should be avoided.
[2]) Note that $\Delta(\alpha_o V^{\circ}_{298})$ is not identical to $\Delta\alpha_o\Delta V^{\circ}_{298}$; therefore, ΔV°_{298} cannot be bracketed out as a common factor in Eq.(4.14).

$$= \left[\Delta V^\circ_{298} + \Delta(\alpha_\circ V^\circ_{298})(T - 298) + \frac{1}{2}\Delta(\alpha_1 V^\circ_{298})(T^2 - 298^2) \right.$$
$$\left. - \left\{ \frac{1}{2}\Delta(\beta_0 V^\circ_{298}) + \frac{1}{6}\Delta(\beta_1 V^\circ_{298})(P+2) \right\}(P-1) \right](P-1), ^{2)}$$

$$(4.14)$$

with v_i denoting the i-th stoichiometric coefficient in Eq.(4.1). Substituting Eqs.(4.4) and (4.14) into Eq.(4.2), the general condition of equilibrium among end-member solids may be summarized as

$$\Delta G^* = 0 = \Delta H^\circ_{298} + \int_{298}^{T} \Delta C^\circ_P(T)dT - T\Delta S^\circ_{298} - T \int_{298}^{T} \left[\frac{\Delta C^\circ_P(T)}{T} \right] dT$$

$$+ \left[\Delta V^\circ_{298} + \Delta(\alpha_\circ V^\circ_{298})(T - 298) + \frac{1}{2}\Delta(\alpha_1 V^\circ_{298})(T^2 - 298^2) \right.$$
$$\left. - \left\{ \frac{1}{2}\Delta(\beta_0 V^\circ_{298}) + \frac{1}{6}\Delta(\beta_1 V^\circ_{298})(P+2) \right\}(P-1) \right](P-1).$$

$$(4.15)$$

In the eventuality that $\alpha(T)$ and $\beta(P)$ are unknown, both may be set equal to zero. The extent to which such an approximation is permissible depends mainly upon the relative contribution of these terms to the total Gibbs energy difference of the reaction. As a rule of thumb, it may be stated that ignoring α and β in a solid-solid reaction is likely to lead to significant error. The smaller the ΔG°_T of the reaction and the higher the equilibrium pressure, the larger is this error. We shall demonstrate this in the worked examples given in Section 4.4. Before proceeding to that, however, the error propagation formalism must be handled. Error propagation should be an integral part of all phase equilibria calculations.

4.3 Error Propagation Calculation

4.3.1 Basic Formalism

All thermodynamic data (like $H^\circ_{f,298}$, S°_{298}, V°_{298}, etc.) have uncertainties inherent to them. Combining them for computing phase equilibria always leads to the propagation of uncertainties into the computed phase diagrams. Error propagation calculation permits assessment of the uncertainties of a computed phase diagram. Only the salient features of error propagation will be outlined in this section, following Anderson (1977a,b) in a general way.

For a function $x = f(u,v)$, the error propagation equation may be written as

$$\sigma_x^2 = \left(\frac{\partial x}{\partial u}\right)_v^2 \sigma_u^2 + \left(\frac{\partial x}{\partial v}\right)_u^2 \sigma_v^2 + 2\rho_{uv}\sigma_u\sigma_v \left(\frac{\partial x}{\partial u}\right)_v \left(\frac{\partial x}{\partial v}\right)_u. \qquad (4.16)$$

In this equation, σ^2_u is the variance of u (σ_u being the standard deviation), whereas ρ_{uv} denotes the correlation coefficient. Thus, the variance of x is the weighted sum of the variances of u and v, the weighting factors being squares of the partial derivatives of x with respect to u and v. The third term on the RHS corrects for the covariance (or non-independence) of the variables, u and v.

Observing that at equilibrium $\Delta G^* = 0$ [see Eq.(4.15)], solving for the uncertainties of a calculated phase diagram in terms of T or P is tantamount to applying Eq.(4.16) to $\Delta G^* = f(T,P)$. Thus,

$$\sigma^2_{\Delta G^*} = \left(\frac{\partial \Delta G^*}{\partial T}\right)^2_P \sigma^2_T + \left(\frac{\partial \Delta G^*}{\partial P}\right)^2_T \sigma^2_P$$

$$+ 2\rho_{TP}\sigma_T\sigma_P \left(\frac{\partial \Delta G^*}{\partial T}\right)_P \left(\frac{\partial \Delta G^*}{\partial P}\right)_T . \tag{4.17a}$$

Note that σ^2_T and σ^2_P, which we wish to solve for, appear on the RHS of this equation, while the LHS, $\sigma^2_{\Delta G^*}$, combines the variances of all other terms contributing to ΔG^*, whose uncertainties must be propagated to evaluate $\sigma^2_{\Delta G^*}$. Disregarding the uncertainties of $\alpha(T)$ and $\beta(P)$, which are unknown, and certainly vanishingly small, we recast (4.15) to a condensed form *for use in error propagation*,

$$\Delta G^* = \Delta H^\circ_T - T\Delta S^\circ_T + \Delta V^\circ_{298}(P-1) + C. \tag{4.18}$$

In this equation, C sums the remaining terms of (4.15), comprising $\Delta[\alpha(T)V^\circ_{298}]$ and $\Delta[\beta(P)V^\circ_{298}]$, whose uncertainties remain unknown, and cannot be propagated.

Considering that T and P are in reality always non-covariant, i.e., $\rho_{TP} = 0$, and taking the derivatives of (4.17a) with respect to the RHS of (4.18), we have[3]

$$\sigma^2_{\Delta G^*} = (-\Delta S^\circ_T)^2\sigma^2_T + (\Delta V^\circ_{298})^2\sigma^2_P. \tag{4.17}$$

Now, if we wish to solve for σ_T at any equilibrium P (for which $\sigma_P = 0$), from (4.17), it follows that

$$(-\Delta S^\circ_T)^2\sigma^2_T = \sigma^2_{\Delta G^*}, \tag{4.19a}$$

such that

$$\sigma_T = \frac{\sigma_{\Delta G^*}}{|-\Delta S^\circ_T|} . \tag{4.19}$$

---

[3] To obtain $(\partial \Delta G^*/\partial T)_P$ from Eq.(4.18), we may proceed as follows:

$(\partial \Delta G^*/\partial T)_P = \partial/\partial T[\Delta H^\circ_T - T\Delta S^\circ_T] = (\partial \Delta H^\circ_T/\partial T)_P - \partial/\partial T(T\Delta S^\circ_T)$
$\qquad = \Delta C^\circ_P - T(\partial \Delta S^\circ_T/\partial T)_P - 1\Delta S^\circ_T = \Delta C^\circ_P - T(\Delta C^\circ_P/T) - \Delta S^\circ_T = -\Delta S^\circ_T$

Likewise, to solve for σ_P at a specified equilibrium T, for which $\sigma_T = 0$, we obtain

$$\sigma_P = \frac{\sigma_{\Delta G^*}}{|\Delta V^\circ_{298}|} . \tag{4.20}$$

Thus, given ΔS°_T and ΔV°_{298}, one may obtain σ_T and σ_P, if $\sigma_{\Delta G^*}$ can be computed by propagating the uncertainties of the tabulated thermodynamic data. Referring to (4.18), the computation of $\sigma_{\Delta G^*}$ is achieved by considering the function $\Delta G^* = f(\Delta H^\circ_T, \Delta S^\circ_T, \Delta V^\circ_{298})$, and applying (4.16) to it. In so doing, two alternative cases must be distinguished. If a calorimetric database is used for ΔH°_T and ΔS°_T, $\rho_{\Delta H \Delta S}$, $\rho_{\Delta S \Delta V}$, and $\rho_{\Delta H \Delta V}$ are zero, because $H^\circ_{f,T}$, S°_T, and V°_{298} are independently measured quantities. On the other hand, if we wish to employ $\Delta H^\circ_{<T>}$ and $\Delta S^\circ_{<T>}$ (see Chap. 5) extracted from reaction reversal experiments, $\rho_{\Delta H \Delta S}$ will be nonzero, because the extracted $\Delta H^\circ_{<T>}$ and $\Delta S^\circ_{<T>}$ data are always correlated. For now, we use a calorimetric database for phase equilibria calculations, such that $\rho_{\Delta H \Delta S}$, $\rho_{\Delta S \Delta V}$, and $\rho_{\Delta H \Delta V}$ are zero. Accordingly,

$$\sigma^2_{\Delta G^*} = \left(\frac{\partial \Delta G^*}{\partial \Delta H^\circ_T}\right)^2 \sigma^2_{\Delta H^\circ_T} + \left(\frac{\partial \Delta G^*}{\partial \Delta S^\circ_T}\right)^2 \sigma^2_{\Delta S^\circ_T} + \left(\frac{\partial \Delta G^*}{\partial \Delta V^\circ_{298}}\right)^2 \sigma^2_{\Delta V^\circ_{298}} . \tag{4.21}$$

Taking its partial derivatives with respect to the RHS of (4.18),

$$\sigma^2_{\Delta G^*} = 1^2 \sigma^2_{\Delta H^\circ_T} + (-T)^2 \sigma^2_{\Delta S^\circ_T} + (P-1)^2 \sigma^2_{\Delta V^\circ_{298}} , \tag{4.22}$$

whence an expression for $\sigma_{\Delta G^*}$ is ultimately obtained,

$$\sigma_{\Delta G^*} = [\sigma^2_{\Delta H^\circ_T} + T^2 \sigma^2_{\Delta S^\circ_T} + (P-1)^2 \sigma^2_{\Delta V^\circ_{298}}]^{0.5}. \tag{4.23}$$

The next step in computing $\sigma_{\Delta G^*}$ is to evaluate the individual terms on the RHS of Eq. (4.23). Employing $\sigma_{H^\circ f,T}$, $\sigma_{S^\circ T}$, and $\sigma_{V^\circ 298}$ from the thermodynamic data tables, they are put together from the general equation (see Anderson 1977b, Eq.14.2),

$$\sigma^2_{\Delta Y^\circ_T} = \sum_i^n \nu_i^2 \sigma^2_{Y_i} + 2 \sum_{i=1}^{n-1} \sum_{j=i+1}^{n} \rho_{Y_i Y_j} \nu_i \nu_j \sigma_{Y_i} \sigma_{Y_j} , \tag{4.24}$$

Y being any of the quantities ($H^\circ_{f,T}$, S°_T, V°_{298}, etc.), and ν_i the stoichiometric coefficient for the phase i. Before Eq.(4.24) can be put to use, however, we must worry about the covariance term, ρ_{YiYj}.

4.3.2 *Covariance of the Tabulated Thermodynamic Data*

$S°_T$ and $V°_{298}$ are independently measured quantities for every phase; consequently, they are never correlated. As such, $\sigma^2_{\Delta S°_T}$ and $\sigma^2_{\Delta V°_{298}}$ are computed from (4.24), setting $\rho_{Y_iY_j} = 0$. However, $H°_{f,T}$ of a group of phases participating in a reaction *need not* be independent. In such cases, the covariance problem will have to be sorted out prior to computing its $\sigma^2_{\Delta H°_T}$. A case in point is the reaction

$$1 \ Ca_3Al_2[SiO_4]_3 + 1 \ SiO_2 = 1 \ Ca[Al_2Si_2O_8] + 2 \ CaSiO_3. \qquad (4.25)$$
$$\quad (Gr) \qquad\qquad (Q) \qquad\qquad (An) \qquad\quad (Woll)$$

As indicated in Chapter 2, $H°_{f,T}$ of many minerals follows from the linear summation of a number of independent enthalpy measurements by solution and combustion calorimetry. Consider, for example, the $H°_{soln,970}$ and $H°_{f,970}$ data for CaO, Al_2O_3, SiO_2, An, Woll, Q, and Gr, listed in Table 4.1. The $H°_{soln,970}$ data were obtained by Charlu et al. (1975, 1978) by high-temperature lead borate melt solution calorimetry at 970 K, whereas the $H°_{f,970}$ data (measured originally by combustion calorimetry) were interpolated from Robie et al. (1979). The general procedure for computing the $H°_{f,T}$ of any phase from such data has been outlined in Chap. 2. To elucidate the problem of covariance in (4.25), let us calculate $H°_{f,970}$ and $\sigma_{H°f,970}$ of the phases involved in that reaction. Taking grossular as an example, and utilizing the formalisms outlined in Chap. 2, along with the

Table 4.1. Independent calorimetric data for deriving $H°_{f,970}$ and $\sigma_{H°_{f,970}}$ of grossular (Gr), anorthite (An), wollastonite (Woll), and quartz (Q)

Phase	$H°_{soln,970} \pm \sigma$ (J mol^{-1})	Abbreviation	Source of data
CaO	-55145 ± 607	H_1	Charlu et al. (1978)
Al_2O_3	32342 ± 168	H_2	Charlu et al. (1975)
SiO_2	-5146 ± 147	H_3	Charlu et al. (1975)
Grossular	177443 ± 858	H_4	Charlu et al. (1978)
Anorthite	63722 ± 691	H_5	Charlu et al. (1978)
Wollastonite	29581 ± 419	H_6	Charlu et al. (1978)

Phase	$H°_{f,970} \pm \sigma$ (J mol^{-1})	Abbreviation	Source of data
Quartz	-905729 ± 500	H_7	Robie et al. (1979)
Corundum	-1693992 ± 650	H_8	Robie et al. (1979)
CaO	-633757 ± 440	H_9	Robie et al. (1979)

abbreviations in Table 4.1 (H_1 for $H°_{soln,970,CaO}$, etc.), we obtain $H°_{f,970,Gr}$ [$\equiv \Delta H°_{970}$ of the reaction $3\ Ca + 2\ Al + 3\ Si + 6\ O_2 = Ca_3Al_2(SiO_4)_3$] and $\pm [\sigma_{H°f,970,Gr}]$ as follows:

$$
\begin{aligned}
H°_{f,970,Gr} &= 3\ H_1 + H_2 + 3\ H_3 - H_4 + 3\ H_7 + H_8 + 3\ H_9 \\
&= 3 \cdot (-55145) + 32342 + 3 \cdot (-5146) - 177443 \\
&\quad + 3 \cdot (-905729) + (-1693992) + 3 \cdot (-633757) \\
&\quad \pm [3^2 \cdot 607^2 + 168^2 + 3^2 \cdot 147^2 + 858^2 + 3^2 \cdot 500^2 + 650^2 + 3^2 \cdot 440^2]^{0.5} \\
&= -6638424 \pm 2948\ \text{J mol}^{-1}. \quad\quad (4.26)
\end{aligned}
$$

Likewise,

$$
\begin{aligned}
H°_{f,970,An} &= H_1 + H_2 + 2\ H_3 - H_5 + 2\ H_7 + H_8 + H_9 \\
&= -4236024 \pm 1605\ \text{J mol}^{-1}, \quad\quad (4.27)
\end{aligned}
$$

$$
H°_{f,970,Woll} = H_1 + H_3 - H_6 + H_7 + H_9 = -1629358 \pm 1005\ \text{J mol}^{-1}, \quad\quad (4.28)
$$

and from Table 4.1,

$$
H°_{f,970,Q} = H_7 = -905729 \pm 500\ \text{J mol}^{-1}. \quad\quad (4.29)
$$

It is important to note that in all these cases, $\sigma_{H°f,970}$ has been evaluated from (4.24), setting $\rho_{Y_iY_j} = 0$. This is permissible, since all the data listed in Table 4.1 are independent; for such a case, $\rho_{Y_iY_j} = 0$, and (4.24) assumes the form

$$
\sigma^2_{\Delta H°_T} = \sum_i \nu_i^2 \sigma^2_{H°_{f,T,i}}, \quad\quad (4.30)
$$

with the subscript T indicating the temperature of interest.

The next step is to calculate $\Delta H°_{970}$ of the reaction (4.25), and its *true* $\sigma_{\Delta H°970}$. Computing $\Delta H°_{970}$ is straightforward:

$$
\Delta H°_{970}(4.25) = H°_{f,970,An} + 2H°_{f,970,Woll} - H°_{f,970,Gr} - H°_{f,970,Q} = 49413\ \text{J}. \quad\quad (4.31)
$$

If $H°_{f,970}$ of all the four phases were independent of each other, $\sigma_{\Delta H°970}$ would follow from (4.30). Assuming, for the moment, that this is true, and utilizing the values of $\sigma_{H°f,970}$ derived above,

$$
\sigma_{\Delta H°_{970}}(4.25) = [\sigma^2_{H°_{f,970,An}} + 2^2\sigma^2_{H°_{f,970,Woll}} + \sigma^2_{H°_{f,970,Gr}} + \sigma^2_{H°_{f,970,Q}}]^{0.5}
$$

$$
= 3944\ \text{J}. \quad\quad (4.32)
$$

However, $H°_{f,970}$ given in Eqs.(4.26) to (4.29), although correct individually, are by no means mutually independent. That they are correlated can be demonstrated by inserting the explicit reaction schemes of (4.26) through (4.29) into Eq.(4.31):

$$\Delta H^\circ_{970}(4.25) = H^\circ_{f,970,An} + 2H^\circ_{f,970,Woll} - H^\circ_{f,970,Gr} - H^\circ_{f,970,Q}$$

$$= (H_1 + H_2 + 2\ H_3 - H_5 + 2\ H_7 + H_8 + H_9)$$

$$+ 2(H_1 + H_3 - H_6 + H_7 + H_9)$$

$$- (3\ H_1 + H_2 + 3\ H_3 - H_4 + 3\ H_7 + H_8 + 3\ H_9) - H_7$$

$$= H_3 + H_4 - H_5 - 2\ H_6.$$

$$= 49413\ \text{J.} \tag{4.33}$$

Clearly, H°_{soln} data for Q, Gr, An, and Woll only are independent; all other quantities cancel out when evaluating ΔH°_{970}. Therefore, the sigmas of the reactions eliminated in the process of obtaining ΔH°_{970} must also be pruned prior to calculating the *true* $\sigma_{\Delta H^\circ 970}$ of reaction (4.25). Having done this, $\sigma_{\Delta H^\circ 970(4.25)}$ is obtained from Eq.(4.30)

$$\sigma_{\Delta H^\circ_{970}(4.25)} = [\sigma^2_{H_3} + \sigma^2_{H_4} + \sigma^2_{H_5} + 2^2\sigma^2_{H_6}]^{0.5} = 1392\ \text{J.} \tag{4.34}$$

In comparing the results obtained in Eqs.(4.32) and (4.34), it is evident that a failure to recognize the covariance of $H^\circ_{f,T}$ among the phases involved in a reaction will inflate the uncertainty of the calculated equilibrium perceptibly. Unfortunately, sorting out the covariances, simple though it is in principle, may be highly complicated as a routine procedure. This is particularly true for phases involving oxides like Na_2O, K_2O and others, which defy easy handling in solution calorimetry. For them, very complex reaction schemes, comprising a large number of independent reactions, have to be adhered to (cf. Hemingway and Robie 1977). The problem may become even more intractable, when the reaction schemes are not meticulously documented (see Hemingway and Robie 1977, p.416).

Regarding the following, a word of caution is warranted. The worked examples of phase equilibria, documented in this book, are geared to familiarize the reader with the computational procedure. No attempt has been made to resolve the covariance problems prior to evaluating σ_T or σ_P; they may be too large in some cases.

4.4 Worked Examples

The objective of this section is to apply the thermodynamic theory developed above to compute phase relations among end-member minerals, and to estimate the uncertainties of the computed phase diagrams. The numerical examples are reproduced in detail as an aid to the beginners. They are meant as problem-solving guides for the exercises in the subsequent chapters.

Table 4.2. Summary of thermodynamic data for computing the diamond = graphite P-T curve

	Diamond	Graphite
$H^o_{f,298}$ (J mol^{-1})	1895.00 (42)[a]	0.00
S^o_{298} (J K^{-1} mol^{-1})	2.38(.01)	5.74(.01)
V^o_{298} (J bar^{-1} mol^{-1})	0.3417(.0001)	0.5298(.0001)
$C^o_P(T)$ (J K^{-1} mol^{-1}) = a + b·T + c·T^{-2} + d·T$^{-0.5}$ + e·T^2, with		
a	98.445	63.160
b	-3.6554E-2	-1.1468E-2
c	1.2166E6	7.4807E5
d	-1.6590E3	-1.0323E3
e	1.0977E-5	1.8079E-6
Validity (T in K)	298-1800	298-1800
β_o (bar^{-1})	0.18E-6	3.0E-6
β_1 (bar^{-2})	0.00	0.00

[a] Quantities in parentheses are 2σ.

Problem Set 4.1, with Solutions

1. The V = f(T) data for diamond and graphite (Clark 1966) can be fitted to the following polynomials by the least squares method:

V^o_T,diamond (J bar^{-1}mol^{-1}) = 0.3409 + 0.2015E-5 T + 0.984E-9 T^2,
V^o_T,graphite(J bar^{-1}mol^{-1}) = 0.5259 + 1.2284E-5 T + 2.165E-9 T^2.

Derive α_o and α_1 for both phases.
2. Using the thermodynamic data (Robie et al. 1979, Clark 1966) summarized in Table 4.2, along with the α_o and α_1 derived earlier, calculate the univariant diamond = graphite P-T curve at 100°C intervals in the temperature range 800-1500°C.
3. Repeat these calculations, ignoring $\alpha(T)$ and $\beta(P)$, and compare the results with those obtained above.
4. Compare the computed P-T diagram with the experimental reversal of the diamond = graphite P-T curve by Kennedy and Kennedy (1976).
5. Compute the $2\sigma_T$ uncertainty of the P-T curve employing the data from Table 4.2.

Problem 1. To derive α_o and α_1 from the V = f(T) data, recall the definition:

$$\alpha \equiv \frac{1}{V^o_{298}} \left(\frac{\partial V}{\partial T} \right)_P . \qquad (4.6)$$

Fitting the molar volume data to a second order polynomial in T,

$$V_T^\circ = p + qT + rT^2, \tag{4.35}$$

and taking its derivative with respect to T, we have

$$\left(\frac{\partial V}{\partial T}\right)_P = q + 2rT. \tag{4.36}$$

Substituting (4.36) into (4.6),

$$\alpha(T) = \frac{q}{V_{298}^\circ} + \frac{2r}{V_{298}^\circ}T = \alpha_o + \alpha_1 T. \tag{4.37}$$

Using V°_{298} (Table 4.2) and V°_T data given earlier for diamond (D) and graphite (G), and employing Eq.(4.37), we obtain

$\alpha_{o,D}$ (K^{-1}) = 0.2015E-5/0.3417 = 0.5897E-5,
$\alpha_{1,D}$ (K^{-2}) = 2·(0.984E-9)/0.3417 = 5.759E-9,

$\alpha_{o,G}$(K^{-1}) = 1.2284E-5/0.5298 = 2.3186E-5, and
$\alpha_{1,G}$(K^{-2}) = 2·(2.165E-9)/0.5298 = 8.173E-9.[4]

Problem 2. Calculation of the diamond (D) = graphite (G) P-T curve between 800 and 1500°C.

The condition for equilibrium of D and G is given by Eq. (4.2), or its more explicit equivalent, Eq.(4.15). We start by consecutively computing the ΔG°_T and $\int_1^P \Delta V_T(P)dP$ terms of Eq.(4.2). To calculate ΔG°_T, the quantities needed are ΔH°_{298}, ΔS°_{298}, and $\Delta C^\circ_p(T)$. These follow from $H^\circ_{f,298}$, S°_{298}, and C°_p (T) data of Table 4.2, bearing in mind that for the D = G reaction, the product and reactant are 1 mol each of graphite and diamond, respectively. Therefore,

ΔH°_{298} = -1895 J and
ΔS°_{298} = 3.36 J K^{-1},

and similarly, the coefficients of the $\Delta C^\circ_p(T)$ function are:

Δa	= -35.285
Δb	= 2.5086E-2
Δc	= -4.6853E5
Δd	= 626.7
Δe	= -9.1691E-6

---

[4] Note that the dimension of α is K^{-1}. To satisfy that requirement both α_o and $(\alpha_1 T)$ must have the same dimension. That explains why the dimension of α_1 is K^{-2}.

Employing these deltas, ΔG°_T can be obtained from

$$\Delta G^\circ_T = \Delta H^\circ_{298} - T\Delta S^\circ_{298} + \int_{298}^{T} \Delta C^\circ_P(T)dT - T\int_{298}^{T} \left[\frac{\Delta C^\circ_P(T)}{T}\right]dT. \quad (4.4)$$

The integrals in (4.4) may be evaluated individually, as in Eqs. (2.23) and (2.27), or collectively. The latter is a somewhat less tedious procedure, and will be adopted here. Substituting T_0 for the lower limit of integration, $(298.15\ K \equiv T_0)$, we have

$$\int_{T_0}^{T} \Delta C^\circ_P(T)dT - T\int_{T_0}^{T}\left[\frac{\Delta C^\circ_P(T)}{T}\right]dT$$

$$= \Delta a\left[T\left(1 - \ln\frac{T}{T_0}\right) - T_0\right] - \frac{\Delta b}{2}(T - T_0)^2 - \Delta C\frac{(T - T_0)^2}{2T \cdot T_0^2}$$

$$- 2\Delta d\left[\frac{(T^{0.5} - T_0^{0.5})^2}{T_0^{0.5}}\right] - \frac{\Delta e}{6}[(T - T_0)^2(T + 2T_0)]$$

$$- \frac{\Delta f}{6}\left[\frac{(T - T_0)^2(T_0 + 2T)}{T_0^3T^2}\right]. \quad (4.38)$$

To give an example of evaluation of ΔG°_T, let us compute ΔG°_{1373}. Inserting Δa through Δe given above into Eq.(4.4), and utilizing the format of integration in (4.38), we have

ΔG°_{1373} (J) = -1895 - 1373.15 (3.36) + [36067.10 - 14495.00
$\qquad$ + 2217.87 - 28426.33 + 3478.06] = -7667.08.

This value of ΔG°_T will be used later to evaluate the equilibrium pressure of the diamond = graphite reaction at 1100°C (1373.15 K). Considering the uncertainties of $H^\circ_{f,298}$ and S°_{298} listed in Table 4.2, the reader may wonder why ΔG°_T is not being rounded off prior to using it any further. By *not* rounding off ΔG°_T, we utilize the full accuracy of the $C^\circ_P(T)$ data and at the same time preserve the internal consistency between $H^\circ_{f,298}$ and $(G^\circ_T - H^\circ_{298})/T$ (Robie et al. 1979, p.10; Garvin et al. 1987, p.14). We shall follow their recommendation throughout this book.

$\qquad$ Next, let us evaluate the $_1\int^P \Delta V_T(P)dP$ term of Eq.(4.2), utilizing the format of Eq.(4.14). A sample calculation follows:

ΔV°_{298} $\qquad$ = 0.1881 J bar^{-1}
$\Delta(\alpha_0 V^\circ_{298})$ $\qquad$ = 1.0269E-5 J bar^{-1}K^{-1}
$\Delta(\alpha_1 V^\circ_{298})$ $\qquad$ = 2.3622E-9 J bar^{-1}K^{-2}
$\Delta(\beta_0 V^\circ_{298})$ $\qquad$ = 1.5279E-6 J bar^{-2}
$\Delta(\beta_1 V^\circ_{298})$ $\qquad$ = 0 J bar^{-3}.

Inserting these into the RHS of (4.14) and putting T = 1373.15 K, the integration of $_1\!\int^P\!\Delta V_T(P)dP$ from 1 bar to an arbitrarily chosen pressure of 40000 bar gives

$$_1\!\int^{40000}\!\Delta V_{1373}(P)dP = [0.1881 + 1.0269E\text{-}5\cdot(1373.15 - 298.15)$$
$$+ (0.5)\cdot2.3622E\text{-}9\cdot(1373.15^2 - 298.15^2)$$
$$- \{(0.5)\cdot1.5279E\text{-}6\}\cdot(40000 - 1)]\cdot(40000 - 1)$$
$$= 6827.99 \text{ J.}$$

Having illustrated this step, we may proceed to calculate the equilibrium pressure, P, of the D = G reaction at any specified T. At equilibrium, the sum of $\Delta G°_T$ and $_1\!\int^P\!\Delta V_T(P)dP$ is zero (4.2). For any T, (4.2) is easily solved for P by iteration. Alternatively, P may be derived graphically. To illustrate the latter option first, ΔG^*, the sum of $\Delta G°_T$ and $_1\!\int^P\!\Delta V_T(P)dP$, was computed over a range of P in steps of 1000 bar at T = 1373.15 K, and plotted against P in Fig. 4.1. The equilibrium pressure can be read off from it where the condition $\Delta G^* = 0$ is fulfilled. At 1373.15 K, that is the case at 46200 bar. Note that in Fig. 4.1, the P-T stability field of diamond, the reactant, is identified by $\Delta G^* > 0$, that of graphite, the product, by $\Delta G^* < 0$.

The alternative procedure of solving for the equilibrium P by iteration involves calculation of ΔG^* at specified intervals up or down the P axis, until the sign of ΔG^* changes. Once that happens, we backtrack along the P-axis at

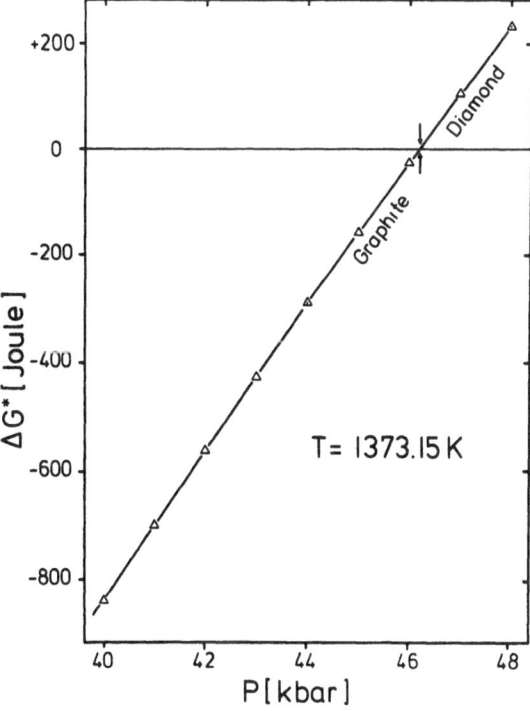

Fig. 4.1 A plot of ΔG^* vs P for the diamond = graphite reaction at 1373.15 K. The *double arrow* highlights the pressure for which $\Delta G^* = 0$. The stability field of diamond is identified by $\Delta G^* > 0$, that of graphite by $\Delta G^* < 0$

Table 4.3. Iterative solution for the equilibrium pressure of the diamond = graphite reaction at 1373.15 K (1100°C)

T (K)	ΔG_T^o (J)	$_1\int^P \Delta V_T(P)dP$ (J)	ΔG^* (J)	P (bar)
1373.15	-7667.08	6827.99	-839.09	40000
1373.15	-7667.08	7376.35	-290.73	44000
1373.15	-7667.08	7900.27	233.14	48000
1373.15	-7667.08	7641.37	-25.71	46000
1373.15	-7667.08	7667.53	0.45	46200
1373.15	-7667.08	7666.22	-0.85	46190
1373.15	-7667.08	7666.75	-0.33	46194
1373.15	-7667.08	7667.27	0.19	46198
1373.15	-7667.08	7667.01	-0.07	46196
1373.15	-7667.08	7667.14	0.06	46197

smaller intervals, until the sign of ΔG^* reverses again. This process of iteration is continued till the equilibrium pressure has been approached to a desired cut-off limit. Table 4.3 reproduces the results obtained at 1100°C by this technique. The equilibrium P is 46196 bar, in agreement with that obtained by the graphical technique. The results of calculation of the D = G equilibrium curve in the range 800-1500°C are listed in Table 4.4, and plotted in Fig. 4.2 (see later).

Problem 3. The D = G P-T curve is to be computed again to see how much difference it makes when $\alpha(T)$ and $\beta(P)$ are ignored.

If both $\alpha(T)$ and $\beta(P)$ in Eq.(4.15) equal zero, the condition for equilibrium of the D = G reaction reduces to

$$\Delta G^* = 0 = \Delta G_T^o + \Delta V_{298}^o(P-1), \qquad (4.39)$$

which may be directly solved for P at any T,

$$P(bar) = \frac{-\Delta G_T^o}{\Delta V_{298}^o} + 1. \qquad (4.40)$$

For T = 1373.15 K, application of (4.40) yields

P(bar) = [- (-7667.08)/0.1881] + 1 = 40761.7.

To facilitate a comparison with the results obtained earlier for nonzero $\alpha(T)$ and $\beta(P)$, both sets of results have been juxtaposed in Table 4.4 (see below). As was to be anticipated, the two differ considerably. Figure 4.2 puts across this message.

Table 4.4. Comparison of the computed and the experimental data on the diamond = graphite equilibrium

T(°C)	T(K)	α,β ignored P(bar)	α,β considered (2σ_T) P(bar)	(K)	Experimental P(bar)
800	1073.15	33479	37303	(9.7)	-
900	1173.15	35935	40273	(9.9)	-
1000	1273.15	38368	43245	(10.2)	-
1100	1373.15	40762	46196	(10.5)	46900
1200	1473.15	43098	49097	(10.9)	49400
1300	1573.15	45358	51912	(11.5)	51900
1400	1673.15	47520	54600	(12.3)	54400
1500	1773.15	49560	57114	(13.4)	56900

Problem 4. Comparison with the experimental P-T curve of the D = G equilibrium.

Kennedy and Kennedy (1976) used a solid-media apparatus, equipped with a zero-friction cell, to reverse the D = G reaction between 1100-1625°C. Their experimental P-T curve was given as

$$P(kbar) = 19.4 + 0.025 \; T(°C). \tag{4.41}$$

These experimental results have also been entered in Table 4.4 and displayed in Fig. 4.2. Kennedy and Kennedy's (1976) experiments vindicate the results obtained by thermodynamic computations, and emphasize the necessity of including $\alpha(T)$ and $\beta(P)$ for calculating the P-T curves of solid-solid reactions.

Problem 5. Estimation of $2\sigma_T$ of the computed P-T curve for the D = G phase transition.

Prior to embarking on the estimation of the uncertainties of a computed P-T curve, two points are made. First, the uncertainties of standard state data given in thermodynamic data tables apply to all temperatures, although they are generally stated for 298.15 K. For example, $\sigma_{H°f,298}$ is the same as $\sigma_{H°f,T}$. Secondly, they are by convention two times the standard deviation. That is, the numbers appearing in Table 4.2 are for $2\sigma_{H°f,T}$, $2\sigma_{S°T}$, etc. Accordingly, we must use one-half of those values to come up with σ_T from Eq.(4.19). Because we need to compute $2\sigma_T$, however, we may as well cut corners and use the values listed in Table 4.2 unmodified, keeping in mind that (4.19) will now yield $2\sigma_T$. As an example, let us compute $2\sigma_T$ at one point (1100°C and

46196 bar) on the D = G P-T curve. Using the data from Table 4.2, we have

$$(2\sigma_{\Delta H_T^\circ})^2 = 1^2(2\sigma_{H_{f,T,D}^\circ})^2 + 1^2(2\sigma_{H_{f,T,D}^\circ})^2 = (42 \text{ J})^2 = 1764 \text{ J}^2;$$

$$(2\sigma_{\Delta S_T^\circ})^2 = 1^2(2\sigma_{S_{T,D}^\circ})^2 + 1^2(2\sigma_{S_{T,G}^\circ})^2$$
$$= (0.01)^2 + (0.01)^2 \text{ (J K}^{-1})^2 = 2\cdot10^{-4} \text{ (J K}^{-1})^2;$$

$$(2\sigma_{\Delta V_{298}^\circ})^2 = 1^2(2\sigma_{V_{298,D}^\circ})^2 + 1^2(2\sigma_{V_{298,G}^\circ})^2$$
$$= (0.0001)^2 + (0.0001)^2 \text{ (J bar}^{-1})^2 = 2\cdot10^{-8} \text{ (J bar}^{-1})^2$$

Inserting these values into Eq.(4.23), we have for T = 1373.15 K and P = 46196 bar,

$$2\sigma_{\Delta G*} = [(2\sigma_{\Delta H_T^\circ})^2 + T^2(2\sigma_{\Delta S_T^\circ})^2 + (P-1)^2(2\sigma_{\Delta V_{298}^\circ})^2]^{0.5}$$
$$= \{1764 \text{ J}^2 + (1373.15 \text{ K})^2[2\cdot10^{-4} \text{ (J K}^{-1})^2]$$
$$+ [(46196-1)\text{bar}]^2[2\cdot10^{-8} \text{ (J bar}^{-1})^2]\}^{0.5}$$
$$= 46.73 \text{ J.}$$

A very important lesson emerges from the calculation of $\sigma_{\Delta G*}$; the $\sigma_{\Delta H^\circ T}$ term makes the bulk of $\sigma_{\Delta G*}$. This will apply to all our subsequent examples. To employ Eq. (4.19), we next need to compute ΔS°_T at 1373.15 K. Entering ΔS°_{298} and Δa through Δe, given above, into Eq.(2.27) -

$$\Delta S_T^\circ = 3.36 - 35.285 \ln(1373.15/298.15) + 2.5085\text{E-}2\cdot(1373.15-298.15)$$
$$- \{(4.6853\text{E}5)/2\}\cdot[\{1/(298.15)^2\} - \{1/(1373.15)^2\}]$$
$$+ 2\cdot626.7\cdot[\{1/(298.15)^{0.5}\} - \{1/(1373.15)^{0.5}\}]$$
$$- \{(9.1691\text{E-}6)/2\}\cdot[(1373.15)^2 - (298.15)^2]$$
$$= 4.4537 \text{ J K}^{-1}.$$

Inserting $2\sigma_{\Delta G*}$ and ΔS°_T into Eq.(4.19), the $2\sigma_T$ uncertainty is

$$2\sigma_T = 2\sigma_{\Delta G*}/|\Delta S_T^\circ| = 46.73 \text{ (J)}/4.4537 \text{ (J K}^{-1}) = 10.5 \text{ K.}$$

Proceeding in this fashion, the $2\sigma_T$ values were obtained all along the P-T curve (Table 4.4). The hatched envelope shown in Fig. 4.2 depicts the $2\sigma_T$ range. It corresponds to a $2\sigma_P$ of about ± 250 bar, in agreement with that obtained using Eq. (4.20).

The result of the calculation of the P-T curve for the diamond = graphite reaction appears very promising indeed, when considered in terms of its agreement with direct experimental reversal or for its surprisingly small uncertainty. For most geologically relevant solid-solid reactions, however, the general picture is by far less encouraging, *if* the computations are executed *exclusively* on the basis of a *calorimetrically derived* thermodynamic database. The next two examples will emphasize this.

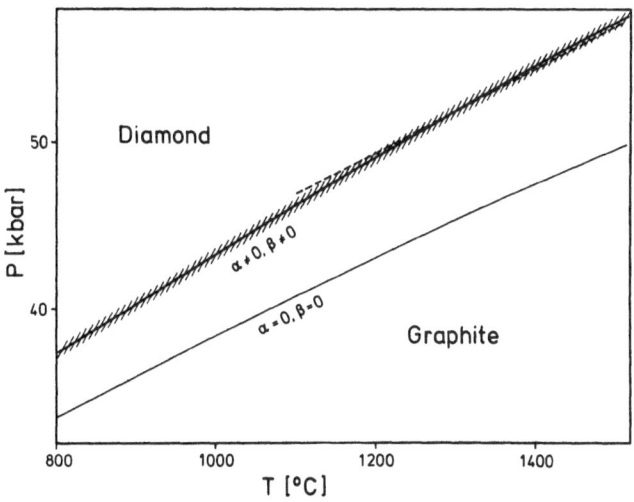

Fig. 4.2 A comparison of calculated and experimental P-T curves of the reaction diamond = graphite. The *heavier solid line* shows the results of calculation with $\alpha(T)$ and $\beta(P)$, while the *thinner line* represents those obtained ignoring $\alpha(T)$ and $\beta(P)$. The experimental reversal data by Kennedy and Kennedy (1976) are reproduced by the *dashed straight line*. Note that the experiments confirm the result of computation with nonzero $\alpha(T)$ and $\beta(P)$. The exceptionally narrow $2\sigma_T$ uncertainty envelope has been *hatched* for clarity

Problem Set 4.2, with Solutions

1. Compute the P-T curve of the reaction kyanite (Ky) = andalusite (And) between 300 and 900 °C, using the data from Table 4.5.
2. Estimate $2\sigma_T$ of the Ky = And reaction from (4.19), disregarding any covariance in the database. How do the reversal experiments on the Ky = And reaction (Newton 1966a; Richardson et al. 1969; Holdaway 1971) compare with the calculations?

Problem 1. The calculation of the P-T curve of the reaction Ky = And followed the algorithm outlined above. The results are given in Table 4.6. Contrary to our experience with the D = G reaction, the P-T curves for the Ky = And reaction, computed with or without $\alpha(T)$ and $\beta(P)$ differ by less than 500 bar. As emphasized earlier, the extent to which the two curves are likely to deviate from each other depends upon the relative contributions of $\alpha(T)$ and $\beta(P)$ to ΔG^* of the reaction. In the last two examples, ΔG°_T is of the same order of magnitude. Thus, the difference in the results must be attributed to the large difference in the equilibrium pressure.

Problem 2. The estimation of $2\sigma_T$ uses Eq.(4.19). For 873.15 K and 4837 bar (a point on the P-T curve, see Table 4.6), we have

$2\sigma_{\Delta G^*} = 2837.1$ J, $\Delta S^\circ_T = 8.49$ J K^{-1}, and $2\sigma_T = 334$ K.

Table 4.5. Thermodynamic data for calculating the kyanite = andalusite P-T curve

	Kyanite	Andalusite
$H_{f,298}^{\circ}$ (J mol^{-1})[a]	-2591730.00 (1900)[d]	-2587525.00 (2100)
S_{298}° (J K^{-1} mol^{-1})[b]	82.30 (0.13)	91.39 (0.14)
V_{298}° (J bar^{-1} mol^{-1})[b]	4.415 (0.005)	5.152 (0.005)
$C_P^{\circ}(T)$ (J K^{-1} mol^{-1})[b] = $a + b \cdot T + c \cdot T^{-2} + d \cdot T^{-0.5}$, with		
a	3.039E2	2.904E2
b	-1.339E-2	-1.052E-2
c	-8.952E5	-1.109E6
d	-2.9043E3	-2.6278E3
Validity (T in K)	250-1600	250-1600
α_o (K^{-1})[c]	6.5593E-6	5.8041E-6
α_1 (K^{-2})[c]	2.4189E-8	3.4619E-8
β_o (bar^{-1})[c]	0.70E-6	0.67E-6
β_1 (bar^{-2})[c]	0.0	0.0

[a] Robie et al. (1979).
[b] Robie and Hemingway (1984a).
[c] Clark (1966).
[d] Quantities in parentheses are 2σ.

Table 4.6. Results of calculation of the kyanite = andalusite P-T curve

T(°C)	T(K)	α,β ignored P(bar)	α,β considered P(bar)	$2\sigma_T$(K)
300	573.15	1408	1392	(311)
400	673.15	2631	2586	(318)
500	773.15	3827	3736	(326)
600	873.15	4993	4837	(334)
700	973.15	6129	5887	(344)
800	1073.15	7237	6886	(352)
900	1173.15	8317	7833	(361)

Figure 4.3 depicts the calculated P-T curve with nonzero $\alpha(T)$ and $\beta(P)$, along with its $2\sigma_T$ uncertainty envelope. For comparison, the reversal brackets by Newton (1966a), Richardson et al. (1969), and Holdaway (1971) are also displayed. Note that only some of the brackets cut the computed P-T curve, though all are located within the $2\sigma_T$ envelope. This type of agreement is quite typical for the majority of the geologically relevant solid-solid reactions.

Table 4.7. Thermodynamic data for calculating the P-T curve of the reaction 2 jadeite = 1 nepheline + 1 low albite. Unless otherwise stated, all data are from Robie et al. (1979)

T(K)	600	700	800	900	1000	1100
ΔG°_T(J)	-7747	-14445	-21121	-27797	-34373	-40998

Phases	$2\sigma_{H^\circ_{f,298}}$ (J mol^{-1})	S°_{298} (J K^{-1} mol^{-1})	$2\sigma_{S^\circ_{298}}$ (J K^{-1} mol^{-1})	V°_{298} (J bar^{-1} mol^{-1})	$2\sigma_{V^\circ_{298}}$ (J bar^{-1} mol^{-1})
Jadeite	4180	133.47	1.25	6.040	0.010
Nepheline	2420	124.35	1.25	5.416	0.006
Low albite	3435	207.40	0.40	10.007	0.013

Phases	α_o[a] (K^{-1})	α_1[a] (K^{-2})	β_o[a] (bar^{-1})	β_1[a] (bar^{-2})
Jadeite	1.3319E-5	2.3072E-8	0.75E-6	0.0
Nepheline	1.6304E-5	5.2520E-8	2.05E-6	5.2E-12
Low albite	1.2955E-5	1.9194E-8	2.02E-6	21.6E-12

[a] Clark (1966).

Problem Set 4.3, with Solutions

1. Compute the P-T curve of the reaction 2 jadeite = 1 nepheline + 1 low albite between 600 and 1100 K, using the thermodynamic data from Robie et al. (1979).
2. Estimate $2\sigma_T$ of the computed P-T curve and compare the results with the reversal brackets by Newton and Kennedy (1968).

Problem 1. This time, we need to do the calculations between 600 and 1100 K. If we choose to execute them in steps of 100 K, we can skip the initial chore of computing ΔG°_T from Eq.(4.4), because it follows directly from $G^\circ_{f,T}$ listed by Robie et al. (1979) for all pertinent temperatures. The values of ΔG°_T, thus derived, and some ancillary input data are summarized in Table 4.7. The results of calculation of the P-T curve are displayed in Table 4.8. Note that the equilibrium pressures calculated with or without $\alpha(T)$ and $\beta(P)$ differ by barely 50 bar. This is owing to two reasons. First, the equilibrium pressures are quite low, and secondly, ΔG°_T is fairly large compared to those of the two reactions considered earlier.

Table 4.8. Results of calculation of the 2 jadeite = 1 nepheline + 1 low-albite P-T curve

T(K)	α,β ignored P(bar)	α,β considered P(bar)	$2\sigma_T$(K)
600	2318	2306	(141)
700	4322	4304	(143)
800	6319	6297	(145)
900	8316	8290	(146)
1000	10283	10247	(148)
1100	12265	12211	(149)

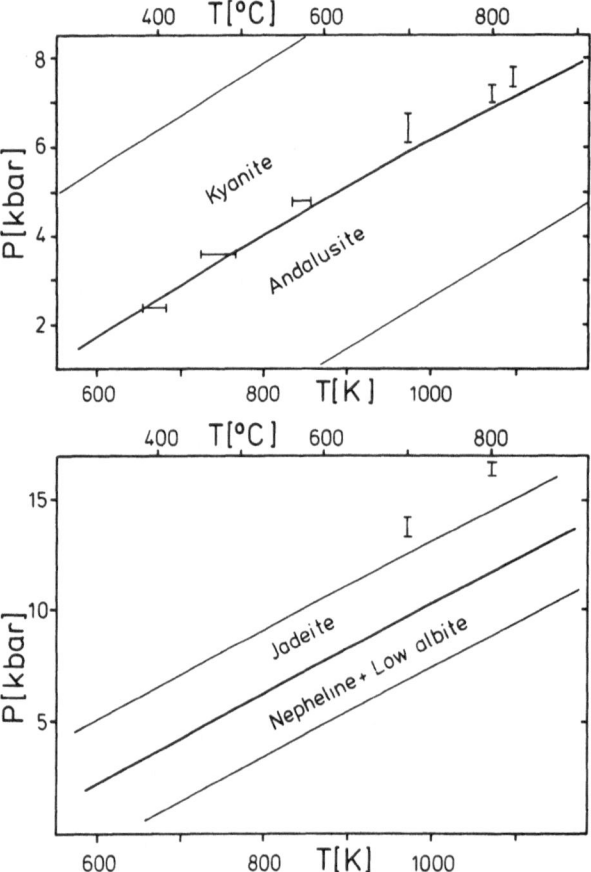

Fig. 4.3 Comparison of experimental reversal brackets of the reactions kyanite = andalusite (*above*) and 2 jadeite = 1 nepheline + 1 low albite (*below*) with the P-T curves calculated on the basis of calorimetric data. Note the large $2\sigma_T$ ranges shown by the pair of thinner lines

Problem 2. To calculate $2\sigma_T$, the 2σ data tabulated in Robie et al. (1979), and reproduced in Table 4.7, are used. The requisite data on $\Delta S°_T$ are also taken from Robie et al. (1979). For one point on the equilibrium curve, say 800 K and 6297 bar, we obtain

$$2\sigma_{\Delta G^*} = 9626.6 \text{ J, and } \Delta S°_{800} = 66.59 \text{ J K}^{-1},$$

such that $2\sigma_T$ amounts to $\pm$ 145 K (Table 4.8). Figure 4.3 displays these results and compares them with the reversal data (Newton and Kennedy 1968). Granting that the reversal brackets are correct, there is no denying that calorimetry and phase equilibria data are inconsistent on a $2\sigma_T$ level. This is striking, especially because no effort has been made to sort out any possible covariance prior to estimating $2\sigma_T$. The removal of covariance, if any, would have exacerbated this inconsistency.

4.5 Concluding Remarks

Calculation of solid-solid phase equilibria is best performed employing data on expansivity and compressibility of all pertinent phases. Disregarding these might lead to serious error, unless the equilibrium curve is located at a pressure well below 10 kbar, and $\Delta G°_T$ of the reaction is large.

Comparison of experimental reversal brackets with P-T curves of solid-solid reactions computed exclusively from calorimetric data warrants the following conclusions:

1. Although the reversal brackets seldom straddle the computed P-T curves, they generally lie within their $2\sigma_T$ envelope. That is, the two categories of data are mutually compatible on a 95% confidence level.

2. The $2\sigma_T$ of P-T curves of the solid-solid reactions calculated from calorimetric data is generally on the order of a few hundreds of degrees, corresponding to $2\sigma_P$ of several kbar. In this respect, the D = G reaction is an exception. Because the reversal brackets are very narrow by comparison, we may use them to put restrictions on the calorimetric data.

An immediate consequence of these facts is that geologically meaningful phase diagrams can hardly be expected from calorimetric data *alone*. Nevertheless, calorimetry is, and will remain, the most fundamental source of thermodynamic data. In Chapter 7, it will be demonstrated that a simultaneous treatment of mutually compatible calorimetric and phase equilibria data provides what is known as an internally consistent thermodynamic dataset. This type of a database is capable of generating reliable phase diagrams.

Chapter 5

Phase Relations Among End-Member Solids and a Pure Fluid

5.1 Introduction

Phase equilibria between mineral end-members and a pure fluid is a limiting case of solid-fluid equilibria in nature. Nevertheless, in the step by step approach adopted in this book, we shall handle this simple case as a prelude to elucidating the more demanding topic of mineral-fluid equilibria in multicomponent systems (Chap. 9). The objective of this chapter is to calculate phase relations between mineral end-members and a pure fluid, such as water or oxygen. Other fluids will be amenable to treatment in an analogous manner.

5.2 Theoretical Background

The theoretical framework required to deal with the problem will be developed from the *general case* of a solid-fluid reaction, where the solids are not end-members, nor is the fluid a pure one. Having achieved that goal, we shall switch to the *special case* of a reaction between mineral end-members and a pure fluid. In either case, the starting point is the mass balance equation written in terms of the components of the phases participating in a reaction. Such a reaction may be generalized as,

$$bB = cC + dD + \nu H_2O, \tag{5.1}$$

in which the reactant, b moles of the solid B, decomposes to yield the reaction products, c and d moles of the solids C and D, plus ν moles of H_2O. The condition of equilibrium for (5.1) is expressed as [cf. Eq.(1.100a)],

$$\Delta\mu = 0 = c\mu_C + d\mu_D - b\mu_B + \nu\mu_{H_2O}. \tag{5.2}$$

Manipulation of this energy balance equation requires, first of all, specification of standard states. For the present purpose, the standard state for the solids is chosen as the pure substance at 1 bar pressure and T of interest. For the fluid, the preferred choice of standard state (see Chap. 3) is the hypothetical ideal gas at 1 bar pressure and at a specified T. The chemical potential of the solid component C, μ_C, at any T and P must then be written in analogy to Eq.(3.1),

$$\mu_C = G^\circ_{f,T,C} + \int_1^P V_{T,C}(P)dP + RT\ln a_C .$$ (5.3)

The chemical potential of H_2O follows from Eq.(3.9a),

$$\mu_{H_2O} = G^\circ_{f,T,H_2O} + RT\ln f^{\nabla}_{H_2O} ,$$ (5.4)

f^{∇}_{H2O} being a dimensionless quantity, equal to f''_{H2O}/f°_{H2O}.
    Inserting the expressions for the relevant chemical potentials into (5.2), and collecting terms, the condition of equilibrium for the reaction (5.1) becomes

$$\Delta\mu = 0 = [cG^\circ_{f,T,C} + dG^\circ_{f,T,D} - bG^\circ_{f,T,B} + \nu G^\circ_{f,T,H_2O}]$$

$$+ \int_1^P [cV_C + dV_D - bV_B]_T(P)dP + RT\ln\frac{a_C^c a_D^d (f^{\nabla}_{H_2O})^\nu}{a_B^b} .$$ (5.5)

The last term on the RHS of (5.5) includes the equilibrium constant K,

$$K = \frac{a_C^c a_D^d (f^{\nabla}_{H_2O})^\nu}{a_B^b} .$$ (5.6)

Note that K now comprises both the activities and the fugacities, no longer just the activities, as in Eq.(1.104). That is owing to our using the 1 bar ideal gas standard state for H_2O, rather than its unit activity standard state at P and T. The ultimate result is the same, however, no matter which standard state is employed.
    Substituting (5.6) into (5.5), the latter may be condensed to

$$\Delta\mu = 0 = \Delta G^\circ_T + \int_1^P \Delta V_{T,s}(P)dP + RT\ln K ,$$ (5.7)

with ΔV_s denoting the volume difference due to the solids. For the *special case* of a reaction involving mineral end-members only, all activity terms in (5.6) are unity. And from Eq.(3.10) it is also clear that for a pure fluid species i, f^{∇}_i is identical to f^*_i. Accordingly, (5.6) reduces to

$$K = (f^*_{H_2O})^\nu .$$ (5.8)

Recalling that for a reaction involving pure substances, $\Delta\mu = \Delta G^*$, and inserting (5.8) into (5.7), the equilibrium condition for the solid-fluid reaction between the mineral end-members and pure H_2O is recovered,

$$\Delta G^* = 0 = \Delta G^\circ_T + \int_1^P \Delta V_{T,s}(P)dP + \nu RT\ln f^*_{H_2O} .$$ (5.9)

To proceed further with the calculation of the equilibrium for the reaction (5.1), the $\Delta G°_T$ term of (5.9) is expanded following (4.4)

$$\Delta G^* = 0 = \Delta H°_{298} - T\Delta S°_{298} + \int\limits_{298}^{T} \Delta C°_P(T)dT - T\int\limits_{298}^{T}\left[\frac{\Delta C°_P(T)}{T}\right]dT$$

$$+ \int\limits_{1}^{P} \Delta V_{T,s}(P)dP + \nu RT\ln f^*_{H_2O}.\tag{5.10}$$

Thus, given the thermodynamic data for the phases of interest, and the fugacity of H_2O at T and P, (5.10) is easily evaluated.

Two approximations are made from time to time in using Eq.(5.10):
1. If data on $\alpha(T)$ and $\beta(P)$ of the pertinent solids are lacking, the $\int_1^P\Delta V_{T,s}(P)dP$ term is approximated by $\Delta V°_{298,s}(P-1)$. As will be demonstrated below, this might lead to some error, although its magnitude is likely to be small compared to solid-solid reactions. Whenever $\alpha(T)$ and $\beta(P)$ data are available, the $\int_1^P\Delta V_{T,s}(P)dP$ term should be evaluated using the format of Eq.(4.14).
2. To the extent that $C°_P(T)$ data of any of the phases are lacking, $\Delta C°_P$ is approximated as zero. This is a risky procedure, because the $\Delta C°_P(T)$-integrals generally contribute substantially to $\Delta G°_T$ of the reaction and, therefore, ignoring them might cause large error in the computed curve. Problem set 5.2, given later, puts across this message. Owing to the proliferation of $C°_P(T)$ measurements by DSC, this approximation is now seldom required.

5.3 Error Propagation Formalism for Solid-Fluid Equilibria

The basic formalism of error propagation was introduced in the last chapter with special reference to solid-solid reactions. This treatment will be extended now to solid-fluid equilibria, keeping reiterations to a minimum.

First, we rewrite the general condition of equilibrium of a solid-fluid reaction in a condensed form analogous to Eq.(4.18),

$$\Delta G^* = \Delta H°_T - T\Delta S°_T + \Delta V°_{298,s}(P-1) + \nu RT\ln f^*_{H_2O} + C,\tag{5.10a}$$

with C comprising the $\alpha(T)$ and $\beta(P)$ terms due to the solids taking part in a reaction. Applying (4.16) to the function $\Delta G^* = f(T,P)$, T and P being non-covariant,

$$\sigma^2_{\Delta G^*} = \left(\frac{\partial\Delta G^*}{\partial T}\right)^2_P\sigma^2_T + \left(\frac{\partial\Delta G^*}{\partial P}\right)^2_T\sigma^2_P.\tag{5.11a}$$

Taking its derivatives with respect to (5.10a),[1]

$$\sigma_{\Delta G^*}^2 = \left[-\Delta S_T^{\circ} + \nu \left\{ R \ln f_{H_2O}^* + \frac{RT}{f_{H_2O}^*} \left(\frac{\partial f_{H_2O}^*}{\partial T} \right) \right\} \right]^2 \sigma_T^2$$

$$+ [\Delta V_{298,s}^{\circ} + \nu V_{H_2O}]^2 \sigma_P^2 . \tag{5.11}$$

Therefore, at any point on the equilibrium curve, we have

$$\sigma_T = \frac{\sigma_{\Delta G^*}}{\left| -\Delta S_T^{\circ} + \nu \left\{ R \ln f_{H_2O}^* + \frac{RT}{f_{H_2O}^*} \left(\frac{\partial f_{H_2O}^*}{\partial T} \right) \right\} \right|}, \tag{5.12a}$$

$$\sigma_P = \frac{\sigma_{\Delta G^*}}{|\Delta V_{298,s}^{\circ} + \nu V_{H_2O}|} . \tag{5.12b}$$

To calculate $\sigma_{\Delta G^*}$ of any solid-fluid reaction, we apply Eq.(4.16) to the function $\Delta G^* = f(\Delta H^{\circ}_T, \Delta S^{\circ}_T, \Delta V^{\circ}_{298,s}, R T \ln f^*_{H2O})$. Again, two alternative cases must be distinguished. To the extent that we use $\Delta H^{\circ}_{<T>}$ and $\Delta S^{\circ}_{<T>}$ extracted from phase equilibria reversals, these two will be correlated (see Sect.5.5). However, for now, we shall utilize a calorimetric database for phase equilibria calculations. In this case, the independent variables of the above function will be non-covariant, and applying (4.16) we obtain

$$\sigma_{\Delta G^*}^2 = \left(\frac{\partial \Delta G^*}{\partial \Delta H_T^{\circ}} \right)^2 \sigma_{\Delta H_T^{\circ}}^2 + \left(\frac{\partial \Delta G^*}{\partial \Delta S_T^{\circ}} \right)^2 \sigma_{\Delta S_T^{\circ}}^2$$

$$+ \left(\frac{\partial \Delta G^*}{\partial \Delta V_{298,s}^{\circ}} \right)^2 \sigma_{\Delta V_{298,s}^{\circ}}^2 + \left(\frac{\partial \Delta G^*}{\partial R T \ln f_{H_2O}^*} \right)^2 \sigma_{R T \ln f_{H_2O}^*}^2 . \tag{5.13a}$$

Taking the derivatives with respect to the RHS of (5.10a),

$$\sigma_{\Delta G^*} = [\sigma_{\Delta H_T^{\circ}}^2 + T^2 \cdot \sigma_{\Delta S_T^{\circ}}^2 + (P-1)^2 \cdot \sigma_{\Delta V_{298,s}^{\circ}}^2 + \nu^2 \cdot \sigma_{R T \ln f_{H_2O}^*}^2]^{0.5}. \tag{5.13}$$

Since $\sigma_{R T \ln f^* H2O}$ is in reality unknown, in analogy to $\sigma_{V H2O}$, it is estmated as 0.5% of $R T \ln f^*_{H2O}$. It should be noted that, apart from very high T and P, the largest contribution to $\sigma_{\Delta G^*}$ is due to the $\sigma^2_{\Delta H^{\circ} T}$ term. Therefore, any fallacy in our judgement of $\sigma_{R T \ln f^* H2O}$ is not likely to affect the results significantly.

---

[1] Obtaining $(\partial \Delta G^*/\partial T)_P$ with respect to (5.10a) may be achieved in two steps. Recalling the derivation of (4.17), we start by writing

$$
\begin{aligned}
(\partial \Delta G^*/\partial T)_P \ &= -\Delta S^{\circ}_T + \partial/\partial T [\nu R T \ln f^*_{H2O}] \\
&= -\Delta S^{\circ}_T + \nu [1 \cdot R \ln f^*_{H2O} + R T \cdot \partial/\partial T (\ln f^*_{H2O})] \\
&= -\Delta S^{\circ}_T + \nu [R \ln f^*_{H2O} + R T (1/f^*_{H2O}) \cdot \partial/\partial T (f^*_{H2O})] \\
&= -\Delta S^{\circ}_T + \nu [R \ln f^*_{H2O} + (R T/f^*_{H2O})(\partial f^*_{H2O}/\partial T)]
\end{aligned}
$$

5.4 Sample Calculations of Dehydration Equilibria

Using the thermodynamic formalisms derived above and the data listed in Table 5.1, let us calculate the P_{H2O}-T curves for three dehydration reactions. The exercises have been designed primarily to demonstrate the effects of the aforementioned approximations on the calculated equilibria, and a few other features of solid-fluid equilibria.

Problem Set 5.1

1. Employing the data from Table 5.1, list $\Delta G°_T$ for the reaction

$$2 \text{ diaspore (D)} = 1 \text{ corundum (C)} + 1 \text{ } H_2O, \hspace{3em} (5.14)$$

between 380 and 540°C in steps of 20°C. How much do the $\Delta C°_p(T)$-integrals of Eq.(5.10), combined, contribute to the $\Delta G°_T$ of this reaction?
2. Compute the P-T curve of the reaction between 380 and 540°C in steps of 20°C, using the complete set of input data of Table 5.1.
3. Repeat computation of the P-T curve, ignoring $\alpha(T)$ and $\beta(P)$. Do the same calculations over again, setting $\Delta C°_p = 0$, in addition to ignoring $\alpha(T)$ and $\beta(P)$. What difference does that make in terms of temperature?
4. Estimate $2\sigma_T$ of the computed P-T.
5. Compare the results with the experimental reversals obtained by Haas (1972).

Solutions to the Problem Set 5.1

Problem 1. In order to estimate the contribution of the sum of the $\Delta C°_p(T)$-integrals to $\Delta G°_T$ of the reaction (5.14), the relevant quantities have been evaluated and shown in Table 5.2. It is clear that the $\Delta C°_p(T)$-integrals, taken together, account for up to 1% of $\Delta G°_T$. This is only a fraction of $\sigma_{\Delta G*}$ of the reaction, which is around 10 kJ. In such cases, the approximation $\Delta C°_p = 0$ is clearly acceptable. It cannot be overemphasized, however, that this is an exceptional case. Generally, the $\Delta C°_p(T)$-integrals contribute very substantially to $\Delta G°_T$ of the reactions, and thus, $\Delta C°_p = 0$ cannot be regarded as a viable approximation. It is *not* recommended for a routine calculation of P-T curves.

Problem 2. To calculate the P-T curve of reaction (5.14), Eq.(5.10) must be solved for P at a given T. The results of such calculations for the 500°C isotherm, based on the complete set of input data indicated in Table 5.1, are listed in Table 5.3. Note that ΔG^* changes its sign just above 8000 bar, indicating an equilibrium P slightly in excess of that pressure. For an explicit computation of the P of equilibrium, (5.10) may be solved for by a graphical technique or by iteration, in a manner akin to that used for solid-solid reactions in Chap. 4. A graphical solution for P at 500°C is obtained from a

Table 5.1. Thermodynamic data for the problem sets 5.1 to 5.3

Phases		$H^\circ_{f,298}$ (J mol^{-1})	S°_{298} (J K^{-1} mol^{-1})	V°_{298} (J bar^{-1} mol^{-1})
Diaspore (D)	AlO(OH)	-1000585.0(5000)[a]	35.27(0.17)	1.7760(0.0026)
Corundum (C)	Al_2O_3	-1675700.0(1300)	50.92(0.10)	2.5575(0.0007)
Kaolinite (Ka)	$Al_2[Si_2O_5](OH)_4$	-4120114.0(3975)	203.05(1.25)	9.9520(0.0260)
Pyrophyllite (Py)	$Al_2[Si_4O_{10}](OH)_2$	-5639800.0(3950)	239.40(0.40)	12.7820(0.0290)
Brucite (Br)	$Mg(OH)_2$	-924540.0(440)	63.18(0.13)	2.4630(0.0070)
Periclase (P)	MgO	-601490.0(290)	26.94(0.17)	1.1248(0.0004)
H_2O		-241814.0(42)	188.83(0.04)	-

$$C^\circ_p(T)(J\ K^{-1}\ mol^{-1})^{d)} = a + b\cdot T + C\cdot T^{-2} + d\cdot T^{-0.5} + e\cdot T^2 + f\cdot T^{-3}$$

Phases	a	b	c	d	e	f
Diaspore	143.24	0.0	-3.231E5	-1.5404E3	0.0	6.463E7
Corundum	157.36	7.1899E-4	-1.8969E6	-9.8804E2	0.0	0.0
Kaolinite	523.23	0.0	-2.2443E6	-4.4267E3	0.0	9.231E7
Pyrophyllite	665.93	0.0	-4.9799E6	-5.8974E3	0.0	6.6181E8
Brucite	136.84	0.0	-4.3619E6	-5.371E2	0.0	5.5269E8
Periclase	65.211	-1.2699E-3	-4.6185E5	-3.8724E2	0.0	0.0
H_2O	7.368	2.7468E-2	-2.2316E5	3.6174E2	-4.8117E-6	0.0

Phases	α_o(K^{-1})	α_1(K^{-2})	β_o(bar^{-1})	β_1(bar^{-2})
Diaspore[b]	2.9842E-5	0.0	0.6194E-6	0.0
Corundum[c]	2.0445E-5	6.8913E-9	0.363E-6	0.0
Brucite[b]	3.2481E-5	0.0	1.218E-6	0.0
Periclase[c]	3.1224E-5	1.3716E-8	0.596E-6	0.83E-12

[a] Quantities in parentheses are 2σ.
[b] From Holland and Powell (1985).
[c] From Clark (1966).
[d] The high-temperature $C^\circ_p(T)$ measurements of diaspore, kaolinite, pyrophyllite and brucite extend to only 509, 560, 679, and 666 K, respectively. To ensure proper extrapolation to high temperatures, their $C^\circ_p(T)$-functions given by Berman and Brown (1985) are used. All other data are from Robie et al. (1979).

Table 5.2. ΔG_T° of the reaction 2 diaspore = 1 corundum + 1 H_2O

T (°C)	T (K)	$\Delta H_{298}^\circ - T \cdot \Delta S_{298}^\circ$ (J)	$_{298}\!\int^T \Delta C_P^\circ dT - T \cdot_{298}\!\int^T (\Delta C_P^\circ/T) dT$ (J)	ΔG_T° (J)
380	653.15	-26863.5	-61.9	-26925.4
400	673.15	-30247.7	-15.8	-30263.5
420	693.15	-33631.9	38.7	-33593.2
440	713.15	-37016.1	101.6	-36914.5
460	733.15	-40400.3	173.3	-40227.0
480	753.15	-43784.5	253.7	-43530.8
500	773.15	-47168.7	343.0	-46825.7
520	793.15	-50552.9	441.2	-50111.7
540	813.15	-53937.1	548.5	-53388.6

Table 5.3. Database for the graphical solution of the equilibrium pressure of the reaction 2 diaspore = 1 corundum + 1 H_2O at 500°C

T (K)	ΔG_T° (J)	$_1\!\int^P \Delta V_{T,s}(P) dP$ (J)	$RT \ln f_{H_2O}^*$ (J)	ΔG^* (J)	P (bar)
773.15	-46825.7	-3039.9	44693.5	-5172.1	3000
773.15	-46825.7	-4051.0	46933.2	-3943.5	4000
773.15	-46825.7	-5060.9	49043.9	-2842.7	5000
773.15	-46825.7	-6069.4	51066.5	-1828.6	6000
773.15	-46825.7	-7076.7	52994.5	-907.9	7000
773.15	-46825.7	-8082.8	54868.7	-39.8	8000
773.15	-46825.7	-9087.5	56696.5	+783.3	9000
773.15	-46825.7	-10091.0	58478.4	+1561.7	10000

ΔG^* vs P plot of the data in Table 5.3 (solid curve of Fig. 5.1). The equilibrium pressure, P, read off the graph, is about 8050 bar. A more precise solution for P can be obtained iteratively.

Problem 3. To repeat the calculations with $\alpha(T) = 0$ and $\beta(P) = 0$, $\Delta V^\circ_{298,s}(P-1)$ was substituted for $_1\!\int^P \Delta V_{T,s}(P) dP$ in (5.10). The results for the 500°C isotherm are listed in Table 5.4. This time, the sign of ΔG^* changes between 7000 and 8000 bar, indicating an equilibrium pressure below 8000 bar. Again P was read off the ΔG^* vs P graph of Fig. 5.1 (dashed curve), and found to be about 7900 bar. The results of the iterative solutions, including those derived from the model $\Delta C^\circ_P = 0$, have been reproduced in Table 5.5 and displayed as a P_{H2O} vs T plot in Fig. 5.2. It is clear that ignoring $\alpha(T)$ and $\beta(P)$ leads to an

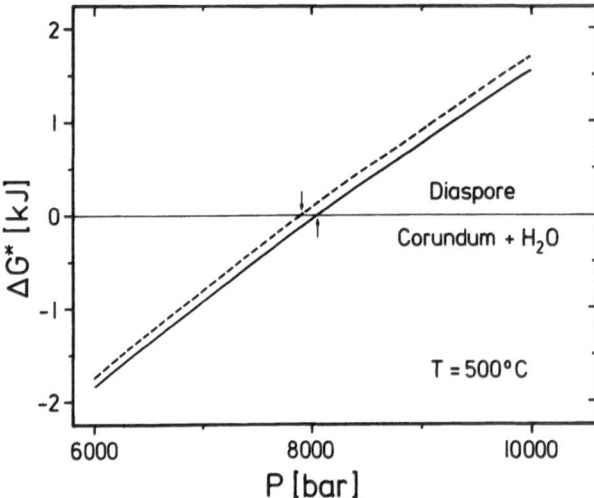

Fig. 5.1 A ΔG^* vs P plot at 500°C for the reaction 2 diaspore = 1 corundum + 1 H_2O. The *solid line* depicts the ΔG^* computed with nonzero $\alpha(T)$ and $\beta(P)$, the *dashed line* with $\alpha(T) = 0$ and $\beta(P) = 0$. The equilibrium pressure corresponding to each set of calculations is indicated by an *arrow*, and can be read from the graph

Table 5.4. Database for the graphical solution of the equilibrium pressure of the reaction 2 diaspore = 1 corundum + 1 H_2O at 500°C, ignoring $\alpha(T)$ and $\beta(P)$

T	ΔG°_T	$\Delta V^\circ_{298,s}(P-1)$	$RTln f^*_{H_2O}$	ΔG^*	P
(K)	(J)	(J)	(J)	(J)	(bar)
773.15	-46825.7	-2982.5	44693.5	-5114.7	3000
773.15	-46825.7	-3977.0	46933.2	-3869.5	4000
773.15	-46825.7	-4971.5	49043.9	-2753.3	5000
773.15	-46825.7	-5966.0	51066.5	-1725.2	6000
773.15	-46825.7	-6960.5	52994.5	-791.7	7000
773.15	-46825.7	-7955.0	54868.7	+88.0	8000
773.15	-46825.7	-8949.5	56696.5	+921.3	9000
773.15	-46825.7	-9944.0	58478.4	+1708.7	10000

error of up to only 2°C in this equilibrium curve. Even the approximation $\Delta C^\circ_P = 0$, over and above ignoring $\alpha(T)$ and $\beta(P)$, causes a maximum error of 5°C. Given the trivial contributions of the two $\Delta C^\circ_P(T)$-integrals (cf. Table 5.2) to ΔG°_T, this is not surprising. It is emphasized again that these results should not be construed to imply that such simplifications are routinely permissible. More often, these terms have a dramatic effect on the calculated P-T curve, as demonstrated by the problem set 5.2 (see below).

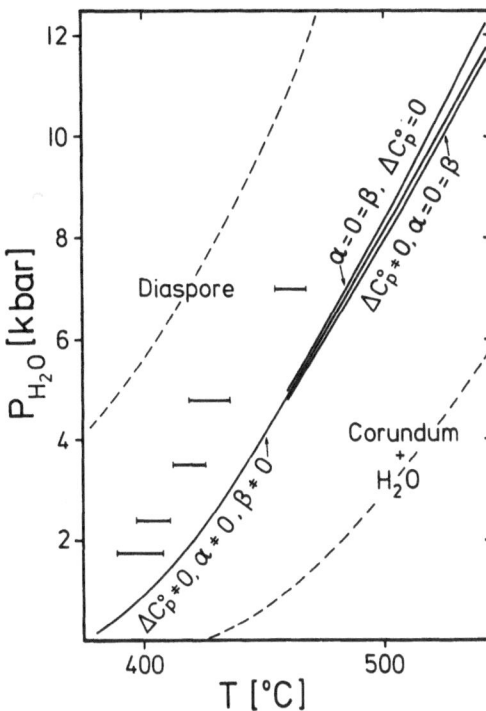

Fig. 5.2 A P_{H_2O}-T   diagram showing the computed univariant curve for the reaction 2 diaspore = 1 corundum + 1 H_2O. The reversals by Haas (1972) are compatible with the computed P-T curve within the range of one σ_T, shown by the *dashed lines*

Table 5.5. The calculated equilibrium pressures for the reaction 2 diaspore = 1 corundum + 1 H_2O between 380 - 540°C, with or without approximations. The $2\sigma_T$ envelope is given only for the first set of calculations

		$\alpha\neq0$, $\beta\neq0$, $\Delta C_p^\circ\neq0$		$\alpha=0$, $\beta=0$, $\Delta C_p^\circ\neq0$	$\alpha=0$, $\beta=0$, $\Delta C_p^\circ=0$
T(°C)	T(K)	P(bar)	$2\sigma_T$(K)	P(bar)	P(bar)
380	653.15	224	94	223	217
400	673.15	1071		1060	1050
420	693.15	2220	126	2192	2222
440	713.15	3531		3480	3570
460	733.15	5032	142	4950	5127
480	753.15	6508		6394	6670
500	773.15	8044	152	7897	8298
520	793.15	9673		9484	10044
540	813.15	11409	153	11178	11919

Problem 4. The estimation of the uncertainty of the calculated P-T curve utilized Eq.(5.12a). The results are listed in Table 5.5, and depicted in Fig. 5.2. Note that due to the large temperature scale chosen to distinguish the computed P-T curves in Fig. 5.2, only one σ_T envelope (pair of thin lines) could be displayed.

Problem 5. The reversal brackets, obtained by Haas (1972), for the reaction (5.14) have been entered in Fig. 5.2. None of them straddle the computed P-T curve; however, they agree within $\pm 1\sigma_T$.

Problem Set 5.2

1. Using the data from Table 5.1, calculate $\Delta G°_T$ for the reaction

$$1 \text{ brucite (Br)} = 1 \text{ periclase (P)} + 1 \text{ H}_2\text{O} \tag{5.15}$$

between 540-780°C at intervals of 40°C. How much does the sum of the $\Delta C°_P$-integrals contribute this time to $\Delta G°_T$?

2. Compute the P-T curve for this reaction in the range 540-780°C with and without the $\alpha(T)$ and $\beta(P)$ terms. Repeat the calculations, ignoring $\alpha(T)$ and $\beta(P)$ and assuming $\Delta C°_P = 0$. Compare the results on a synoptic P-T diagram, and contrast them with the experimental counterpart (Barnes and Ernst 1963; Schramke et al. 1982). Also, estimate its $2\sigma_T$ at several points on the calculated curve.

Solutions to the Problem Set 5.2

Problem 1. The $\Delta G°_T$ of the reaction (5.15) and the contribution of the two $\Delta C°_P(T)$-integrals to $\Delta G°_T$ have been itemized in Table 5.6. This time, the sum of the two $\Delta C°_P(T)$-integrals contributes more than 10% to $\Delta G°_T$. That is an order of magnitude larger than its $\sigma_{\Delta G^*}$ (which is around 500 to 700 J). We thus anticipate that ignoring the $\Delta C°_P(T)$-integrals will lead to an erroneous P-T curve (see below).

Problem 2. The results of calculation of the P_{H2O} - T curve of the reaction (5.15) according to three different models: (1) $\Delta C°_P \neq 0$, $\alpha(T) \neq 0$, $\beta(P) \neq 0$; (2) $\Delta C°_P \neq 0$, but with $\alpha(T) = 0$ and $\beta(P) = 0$; and (3) both $\Delta C°_P = 0$ as well as

Table 5.6. $\Delta G°_T$ of the reaction 1 brucite = 1 periclase + 1 H_2O

T (°C)	T (K)	$\Delta H°_{298} - T \cdot \Delta S°_{298}$ (J)	$_{298}\int^T \Delta C°_P dT - T \cdot_{298}\int^T (\Delta C°_P/T) dT$ (J)	$\Delta G°_T$ (J)
540	813.15	-42842.6	4597.6	-38245.0
580	853.15	-48946.2	5328.1	-43618.0
620	893.15	-55049.8	6103.2	-48946.6
660	933.15	-61153.4	6920.7	-54232.7
700	973.15	-67257.0	7779.2	-59477.8
740	1013.15	-73360.6	8676.4	-64684.2
780	1053.15	-79464.2	9610.8	-69853.4

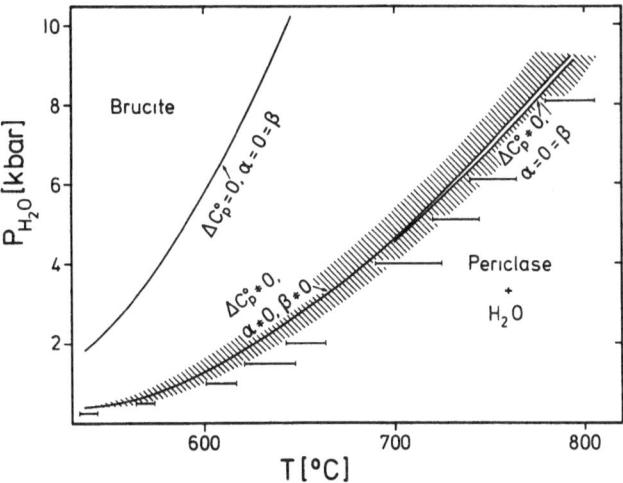

Fig. 5.3 A P_{H_2O}-T diagram showing the computed univariant curve for the reaction 1 brucite = 1 periclase + 1 H_2O. In this pressure range, $\alpha(T)$ and $\beta(P)$ barely affect the equilibrium. Ignoring the ΔC_P°-integrals, by contrast, leads to an erroneous result, because of their large contribution to the ΔG_T° of this reaction (see Table 5.6). Also shown are the reversals to 2 kbar by Barnes and Ernst (1963), and those between 4 and 8.1 kbar by Schramke et al. (1982); they only marginally agree with the $2\sigma_T$ range (*hatched*) of the calculated P-T curve

Table 5.7. Results of calculation of the P_{H_2O} - T curve of the reaction 1 brucite = 1 periclase + 1 H_2O. The $2\sigma_T$ envelope has been obtained only for the first set of calculations

	(a) $\alpha\neq0$, $\beta\neq0$, $\Delta C_P^\circ\neq0$			(b) $\alpha=0$, $\beta=0$, $\Delta C_P^\circ\neq0$	(c) $\alpha=0$, $\beta=0$, $\Delta C_P^\circ=0$
T(°C)	T(K)	P(bar)	$2\sigma_T$(K)	P(bar)	P(bar)
540	813.15	404	7.8	404	1891
580	853.15	876		862	4214
620	893.15	1796		1779	7553
660	933.15	3075	12.7	3034	11990
700	973.15	4646		4570	18182
740	1013.15	6479		6355	28337
780	1053.15	8561	17.4	8379	-

$\alpha(T) = 0 = \beta(P)$ are indicated in Table 5.7. Also listed is its $2\sigma_T$ at 540, 660, and 780°C.

Figure 5.3 depicts the data from Table 5.7, and compares them with the reversal brackets reported by Barnes and Ernst (1963) up to 2 kbar, and by Schramke et al. (1982) from 4 to 8.1 kbar. These only marginally overlap with the $2\sigma_T$ envelope (hatched in Fig. 5.3). Ignoring $\alpha(T)$ and $\beta(P)$ has only a trivial effect at low pressures. However, the curve calculated for $\Delta C°_P = 0$ is dramatically off the mark, as anticipated above from the large contribution of the two $\Delta C°_P(T)$-integrals to $\Delta G°_T$.

Problem Set 5.3

1. Calculate the univariant equilibrium curve of the reaction

$$2 \text{ kaolinite} = 1 \text{ pyrophyllite} + 2 \text{ diaspore} + 2 \text{ H}_2\text{O} \qquad (5.16)$$
$$\text{(Ka)} \qquad \text{(Py)} \qquad \text{(D)}$$

in the range 240- 290°C, using the data summarized in Table 5.1.
2. The calculated P-T curve will differ from the previous examples in two respects. First, dual solutions for pressure will emerge at all temperatures, rather than a unique one. And secondly, a sudden change in the slope of the P-T curve will appear at a very low P. Give explanations for both.

Solutions to the Problem Set 5.3

Problem 1. The results of iterative solutions for equilibrium pressures of the reaction (5.16) at several temperatures have been indicated in Table 5.8. Since no solution was found at 290°C, the equilibrium P was sought at shorter intervals between 280 - 290°C.

Problem 2. Two features of the computed P-T curve set it apart from the previous examples. First, a double solution for P emerges at every T. Secondly, an abrupt break in the dP/dT slope is found between 260 and 270°C.
    To understand the former, let us take a look at the results of calculation of the 280°C isotherm, reproduced in Table 5.9. From a perusal of the last two

Table 5.8. The computed P_{H_2O} - T curve of the reaction 2 kaolinite = 1 pyrophyllite + 2 diaspore + 2 H_2O

T(°C)	T(K)	P_1(bar)	P_2(bar)
240	513.15	27	13624
250	523.15	35	12680
260	533.15	46	11602
270	543.15	773	10326
280	553.15	2065	8599
285	558.15	3110	7358
287.5	560.65	4002	6370

Table 5.9. Two solutions for P_{H_2O} of 2 kaolinite = 1 pyrophyllite + 2 diaspore + 2 H_2O reaction at 280°C

ΔG_T° (J)	$\Delta V_{298,s}^\circ$(P-1) (J)	$2RTlnf_{H_2O}^*$ (J)	ΔG^* (J)	P (bar)
-37660.0	-3566.4	40645.5	-580.9	1000
-37660.0	-5351.4	42719.6	-291.8	1500
-37660.0	-7136.4	44772.1	-24.3	2000
-37660.0	-8921.4	46739.3	+157.9	2500
-37660.0	-10706.4	48688.4	+322.0	3000
-37660.0	-12491.4	50563.6	+412.2	3500
-37660.0	-14276.4	52402.5	+466.1	4000
-37660.0	-16061.4	54242.7	+521.3	4500
-37660.0	-17846.4	56028.5	+522.1	5000
-37660.0	-19631.4	57802.2	+510.8	5500
-37660.0	-21416.4	59547.6	+471.2	6000
-37660.0	-23201.4	61276.5	+415.1	6500
-37660.0	-24986.4	62988.7	+342.3	7000
-37660.0	-26771.4	64686.9	+255.5	7500
-37660.0	-28556.4	66366.6	+150.2	8000
-37660.0	-30341.4	68026.7	+25.3	8500
-37660.0	-32126.4	69676.6	-109.8	9000
-37660.0	-33911.4	71310.8	-260.6	9500
-37660.0	-35696.4	72933.0	-423.4	10000

columns of Table 5.9, it is apparent that the sign of ΔG^* changes twice. This happens between 2000 and 2500 bar, and once again between 8500 and 9000 bar, consistent with two solutions of equilibrium pressures at 2065 and 8599 bar (see Table 5.8). To comprehend the reason behind this, recall the condition of equilibrium of the reaction (5.16):

$$0 = \Delta G_T^\circ + \Delta V_{298,s}^\circ(P-1) + 2\,RTlnf_{H_2O}^*, \tag{5.17a}$$

which may be recast to

$$\Delta G_T^\circ + \Delta V_{298,s}^\circ(P-1) = -2\,RTlnf_{H_2O}^*. \tag{5.17b}$$

Note that at any specified T, both sides of (5.17b) are functions of P. Therefore, the equilibrium pressure must also be discernible from a simultaneous plot of both sides of this equation against P. Figure 5.4 is such a plot.

The LHS of Eq. (5.17b) is linear in P; it increases with P when $\Delta V^\circ_{298,s} > 0$, and decreases if $\Delta V^\circ_{298,s} < 0$. By contrast, the RHS invariably decreases

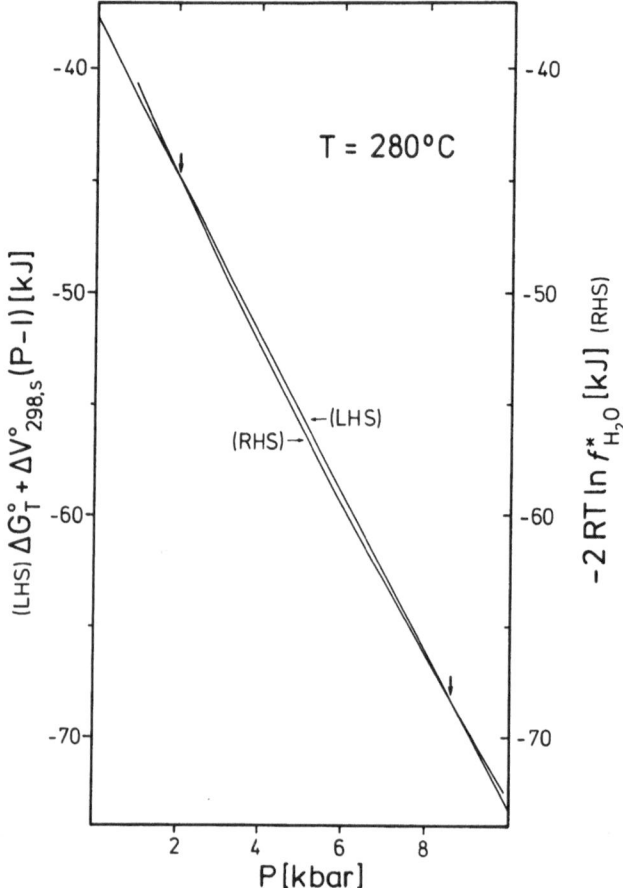

Fig. 5.4 An isothermal (T = 280°C) plot of $\Delta G_T^\circ + \Delta V_{298,s}^\circ(P-1)$ vs P for the reaction 2 kaolinite = 1 pyrophyllite + 2 diaspore + 2 H_2O [curve labeled *LHS* (of 5.17b)]. The other curve, labeled *RHS* (of 5.17b), shows the function $-2RT\ln f_{H_2O}^*$ vs P. The intersections of the two are highlighted by *arrows*; they correspond to the dual solutions for the equilibrium pressures at the stated temperature

with rising pressure, because f_{H2O}^* increases with pressure. For this simple reason, a unique solution for P is expected whenever $\Delta V_{298,s}^\circ > 0$. To the extent $\Delta V_{298,s}^\circ < 0$, one *or* two solutions might result, depending upon the rate of change of LHS and RHS with P. When $\Delta V_{298,s}^\circ$ is negative *and* large enough, as in Eq. (5.16), for which $\Delta V_{298,s}^\circ = -3.570$ J bar^{-1}, the rate of decrease of the LHS of (5.17b) is sufficient to cause a double intersection (highlighted by arrows in Fig. 5.4). In other words, whenever $\Delta V_{298,s}^\circ$ is exceptionally large and negative, the possibility of a dual solution for P must be entertained.

The second extraordinary feature of the P-T curve calling for an explanation is an abrupt change of dP/dT slope between 260 and 270°C,

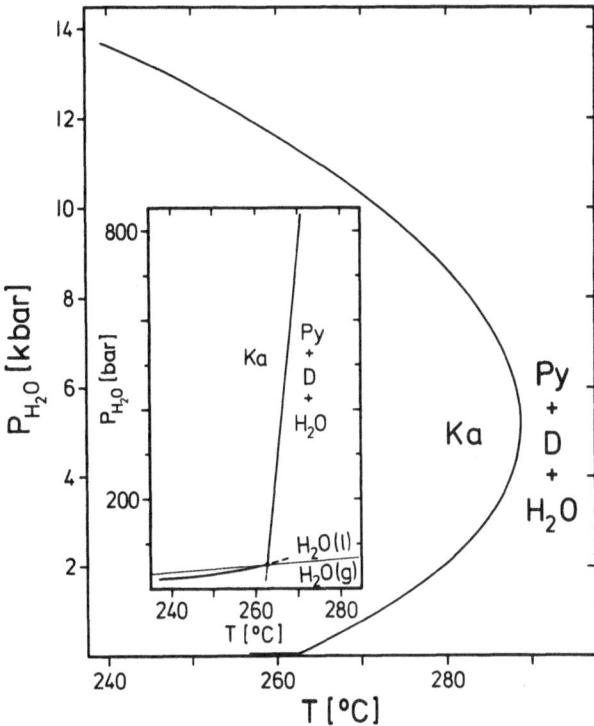

Fig. 5.5 A P_{H_2O}-T plot of the univariant curve for the reaction 2 kaolinite (Ka) = 1 pyrophyllite (Py) + 2 diaspore (D) + 2 H_2O. The dual solutions for P (see Table 5.8) result in "sweeping back" of the curve at high pressures. The *inset* shows the details of the low pressure region, with the abrupt change in the slope of the curve. Indicated in *bold lines* are the equilibrium curves of the above reaction involving liquid H_2O at higher P, and H_2O gas at lower P. The *thinner line* depicts the saturation curve of H_2O. The three curves emanate from the invariant point, shown by a dot

depicted in Fig. 5.5. The inset enlarges the P-T region of interest, and shows the saturation curve of H_2O, in addition to two segments of the P-T curve of reaction (5.16). The three curves intersect around 263°C, 53 bar pressure, generating an invariant point at which Ka, Py, D, H_2O (liquid), and H_2O (gas) coexist. The steeper segment of the curve delineates the decomposition of Ka to Py + D + H_2O (liquid), whereas the flatter one shows the analogous curve involving H_2O (gas). At the P corresponding to the invariant point, H_2O undergoes the first order liquid-gas phase transition, which is accompanied by a huge change in its entropy and volume. This explains the abrupt break in the slope of the P-T curve for reaction (5.16), in keeping with the Clausius-Clapeyron equation:

$$\frac{dP}{dT} = \frac{\Delta S^*}{\Delta V}.$$ (5.18)

In summary then, there are several lessons to be learned from the foregoing exercises, which are of general validity:

1. It is essential to include the two $\Delta C^\circ_p(T)$-integrals for phase equilibria calculations of devolatilization reactions. Practically in all cases, $\Delta C^\circ_p(T) = 0$ is a non-viable approximation.

2. Ignoring $\alpha(T)$ and $\beta(P)$ with impunity is possible only at rather low pressures; they become important with increasing pressure.

3. The dP/dT slopes of devolatilization equilibria are fairly flat at low P. With increasing P, they become progressively steeper and may even become negative at high enough pressures.

4. Although $H^\circ_{f,298}$ and S°_{298} of many minerals have been measured by calorimetry with remarkable accuracy, the $2\sigma_T$ of calculated P-T curves are on the order of several tens to 100°C. In this regard, the equilibrium curve of the reaction Br = P + H_2O is an exception. Stated alternatively, geologically meaningful phase diagrams are seldom obtained from calorimetric data *alone*.

5. The experimental reversal brackets may or may not straddle the P-T curves computed from calorimetric data, but most of them plot within the $\pm 2\sigma_T$ range of the computed curve.

6. The experimental reversal brackets generally put more stringent constraints on the equilibria than calorimetry (Fig. 5.2 is such an example). To put this message across, Zen (1977) coined the term "phase equilibria calorimetry". In Chapter 7, it will be demonstrated how reaction reversals and calorimetric data may be simultaneously handled to derive an internally consistent set of thermodynamic data, which allows the computation of reliable phase diagrams.

5.5 Thermodynamic Interpolation and Extrapolation of Reaction Reversal Data, and Linear Summation of Independent Equilibria

5.5.1 Theoretical Basis

For some geochemical problems, it suffices to interpolate or extrapolate experimental reaction reversal data over a limited range of temperature, rather than going through the more elaborate process of calculating phase diagrams from thermodynamic data. A simple algorithm can be devised for achieving that goal; knowledge of $H^\circ_{f,298}$ and S°_{298} of the participating phases is unneccessary, although their $C^\circ_p(T)$ functions are still very useful. This method has been, and continues to be, enormously popular in geochemistry. Although it is introduced here in the context of solid-fluid equilibria, it is equally applicable to solid-solid reactions. In fact, current calibrations of the majority of geothermometers and geobarometers depend on this technique (see Chap. 9).

Prior to deriving the relevant thermodynamic formalisms, it is essential to comprehend the nature of a reversal experiment. In a reversal experiment, the P-T conditions, at which $\Delta G^* = 0$, are *not* established. Therefore, the term phase *equilibria* measurement is, if anything, a misnomer. In reality, the reversal experiments seek to determine the narrowest possible P-T brackets delimiting the stability field of the reactants (with $\Delta G^* > 0$) and the products ($\Delta G^* < 0$). Since this approach can operate only with the condition of equilibrium (for which $\Delta G^* = 0$)[2]), we are obliged to make some assumptions. The postulate inherent to our treatment is that the equilibrium curve may pass through *any* point between the reversal brackets, including the bracketing points themselves, although, in general, it will have a tendency to pass through the middle of the brackets.

With this background information in mind, let us consider the reaction

$$bB = cC + dD + vH_2O. \tag{5.1}$$

The condition of equilibrium for (5.1) is given by the equation

$$0 = \Delta G_T^\circ + \int_1^P \Delta V_{T,s}(P)dP + RT\ln K . \tag{5.7}$$

Depending upon whether or not the $C_P^\circ(T)$ functions for the phases are known, the following two alternative treatments are possible. In either case, it is justified to replace the volume integral on the RHS of (5.7) by the product, $\Delta V^\circ_{298,s}(P-1)$.

Case 1: $C_P^\circ(T)$ data are not available

To handle the experimental reaction reversal data, (5.7) is first rearranged as

$$-\Delta G_T^\circ = \Delta V_{298,s}^\circ(P-1) + RT\ln K. \tag{5.19}$$

If the RHS of (5.19) is evaluated for a series of *extremely narrow* brackets spanning a *huge* range of temperature, and plotted against T, the $-\Delta G_T^\circ$ vs T plot will turn out to be a very slightly curved line. Such a departure from linearity reveals any perceptible, if small, contributions of the two $\Delta C_P^\circ(T)$-integrals to ΔG_T°. In principle, knowledge of $-\Delta G_T^\circ$ vs T suffices to evaluate ΔS_T° and ΔH_T° at a temperature explored experimentally. The former is found from the relation

$$\Delta S_T^\circ = -\left(\frac{\partial \Delta G_T^\circ}{\partial T}\right)_P . \tag{5.20}$$

That is, the slope of the tangent drawn to the $-\Delta G_T^\circ$ vs T curve at the specified T yields ΔS_T°. Substituting that ΔS_T° back into the identity

$$\Delta H_T^\circ = \Delta G_T^\circ + T\Delta S_T^\circ, \tag{5.21}$$

---

[2] An explicit treatment in terms of the inequalities, $\Delta G^* > 0$ and $\Delta G^* < 0$, will be given in Chap. 7.

$\Delta H°_T$ is obtained for that temperature. Evaluation of $\Delta H°_T$ and $\Delta S°_T$ as a function of temperature yields $\Delta C°_P$ [cf. Eqs.(1.14) and (1.15)].[3]

Unfortunately, in the real world of phase equilibria studies, the reversal brackets are never sufficiently narrow, nor do they span a large enough range of T. Most commonly, the bracketing data extend over a range of merely 50-300 K. In this case, not even the narrowest achievable (10-20 K) reversal brackets help resolve the temperature dependencies of $\Delta H°_T$ and $\Delta S°_T$, and consequently, the magnitude of $\Delta C°_P(T)$ remains unknown. In such cases, a plot of the RHS of (5.19) vs T always yields a straight line. This may be due to either of two reasons. First, $\Delta C°_P = 0$. If this is true, the slope and the intercept of the straight line would correspond to $\Delta S°_{298}$ and $-\Delta H°_{298}$ [cf. Eq.(5.27)]. An alternative interpretation is that $\Delta C°_P(T)$ is not resolved by the experiments, even though it is nonzero. The first alternative may be dismissed, because $\Delta C°_P = 0$ is highly improbable. Accordingly, the slope and the intercept of the straight line are best interpreted as $\Delta S°_{<T>}$ and $-\Delta H°_{<T>}$, *average* standard entropy and enthalpy differences of the reaction *for the experimental range of T*. Thus, to manipulate the reversal brackets of a reaction, for which $\Delta C°_P(T)$ is unknown, (5.19) will be recast in terms of $\Delta H°_{<T>}$ and $\Delta S°_{<T>}$,

$$-\Delta H°_{<T>} + T\Delta S°_{<T>} = \Delta V°_{298,s}(P-1) + RT\ln K. \tag{5.22}$$

The RHS of (5.22) is then evaluated for the bracketing data points and $\Delta H°_{<T>}$ and $\Delta S°_{<T>}$ obtained by linear regression.

Given $\Delta H°_{<T>}$ and $\Delta S°_{<T>}$, Eq.(5.22) can be used to interpolate the phase equilibria between the individual brackets. It may also be utilized for an extrapolation over a *short range* of temperature. This restriction stems from the fact that $\Delta H°_{<T>}$ and $\Delta S°_{<T>}$ are no longer valid at temperatures outside the range of the experiments. If an accurate extrapolation is desired, the knowledge of $\Delta H°_T$ and $\Delta S°_T$ is essential; that is, $C°_P(T)$ of all phases are indispensable (see below). Yet another utility of $\Delta H°_{<T>}$ and $\Delta S°_{<T>}$ data should be mentioned. Often, by design or by chance, the reversal data are available for two or more linearly independent reactions. Knowing their $\Delta H°_{<T>}$ and $\Delta S°_{<T>}$, those of the linearly dependent reactions may be obtained. The linear addition of $\Delta H°$ and $\Delta S°$ is permissible because they are state properties, and as such, path-independent.

The literature of the last three decades provides numerous examples of applications of $\Delta H°_{<T>}$ and $\Delta S°_{<T>}$ to solve a variety of geochemical problems. The basic data on $\Delta G°_T$, required for this purpose, have been reported by individual workers in two different formats, one of which is simply

$$\Delta G°_T = \Delta H°_{<T>} - T\Delta S°_{<T>}. \tag{5.23}$$

---

[3] Lee and Ganguly (1988) gave an an example of the evaluation of $\Delta C°_P$ for a solid-solid reaction by *nonlinear* regression of $-\Delta G°_T$ vs T data.

Others prefer to report $\log K$ (rather than $\Delta G°_T$) for the reaction. To derive $\log K$, one may start by rewriting Eq.(5.22) as

$$RT\ln K = -\Delta H°_{\langle T \rangle} + T\Delta S°_{\langle T \rangle} - \Delta V°_{298,s}(P-1). \tag{5.24}$$

Dividing both sides of (5.24) by RT and converting $\ln K$ to $\log K$,

$$\log K = -\left[\frac{\Delta H°_{\langle T \rangle}}{2.3026R}\right] \cdot \frac{1}{T} + \left[\frac{\Delta S°_{\langle T \rangle}}{2.3026R}\right] + \left[\frac{-\Delta V°_{298,s}}{2.3026R}\right] \cdot \frac{(P-1)}{T}. \tag{5.25}$$

Note that all three square-bracketed terms in the above equation are constants, independent of T and P. Defining them as A (in K), B (dimensionless) and C (in K bar^{-1}), respectively, Eq.(5.25) is popularly quoted as

$$\log K = -\frac{A}{T} + B + C \cdot \frac{(P-1)}{T}. \tag{5.26}$$

It is clear now that Eqs.(5.23) and (5.26) are merely alternative ways of reporting the same set of data. In our worked examples, we shall arbitrarily apply (5.23) to dehydration reactions and (5.26) to redox reactions.

Case 2: $C°_P(T)$ data of all phases are available

As emphasized above, for a very accurate extrapolation of the experimental reversal data, $\Delta H°_T$ and $\Delta S°_T$ must be known. In other words, data on $C°_P(T)$ of the participating phases are essential. To handle the reversal experiments, using the available $C°_P(T)$ data, (5.19) is first recast to

$$-\Delta H°_{298} + T\Delta S°_{298} = \int_{298}^{T} \Delta C°_P(T)dT - T\int_{298}^{T}\left[\frac{\Delta C°_P(T)}{T}\right]dT$$

$$+ \Delta V°_{298,s}(P-1) + RT\ln K. \tag{5.27}$$

The RHS of (5.27) is then evaluated for each bracketing point, and the data subjected to a linear least squares regression to come up with $\Delta H°_{298}$ and $\Delta S°_{298}$. This method of data reduction is popularly known in thermochemistry as the "second law method" (see Pankratz et al. 1984).

Given $\Delta H°_{298}$ and $\Delta S°_{298}$, in addition to the quantities on the RHS of (5.27), experimental reversal data may be interpolated and even extrapolated with some confidence. Extrapolation over (5.27), rather than (5.22) is recommended, whenever the appropriate $C°_P(T)$ data are available. Moreover, $\Delta H°_{298}$ and $\Delta S°_{298}$ of the independent reactions may be added to derive those for the linearly dependent equilibria.

5.5.2 Worked Examples of Dehydration Reactions

Let us examine the reaction reversal data for two dehydration reactions in order to familiarize ourselves with the technique of deriving $\Delta H°_{<T>}$ and $\Delta S°_{<T>}$, and of $\Delta H°_{298}$ and $\Delta S°_{298}$. The derived data will then be utilized to extrapolate the equilibrium curves and to construct a phase diagram.

Problem Set 5.4

1. Using the experimental reversal brackets for the reactions

$$1 \text{ paragonite} = 1 \text{ albite} + 1 \text{ corundum} + 1 \text{ H}_2\text{O and} \qquad (5.28)$$
$$\quad (\text{Pg}) \qquad\qquad (\text{Ab}) \qquad\quad (\text{C})$$
$$1 \text{ paragonite} = 1 \text{ jadeite} + 1 \text{ kyanite} + 1 \text{ H}_2\text{O}, \qquad (5.29)$$
$$\quad (\text{Pg}) \qquad\qquad (\text{Jd}) \qquad\quad (\text{Ky})$$

reproduced in Table 5.10, and the standard state molar volumes of the solids (Table 5.11), calculate $\Delta H°_{<T>}$ and $\Delta S°_{<T>}$ for both reactions.

Table 5.10. The reaction reversal data to be employed for solving the problem set 5.4

P_{H_2O}(bar)	T(°C)	Stable assemblage
Reaction (5.28): 1 Pg = 1 Ab + 1 C + 1 H$_2$O (Chatterjee 1970)		
1000	530	Pg
1000	550	Ab + C + H$_2$O
2000	555	Pg
2000	575	Ab + C + H$_2$O
3000	580	Pg
3000	600	Ab + C + H$_2$O
5000	625	Pg
5000	640	Ab + C + H$_2$O
6000	620	Pg
6000	650	Ab + C + H$_2$O
7000	650	Pg
7000	670	Ab + C + H$_2$O
Reaction (5.29): 1 Pg = 1 Jd + 1 Ky + H$_2$O (Holland 1979)		
24000	550	Pg
26000	550	Jd + Ky + H$_2$O
24000	600	Pg
25500	600	Jd + Ky + H$_2$O
24000	650	Pg
25000	650	Jd + Ky + H$_2$O
23000	700	Pg
24500	700	Jd + Ky + H$_2$O

Table 5.11 V°_{298} of solids relevant to the problem set 5.4

Phases	V°_{298}(J bar^{-1} mol^{-1})	Source of data
Paragonite	13.1880	Flux and Chatterjee (1986)
Albite	10.0430	Robie et al. (1979)
Kyanite	4.4150	Robie and Hemingway (1984a)
Jadeite	6.0400	Holland (1979)
Corundum	2.5575	Robie et al. (1979)

Table 5.12. Heat capacity power functions, $C^{\circ}_p(T)$ (in J K^{-1} mol^{-1}) = a + b·T + c·T^{-2} + d·T$^{-0.5}$ + e·T^2 + f·T^{-3}, for phases participating in the reactions (5.28) through (5.30)

$C^{\circ}_p(T)$ coeff.	Phase abbreviations Pg[a]	C[b]	Ab[c]	Jd[c]	Ky[d]	H$_2$O[b]
a	7.4526E2	1.5736E2	3.9187E2	3.1129E2	3.039E2	7.3680
b	0.0	7.1899E-4	0.0	0.0	-1.339E-2	2.7468E-2
c	0.0	-1.8969E6	-9.3975E6	-5.3503E6	-8.952E5	-2.2316E5
d	-7.2731E3	-9.8804E2	-2.2696E3	-2.0051E3	-2.9043E3	3.6174E2
e	0.0	0.0	0.0	0.0	0.0	-4.8117E-6
f	-6.248E7	0.0	1.32604E9	6.6257E8	0.0	0.0
Validity (in K)		298-1800			250-1600	298-1800

[a] $C^{\circ}_p(T)$ measurements to 900 K (Robie and Hemingway 1984b) fitted to the Berman-Brown polynomial.
[b] Robie et al. (1979).
[c] Berman and Brown (1985).
[d] Robie and Hemingway (1984a)

2. Utilize the above data along with the $C^{\circ}_p(T)$ polynomials given in Table 5.12 to compute ΔH°_{298} and ΔS°_{298} of the reactions (5.28) and (5.29).
3. Employing $\Delta H^{\circ}_{<T>}$ and $\Delta S^{\circ}_{<T>}$, as well as ΔH°_{298} and ΔS°_{298}, of (5.28) and (5.29), extrapolate the P-T curves of both reactions to higher temperatures until they generate an invariant point. How do the computed curves behave within and outside the range covered by the original experiments?
4. Calculate σ_T of the P-T curves for the reactions (5.28) and (5.29), utilizing the ΔH°_{298} and ΔS°_{298} data.
5. Predict the P-T curve of the third reaction emanating from the above invariant point,

$$1 \text{ jadeite} + 1 \text{ kyanite} = 1 \text{ albite} + 1 \text{ corundum}, \qquad (5.30)$$
$$\quad \text{(Jd)} \qquad \text{(Ky)} \qquad \text{(Ab)} \qquad \text{(C)}$$

using ΔH°_{298} and ΔS°_{298} data, and estimate its σ_T.

Solutions to the Problem Set 5.4

Problem 1. To derive $\Delta H^\circ_{<T>}$ and $\Delta S^\circ_{<T>}$ of the reactions (5.28) and (5.29), the RHS of Eq.(5.22) must be evaluated first. Considering that in both reactions 1 mol H_2O is liberated, the condition of equilibrium is obtained by recasting (5.22) to

$$-\Delta H^\circ_{<T>} + T\Delta S^\circ_{<T>} = \Delta V^\circ_{298,s}(P\text{-}1) + RT\ln f^*_{H_2O}. \tag{5.31}$$

The results of evaluation of the RHS of (5.31) are listed in Table 5.13. A linear regression of the data in the last column of Table 5.13 against T (in K) gave

$$
\begin{aligned}
\Delta H^\circ_{<T>}(5.28) &= 99550(6854) \text{ J,} \\
\Delta S^\circ_{<T>}(5.28) &= 174.407(7.806) \text{ J K}^{-1}\text{, and} \\
\rho_{\Delta H\Delta S(5.28)} &= -0.9988;
\end{aligned}
\tag{5.32}
$$

Table 5.13. Results of the calculation of the RHS of Eq. (5.31) for the reactions (5.28) and (5.29)

T (K)	P (bar)	f_{H_2O} (bar)	$\Delta V^\circ_{298,s}(P\text{-}1)$ (J)	$RT\ln f^*_{H_2O}$ (J)	$\Delta V^\circ_{298,s}(P\text{-}1)+RT\ln f^*_{H_2O}$ (J)
\multicolumn{6}{c}{Reaction (5.28): 1 Pg = 1 Ab + 1 C + 1 H_2O}					
803.15	1000	517	- 587	41722	41135
823.15	1000	555	- 587	43246	42659
828.15	2000	911	-1174	46921	45747
848.15	2000	976	-1174	48541	47367
853.15	3000	1456	-1762	51664	49902
873.15	3000	1540	-1762	53282	51520
898.15	5000	3175	-2937	60211	57274
913.15	5000	3299	-2937	61507	58570
893.15	6000	4205	-3524	61962	58438
923.15	6000	4519	-3524	64596	61072
923.15	7000	5955	-4112	66714	62602
943.15	7000	6207	-4112	68484	64372
\multicolumn{6}{c}{Reaction (5.29): 1 Pg = 1 Jd + 1 Ky + H_2O}					
823.15	24000	284459	-65589	85948	20359
823.15	26000	437243	-71055	88890	17835
873.15	24000	275336	-65589	90932	25343
873.15	25500	374585	-69689	93167	23478
923.15	24000	266483	-65589	95889	30300
923.15	25000	324396	-68322	97398	29076
973.15	23000	212834	-62856	99263	36407
973.15	24500	283033	-66956	101570	34614

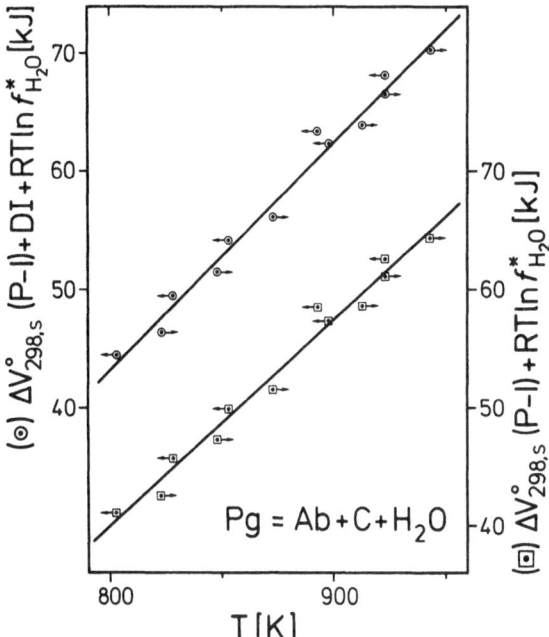

Fig. 5.6 A plot of $\Delta V^{\circ}_{298,s}$(P-1) + $RTln f^*_{H_2O}$(*box symbols*) vs T for the bracketing experiments on the reaction 1 paragonite (Pg) = 1 albite (Ab) + 1 corundum (C) + 1 H_2O. The relative stability of the products is indicated by the *arrows pointing right*, that of the reactant by *arrows pointing left*. The *straight line* through the data points is from the linear regression. The upper part of the diagram depicts $\Delta V^{\circ}_{298,s}$(P-1) + DI + $RTln f^*_{H_2O}$ (*circles*) vs T, with DI denoting $_{298}\int^T \Delta C^{\circ}_P dT$ - T $_{298}\int^T (\Delta C^{\circ}_P/T)dT$. The *arrows* have the same significance as above, as also the straight line

$$\Delta H^{\circ}_{<T>}(5.29) = 70752(6339)\ J,$$
$$\Delta S^{\circ}_{<T>}(5.29) = 109.033(7.045)\ J\ K^{-1},\ and$$
$$\rho_{\Delta H \Delta S(5.29)} = -0.9981. \tag{5.33}$$

The quantities appearing above in parentheses are $\pm 1\sigma$. Note that $\Delta H^{\circ}_{<T>}$ and $\Delta S^{\circ}_{<T>}$ have not been rounded off to preserve the internal consistency between the two. This is essential for using the data to calculate phase relations. The use of the correlation coefficient, $\rho_{\Delta H \Delta S}$, will become clear from the error propagation calculations given later.

The RHS of Eq.(5.31) for each bracketing point of reaction (5.28) are depicted in Fig. 5.6 by box symbols. The relative stability of the reactant is indicated by arrows pointing left, that of the products by arrows pointing to the right. A straight line fitting the data passes through the bracketing points without violating any of those points. This may be regarded as a minimum requirement for applying the method of regression. This condition is, however, not always fulfilled (Day and Kumin 1980, p.269), requiring us to prefer the technique of mathematical programming (see Chap. 7).

Table 5.14. Results of the computation of the RHS of Eq. (5.34) of the reactions (5.28) and (5.29)

T (K)	P (bar)	f_{H_2O} (bar)	DI[a] (J)	$\Delta V^\circ_{298,s}(P-1)$ (J)	$RTlnf^*_{H_2O}$ (J)	$\Delta V^\circ_{298,s}(P-1)+$ $DI+RTlnf^*_{H_2O}$ (J)
		Reaction (5.28): 1 Pg = 1 Ab + 1 C + 1 H$_2$O				
803.15	1000	517	3403	- 587	41722	44538
823.15	1000	555	3715	- 587	43246	46374
828.15	2000	911	3796	-1174	46921	49543
848.15	2000	976	4127	-1174	48541	51494
853.15	3000	1456	4212	-1762	51664	54114
873.15	3000	1540	4562	-1762	53282	56082
898.15	5000	3175	5020	-2937	60211	62294
913.15	5000	3299	5306	-2937	61507	63876
893.15	6000	4205	4926	-3524	61962	63364
923.15	6000	4519	5501	-3524	64596	66573
923.15	7000	5955	5501	-4112	66714	68103
943.15	7000	6207	5903	-4112	68484	70275
		Reaction (5.29): 1 Pg = 1 Jd + 1 Ky + H$_2$O				
823.15	24000	284459	3908	-65589	85948	24267
823.15	26000	437243	3908	-71055	88890	21743
873.15	24000	275336	4743	-65589	90932	30086
873.15	25500	374585	4743	-69689	93167	28221
923.15	24000	266483	5662	-65589	95889	35962
923.15	25000	324396	5662	-68322	97398	34738
973.15	23000	212834	6664	-62856	99263	43071
973.15	24500	283033	6664	-66956	101570	41278

[a] The abbreviation DI denotes $[_{298}\int^T\Delta C^\circ_P dT - T \cdot_{298}\int^T (\Delta C^\circ_P/T)dT]$, the sum of the two ΔC°_P-integrals of Eq. (5.34).

Problem 2. To derive ΔH°_{298} and ΔS°_{298} of the above reactions, we rewrite Eq. (5.27), bearing in mind that 1 mol H$_2$O is released in each case,

$$-\Delta H^\circ_{298} + T\Delta S^\circ_{298} = \int_{298}^T \Delta C^\circ_P(T)dT - T\int_{298}^T \left[\frac{\Delta C^\circ_P(T)}{T}\right]dT$$
$$+ \Delta V^\circ_{298,s}(P-1) + RTlnf^*_{H_2O}. \qquad (5.34)$$

This time, the RHS of Eq. (5.34) must be evaluated, utilizing the $C^\circ_P(T)$ polynomials of the participating phases listed in Table 5.12. The results are summarized in Table 5.14.

Proceeding in a fashion outlined earlier, $\Delta H°_{298}$ and $\Delta S°_{298}$ of the reactions (5.28) and (5.29) have been derived,

$$\Delta H°_{298}(5.28) = 110554(6849) \text{ J},$$
$$\Delta S°_{298}(5.28) = 192.276(7.801) \text{ J K}^{-1}, \text{ and}$$
$$\rho_{\Delta H \Delta S(5.28)} = -0.9987; \tag{5.35}$$

$$\Delta H°_{298}(5.29) = 82009(6381) \text{ J},$$
$$\Delta S°_{298}(5.29) = 127.406(7.091) \text{ J K}^{-1}, \text{ and}$$
$$\rho_{\Delta H \Delta S(5.29)} = -0.9981. \tag{5.36}$$

The RHS of (5.34) for each bracketing point of reaction (5.28) is shown in Fig. 5.6 by small circles. As before, the straight line fitting the data passes through all the bracketing points. The difference in the slopes of the two curves reflects the difference in the magnitudes of $\Delta S°_{<T>}$ and $\Delta S°_{298}$ [Eqs. (5.32) and (5.35)]. Their intercepts at 0 K correspond to the negative of $\Delta H°_{<T>}$ and $\Delta H°_{298}$, respectively. What has been stated here for reaction (5.28) applies also to reaction (5.29), not shown in Fig. 5.6.

Problem 3. The P-T curves of reactions (5.28) and (5.29) will now be extrapolated to higher temperatures. If we use the $\Delta H°_{298}$ and $\Delta S°_{298}$ data for that purpose, the equilibrium curves of the reactions will have to be obtained by solving the equation

$$\Delta G^* = 0 = \Delta H°_{298} - T\Delta S°_{298} + \int_{298}^{T} \Delta C°_P(T)dT - T \int_{298}^{T} \left[\frac{\Delta C°_P(T)}{T}\right]dT$$
$$+ \Delta V°_{298,s}(P-1) + RT\ln f^*_{H_2O} \tag{5.34a}$$

for P at some specified T. The results of computations have been displayed in bold lines in Fig. 5.7. The P-T curves of the two reactions intersect with each other at about 915°C, 22.5 kbar. And to generate those curves from $\Delta H°_{<T>}$ and $\Delta S°_{<T>}$ data,

$$\Delta G^* = 0 = \Delta H°_{<T>} - T\Delta S°_{<T>} + \Delta V°_{298,s}(P-1) + RT\ln f^*_{H_2O} \tag{5.31a}$$

is similarly solved for P. These results are indicated by thinner lines in Fig. 5.7. Within the temperature range of the reversal experiments, they are practically indistinguishable from the P-T curves obtained earlier. They differ only in the temperature range outside of those covered by the reversals. This is also seen from the invariant point, which is now located at 883°C, 21.8 kbar.

Two important lessons emerge clearly from Fig. 5.7. First, the computed P-T curves invariably agree with reversal brackets, irrespective of whether they are based on $\Delta H°_{298}$ and $\Delta S°_{298}$, or $\Delta H°_{<T>}$ and $\Delta S°_{<T>}$. That is, interpolation can be achieved equally well with either set of data. By contrast, the extrapolated parts of those curves indicate appreciably different locations for the invariant points. As argued earlier, this is due to the fact that $\Delta H^{\circ}_{<T>}$ and $\Delta S^{\circ}_{<T>}$ are valid only within the range explored by the experiments. Outside of that range, ΔH°_{298} and ΔS°_{298} data must be used along with $\Delta C^{\circ}_P(T)$.

Problem 4. The next task is to calculate σ_T of the P-T curves obtained from the data summarized in (5.35) and (5.36). For this purpose recall that in extracting ΔH°_{298} and ΔS°_{298} from reversal experiments, it was tacitly assumed that the quantities $\Delta V^{\circ}_{298,s}$, $\Delta C^{\circ}_P(T)$ and $RT\ln f^*_{H2O}$ on the RHS of (5.34) are known exactly. To formalize the error propagation, we may then rewrite (5.34a) as

$$\Delta G^* = \Delta H^{\circ}_{298} - T\Delta S^{\circ}_{298} + C, \tag{5.34b}$$

C comprising all those terms whose errors need not be propagated. Considering $\Delta G^* = f(T)$, and applying Eq.(4.16) to it,

$$\sigma^2_{\Delta G^*} = \left(\frac{\partial \Delta G^*}{\partial T} \right)^2_P \sigma^2_T . \tag{5.37a}$$

Taking its derivative with respect to (5.34b), we have

$$\sigma_T = \frac{\sigma_{\Delta G^*}}{|-\Delta S^{\circ}_{298}|} . \tag{5.37}$$

And to compute $\sigma_{\Delta G^*}$, we apply the basic error propagation equation (4.16) to the function $\Delta G^* = f(\Delta H^{\circ}_{298}, \Delta S^{\circ}_{298})$, paying attention to the fact that *in this case* ΔH°_{298} and ΔS°_{298} *are correlated.* Thus,

$$\sigma^2_{\Delta G^*} = \left(\frac{\partial \Delta G^*}{\partial \Delta H^{\circ}_{298}} \right)^2_P \sigma^2_{\Delta H^{\circ}_{298}} + \left(\frac{\partial \Delta G^*}{\partial \Delta S^{\circ}_{298}} \right)^2_P \sigma^2_{\Delta S^{\circ}_{298}}$$

$$+ 2\rho_{\Delta H\Delta S} \sigma_{\Delta H^{\circ}_{298}} \sigma_{\Delta S^{\circ}_{298}} \left(\frac{\partial \Delta G^*}{\partial \Delta H^{\circ}_{298}} \right)_P \left(\frac{\partial \Delta G^*}{\partial \Delta S^{\circ}_{298}} \right)_P . \tag{5.38a}$$

Again, taking the derivatives with respect to (5.34b), we have

$$\sigma_{\Delta G^*} = [\sigma^2_{\Delta H^{\circ}_{298}} + T^2 \cdot \sigma^2_{\Delta S^{\circ}_{298}} - 2 \cdot T \cdot \rho_{\Delta H\Delta S} \cdot \sigma_{\Delta H^{\circ}_{298}} \cdot \sigma_{\Delta S^{\circ}_{298}}]^{0.5}. \tag{5.38}$$

Knowledge of ΔS°_{298}, in addition to $\sigma_{\Delta G^*}$, enables us to derive σ_T from Eq.(5.37). One example of calculation of σ_T for each of the reactions follows. For Pg = Ab + C + H_2O (5.28), using the set of data given in (5.35), we have at 540°C:

$$\sigma_{\Delta G^*} = [(6849)^2 + (813.15)^2 \cdot (7.801)^2$$
$$- 2 \cdot (813.15) \cdot (-0.9987) \cdot (6849) \cdot (7.801)]^{0.5} \, J = 13188 \, J, \text{ and}$$

$\Delta S_{298}^{\circ} = 192.276$ J K^{-1}, such that

$\sigma_T(5.28) = 68.6$ K.

Likewise, employing the data summarized in Eq.(5.36) for Pg = Jd + Ky + H_2O (5.29) reaction, we obtain at 540°C:

$$\sigma_{\Delta G^*} = [(6381)^2 + (813.15)^2 \cdot (7.091)^2$$
$$- 2 \cdot (813.15) \cdot (-0.9881) \cdot (6381) \cdot (7.091)]^{0.5} \text{ J} = 12111 \text{ J, and}$$

$\Delta S_{298}^{\circ} = 127.406$ J K^{-1}, so that

$\sigma_T(5.29) = 95.1$ K.

The σ_T range for each reaction has been calculated for a series of points along the P-T curve and shown in Fig. 5.7 by the pair of dashed lines.

Problem 5. We now focus on the calculation of the third univariant curve emanating from the invariant point in Fig. 5.7, the P-T curve of the reaction

1 jadeite + 1 kyanite = 1 albite + 1 corundum.   (5.30)
    (Jd)   (Ky)   (Ab)   (C)

To achieve our goal, we need ΔH°_{298} and ΔS°_{298} for this reaction. From the linear-dependency relation

$1 \cdot (5.30) = 1 \cdot (5.28) - 1 \cdot (5.29),$ (5.39)

we immediately obtain

$\Delta H_{298}^{\circ}(5.30) = \Delta H_{298}^{\circ}(5.28) - \Delta H_{298}^{\circ}(5.29) = 28545$ J,   (5.40a)

$\Delta S_{298}^{\circ}(5.30) = \Delta S_{298}^{\circ}(5.28) - \Delta S_{298}^{\circ}(5.29) = 64.870$ J K^{-1}.   (5.40b)

Using the appropriate data for ΔV°_{298} and $\Delta C^{\circ}_P(T)$ from Tables 5.11 and 5.12, the P-T curve of the solid-solid equilibrium (5.30) has been computed and depicted in Fig. 5.7 at temperatures in excess of that of the invariant point. To calculate its σ_T, we utilize the linear additivity relation, $\Delta G^*(5.30) = \Delta G^*(5.28) - \Delta G^*(5.29)$, implying that $\Delta G^*(5.30) = f[\Delta G^*(5.28), \Delta G^*(5.29)]$. Since $\Delta G^*(5.28)$ and $\Delta G^*(5.29)$ were obtained from independent sets of reaction reversals, they are not mutually correlated. Therefore,

$$\sigma^2_{\Delta G^*(5.30)} = \left[\frac{\partial \Delta G^*(5.30)}{\partial \Delta G^*(5.28)}\right]^2 \sigma^2_{\Delta G^*(5.28)}$$

$$+ \left[\frac{\partial \Delta G^*(5.30)}{\partial \Delta G^*(5.29)}\right]^2 \sigma^2_{\Delta G^*(5.29)}.$$ (5.41a)

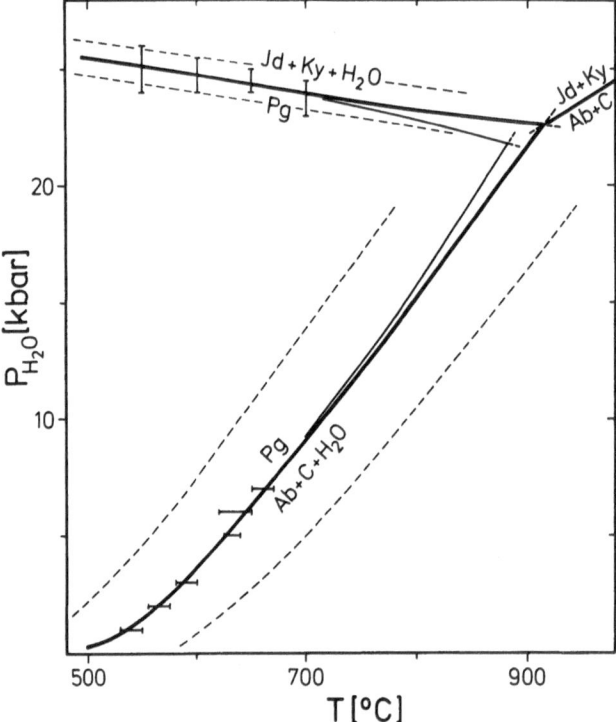

Fig 5.7 A P_{H_2O} - T plot of the equilibria (5.28) through (5.30), computed from ΔH°_{298} and ΔS°_{298} (*bold lines*), and $\Delta H^\circ_{\leq T>}$ and $\Delta S^\circ_{\leq T>}$ (*thinner lines*) data for the reactions (5.28) and (5.29). Note that both sets of curves are identical within the temperature range of the reversal brackets, but their extrapolations to higher temperatures differ perceptibly. The σ_T envelopes of the first two reactions are indicated by *dashed lines*

Taking the derivatives with respect to the above linear additivity equation, Eq.(5.41a) simplifies to

$$\sigma^2_{\Delta G*(5.30)} = (1)^2 \cdot \sigma^2_{\Delta G*(5.28)} + (-1)^2 \cdot \sigma^2_{\Delta G*(5.29)} = \sigma^2_{\Delta G*(5.28)} + \sigma^2_{\Delta G*(5.29)}, \qquad (5.41)$$

such that

$$\sigma_{\Delta G*(5.30)} = [\sigma^2_{\Delta G*(5.28)} + \sigma^2_{\Delta G*(5.29)}]^{0.5}. \qquad (5.42)$$

Knowing $\sigma_{\Delta G*(5.30)}$ for a given temperature and $\Delta S^\circ_{298}(5.30)$, σ_T is obtained from (5.37). One numerical example of the calculation of σ_T will do. Using the data on $\sigma_{\Delta G*(5.28)}$ and $\sigma_{\Delta G*(5.29)}$ at 1000°C, we have

$$\sigma_{\Delta G*(5.30)} = 22774 \text{ J}, \Delta S^\circ_{298}(5.30) = 64.87 \text{ J K}^{-1}, \text{ and } \sigma_T = 351 \text{ K}.$$

The message is sobering. Even though the P-T curve of the linearly dependent equilibria can be predicted from the linear summation of $\Delta H°_{298}$ and $\Delta S°_{298}$, errors keep piling up, and the resulting σ_T is phenomenal. The σ_T of (5.30) has not been shown in Fig. 5.7.

5.5.3 Worked Examples of Redox Reactions

In the redox (reduction-oxidation) equilibria handled in this section, magnetite (Mt), wüstite (W), and iron (I) are regarded as end-member minerals reacting with a pure fluid, oxygen. Note that wüstite, Fe_xO, is strictly speaking not an end-member, but a phase of variable composition. It exhibits cation deficiency, the extent of which depends on temperature in a given assemblage (Darken and Gurry 1953, Fig. 14-5); the value of x in its formula unit varies between 0.85 and 0.95. Fortunately, in the temperature range to be explored here, its composition can be approximated for all intents and purposes as $Fe_{.947}O$. The thermodynamic data for such a wüstite is given by Robie et al. (1979), and will be used below.

   Bearing the above premise in mind, we now treat the following three equilibria:

$$0.5\ Fe_3O_4\ (Mt) = 1.5\ Fe\ (I) + 1\ O_2, \tag{5.43}$$

$$2\ Fe_{.947}O\ (W) = 1.894\ Fe\ (I) + 1\ O_2,\ and \tag{5.44}$$

$$2.4035\ Fe_3O_4\ (Mt) = 7.6140\ Fe_{.947}O\ (W) + 1\ O_2. \tag{5.45}$$

The mass balance of redox reactions is generally written such (Eugster and Wones 1962, p.89) that in each case 1 mole O_2 is released. Although this practice is not compelling, as we shall see below, it has the advantage of simplicity. Note that two of the three reactions are linearly independent. If (5.43) and (5.44) are regarded as independent, (5.45) follows from linear summation,

$$1\cdot(5.45) = 4.807\cdot(5.43) - 3.807\cdot(5.44). \tag{5.46}$$

Besides being of considerable geochemical significance, the redox equilibria are utilized to buffer the oxygen fugacity, f_{O2}, of charges in hydrothermal experiments (Eugster 1957; Eugster and Wones 1962). The question of primary importance in the buffering assemblages is the magnitude of the equilibrium f_{O2} as a function of temperature (at any total pressure, P_{tot}). A convenient way to calculate f_{O2} of such an assemblage is to utilize $G°_{f,T}$ data given by Robie et al. (1979). For the above reactions this procedure is almost a necessity, because Robie et al. (1979) do not provide the $C°_p(T)$ data for iron.

The condition of equilibrium for a redox reaction is derived from Eq.(5.7), replacing the $_1\int^P \Delta V_{T,s}(P)dP$ term by $\Delta V^\circ_{298,s}(P_{tot}-1)$. Ignoring $\alpha(T)$ and $\beta(P)$ is unlikely to introduce a tangible error, unless P_{tot} is extremely high. Recasting (5.7) accordingly,

$$0 = \Delta G^\circ_T + \Delta V^\circ_{298,s}(P_{tot}-1) + RT\ln K. \qquad (5.47)$$

Using the hypothetical ideal gas standard state for the fluid at 1 bar and specified T, and noting that the participating solids have unit activity at T and P, the equilibrium constant for the redox reactions may be expressed as

$$K = (f^*_{O_2})^\nu, \qquad (5.48a)$$

where ν is the number of moles of O_2 liberated by the particular reaction. Because the mass balance equations (5.43) through (5.45) have been formulated such that 1 mol O_2 is set free in every case, (5.48a) reduces to

$$K = f^*_{O_2}. \qquad (5.48)$$

Introducing (5.48) into (5.47), and solving for $\log f^*_{O2}$,

$$\log f^*_{O_2} = \frac{-\Delta G^\circ_T - \Delta V^\circ_{298,s}(P_{tot}-1)}{2.3026RT}. \qquad (5.49)$$

Given the data on $G^\circ_{f,T}$ and V°_{298} of the phases participating in a redox reaction, its f^*_{O2} is computed from (5.49) for any P_{tot} and T. The worked examples given below address some of the problems of computing redox equilibria.

Problem Set 5.5

1. Employing $G^\circ_{f,T}$ and ancillary data from Robie et al. (1979), tabulate the following quantities at 298.15 K, and at 400-1000 K in steps of 100 K for the reactions (5.43) and (5.44): (i) ΔG°_T, and (ii) $\log f_{O2}$ (in bar) at 1 bar and 2000 bar P_{tot}.
2. Rewrite the same data from Robie et al. (1979) in terms of $\log K$ of the reactions (5.43) and (5.44). Are these $\log K$ valid over the entire range 298.15-1000 K?
3. Using the $\log K$ for (5.43) and (5.44), compute the $\log f_{O2}$ vs T relation for the reactions (5.43) through (5.45) at $P_{tot} = 2$ kbar. Summarize the results in a phase diagram, indicating the stability fields for Mt, W, and I.

Solutions to the Problem Set 5.5

Problem 1. ΔG°_T for the reactions (5.43) and (5.44), computed from $G^\circ_{f,T}$ data of Robie et al. (1979), are listed in Table 5.15. To solve for f_{O2} of those

Table 5.15. Calculated ΔG_T° and $\log f_{O_2}$ for the reactions 0.5 Mt = 1.5 I + 1 O_2 [Eq. (5.43)] and 2 W = 1.894 I + 1 O_2 [Eq. (5.44)]

T	Reaction (5.43)			Reaction (5.44)		
	ΔG_T°	$\log f_{O_2}$	$\log f_{O_2}$	ΔG_T°	$\log f_{O_2}$	$\log f_{O_2}$
(K)	(J)	(bar)	(bar)	(J)	(bar)	(bar)
		($P_{tot} = 1$ bar)	($P_{tot} = 2$ kbar)		($P_{tot} = 1$ bar)	($P_{tot} = 2$ kbar)
298.15	506283	-88.70	-88.29	490310	-85.90	-85.53
	(± 1067)[a]			(± 1758)		
400	488846	-63.84	-63.53	476208	-62.19	-61.91
500	472127	-49.32	-49.08	462920	-48.36	-48.14
600	455840	-39.68	-39.48	450022	-39.18	-38.99
700	439998	-32.83	-32.66	437400	-32.64	-32.48
800	424563	-27.72	-27.57	424860	-27.74	-27.60
900	409534	-23.77	-23.63	412288	-23.93	-23.80
1000	394717	-20.62	-20.50	399664	-20.88	-20.77

[a] The quantities in parentheses are $\pm 2\sigma_{\Delta G^\circ}$.

reactions at desired T and P_{tot}, we also need to know $\Delta V^\circ_{298,s}$. Using V°_{298} data from Robie et al. (1979),

$$\Delta V^\circ_{298,s}\ (5.43) = -1.1624 \text{ J bar}^{-1} \text{ and} \tag{5.50}$$

$$\Delta V^\circ_{298,s}\ (5.44) = -1.0648 \text{ J bar}^{-1}. \tag{5.51}$$

Equation (5.49) is then utilized to compute $\log f^*_{O2}$ at given T and P_{tot}. Recalling the definition, $f^*_{O2} \equiv f_{O2}/f^\circ_{O2}$, and that $f^\circ_{O2} = 1$ bar at all T, $\log f_{O2}$ is expressed in bar (Table 5.15).

Problem 2. To recast the ΔG°_T of Table 5.15 to $\log K$, we must know $\Delta H^\circ_{<T>}$, $\Delta S^\circ_{<T>}$, and $\Delta V^\circ_{298,s}$. The requisite $\Delta H^\circ_{<T>}$ and $\Delta S^\circ_{<T>}$ were obtained from a linear least squares fit to ΔG°_T:

$$\Delta G_T^\circ (5.43) = 552182 - 158.815 \text{ T } (\pm 1074) \text{ (J), and} \tag{5.52}$$

$$\Delta G_T^\circ (5.44) = 527697 - 128.490 \text{ T } (\pm 558) \text{ (J).} \tag{5.53}$$

The first term on the RHS of these equations is $\Delta H^\circ_{<T>}$, the second one is $\Delta S^\circ_{<T>} \cdot$ T. The quantities indicated in parentheses are $\sigma_{\Delta G^\circ}$ of the least squares fit. For (5.43), it is twice as large as $\sigma_{\Delta G^\circ}$ of the original data reproduced in Table 5.15[4]; for (5.44), it is about the same. This allows the conclusion that

[4] The huge $\sigma_{\Delta G^\circ}$ implies that $\int_{298}^{T} \Delta C^\circ_p(T)dT - T_{298}\int^{T}[\Delta C^\circ_p(T)/T]dT$ term in Eq.(5.43) is too strongly temperature dependent to permit a linear fit to ΔG°_T data within their calorimetric uncertainties.

the derived values of $\Delta H°_{<T>}$ and $\Delta S°_{<T>}$ are valid over the entire temperature range only for (5.44). Thus, $\log K$ of Eq.(5.44) only will be valid over the 298-1000 K range, not, however, that of (5.43). This must be borne in mind to judge the results of the following calculations.

A first step in rewriting the Robie et al. (1979) data to $\log K$ is to obtain the constants A, B, and C. Inserting the appropriate $\Delta H°_{<T>}$, $\Delta S°_{<T>}$, and $\Delta V°_{298,s}$ into (5.25), the constants A, B, and C [of (5.26)] for the reactions (5.43) and (5.44) are recovered,

$$A(5.43) = 28843 \text{ K}, B(5.43) = 8.296, C(5.43) = 0.0607 \text{ K bar}^{-1}, \text{ and} \qquad (5.54)$$

$$A(5.44) = 27564 \text{ K}, B(5.44) = 6.712, C(5.44) = 0.0556 \text{ K bar}^{-1}. \qquad (5.55)$$

These are then employed to express the $\log K$ of those reactions:

$$\log K (5.43) = -\frac{28843}{T} + 8.296 + 0.0607\frac{(P-1)}{T} \quad \text{and} \qquad (5.56)$$

$$\log K (5.44) = -\frac{27564}{T} + 6.712 + 0.0556\frac{(P-1)}{T}, \qquad (5.57)$$

with T given in K and P in bar.

Problem 3. To obtain a $\log f_{O2}$ vs T diagram, showing the stability fields of Mt, W, and I at 2 kbar P_{tot}, we start with $\log K$ for the reactions (5.43) and (5.44). To complete these phase relations, however, we need the $\log K$ of the linearly dependent reaction (5.45). Using the linear additivity relationship given in (5.46), we have

$$\log K(5.45) = 4.807 \cdot \log K(5.43) - 3.807 \cdot \log K(5.44). \qquad (5.58)$$

The linear addition of $\log K$ is permissible, because it sums state quantities [see RHS of (5.25)]. Entering $\log K(5.43)$ and $\log K(5.44)$ given above into (5.58), $\log K$ of the reaction (5.45) is recovered,

$$\log K (5.45) = -\frac{33712}{T} + 14.326 + 0.0801\frac{(P-1)}{T}, \qquad (5.59)$$

with T given in K and P in bar.

The calculation of the $\log f_{O2}$ vs T diagram at T and P requires us to convert $\log K$ to $\log f^*_{O2}$. Due to the formulation of the mass balance of the redox reactions in terms of 1 mole of O_2, we have

$$\log K = \log f^*_{O_2}. \qquad (5.48)$$

Recalling that f^*_{O2} is numerically equal to f_{O2} (in bar), we may obtain $\log f_{O2}$ of the reactions (5.45) to (5.47) at given T and P, using the above $\log K$ data.

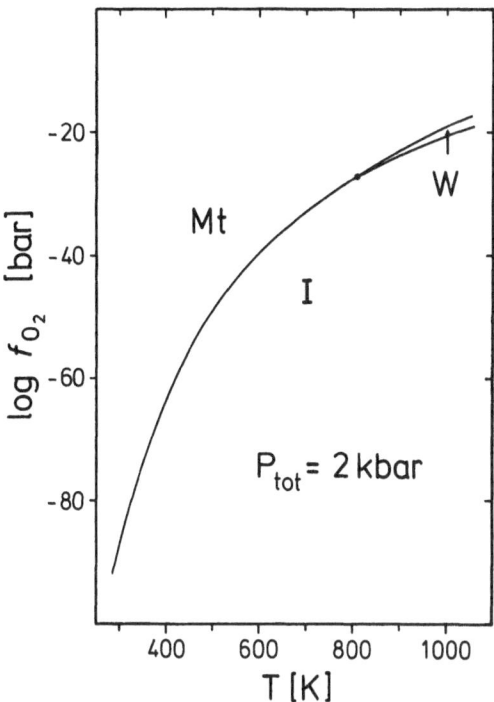

Fig. 5.8 A calculated isobaric $\log f_{O_2}$ - T diagram (for P_{tot} = 2000 bar) delineating the stability fields of magnetite (Mt), wüstite (W), and iron (I)

Table 5.16. The $\log f_{O_2}$ calculated from $\log K$ data for the reactions 0.5 Mt = 1.5 I + 1 O_2 [Eq. (5.43)], 2 W = 1.894 I + 1 O_2 [Eq. (5.44)], and 2.4035 Mt = 7.6140 W + 1 O_2 [Eq. (5.45)] for P_{tot} = 2 kbar

T(K)	Reaction (5.43) $\log f_{O_2}$(bar)	Reaction (5.44) $\log f_{O_2}$(bar)	Reaction (5.45) $\log f_{O_2}$(bar)
298.15	-88.04	-85.37	-98.21
400	-63.51	-61.92	-69.55
500	-49.15	-48.19	-52.78
600	-39.57	-39.04	-41.59
700	-32.73	-32.51	-33.61
800	-27.61	-27.60	-27.61
900	-23.62	-23.79	-22.95
1000	-20.43	-20.74	-19.23

Eugster and Wones (1962) pioneered in these calculations, and popularized them among geochemists. The results of calculation of $\log f_{O2}$ vs T relations for reactions (5.43), (5.44) and (5.45) at a P_{tot} of 2 kbar are listed in Table 5.16, and depicted in Fig. 5.8. The $\log f_{O2}$ data reported in Tables 5.15 and 5.16 agree well.

Two general features of the $\log f_{O2}$ - T relation emerge from the foregoing exercises. First, $\log f_{O2}$ is vanishingly small for *all* temperatures and pressures, and secondly, it is barely dependent on the total pressure. Accordingly, under no conditions can P_{fluid} equal P_{tot}. This adds a degree of freedom to the redox equilibria. Any two of the three variables, T, P_{tot}, and f_{O2} are arbitrarily chosen to uniquely specify the state of the system.

Chapter 6

Phase Relations Among End-Member Solids and a Binary Fluid Mixture

6.1 Introduction

Phase equilibria calculation involving end-member solids and a binary fluid mixture is a logical extension of the topic handled in the last chapter. It is also geologically relevant, because the natural fluids are mixtures of a number of species including H_2O, CO_2, CH_4, etc. The computation of the mixed-volatile equilibria demands that the thermodynamic properties of the fluid mixtures be known. Accordingly, research work on the equations of state of the fluid mixtures of relevance to geochemistry has been a centerpiece of activity for about a decade (Holloway 1977; Jacobs and Kerrick 1981a,b; Kerrick and Jacobs 1981; Bowers and Helgeson 1983; Saxena and Fei 1988). The availability of these data has promoted our ability to predict mixed-volatile reactions with confidence.

The objective of this chapter is to treat mixed-volatile phase equilibria. The thermodynamic background will be provided briefly, following the developments in the preceding chapters. Next, worked examples will be given for the computation of some mixed-volatile reactions involving H_2O-CO_2 mixtures. A lack of equation of state for this fluid mixture had forced earlier workers (like Greenwood 1967; Skippen 1971) to assume that the fluids mix ideally. Given an explicit equation of state for the H_2O-CO_2 binary (Kerrick and Jacobs 1981; Jacobs and Kerrick 1981a), such a simplification is no longer necessary. Because the equation of state for the H_2O-CO_2 mixture is valid above 400°C, however, the worked examples will be restricted to temperatures higher than that.

6.2 Thermodynamic Background

Consider the mass balance equation

$$bB = cC + dD + v_1F_1 + v_2F_2, \tag{6.1}$$

written in terms of the components of phases participating in the reaction. Let B, C, and D denote the solids, F_1 and F_2 the fluid components. The variables describing the equilibrium are P_{total} (abbreviated below as P_{tot}), T, and the mole fraction of the fluid component 2, X_2 (an abbreviation for X_{F2}). Since

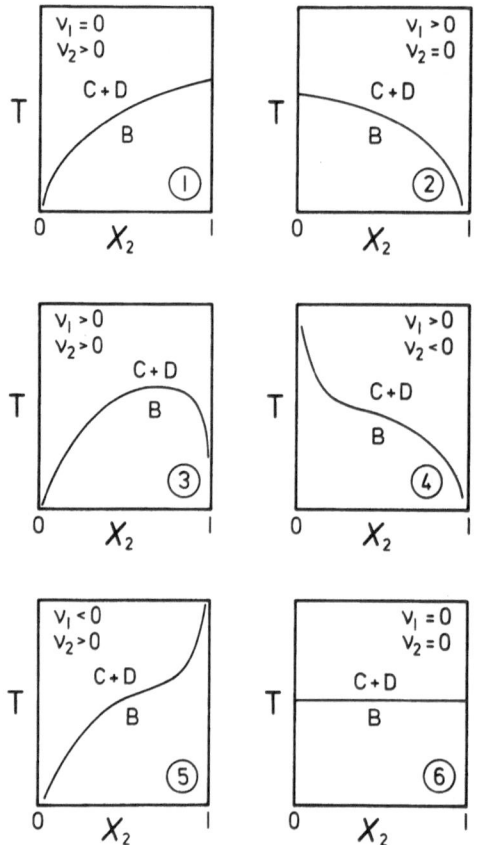

Fig. 6.1 Schematic isobaric (P_{tot} = $P_{F1} + P_{F2}$) T-X_2 curves for solid-fluid equilibria of mixed-volatile reactions showing the six types of reactions discussed in the text

there is only one independent composition variable in a binary mixture, an arbitrary specification of X_2 as the master variable implies

$$X_1 = 1 - X_2. \tag{6.2}$$

As indicated by Greenwood (1967, Fig.1), these equilibria are best handled in terms of isobaric ($P_{tot} = P_{F1} + P_{F2}$ = constant) T - X_2 sections, on which they appear as univariant curves. If the $\Delta H°$ of such a reaction is positive, which is generally the case, he also showed that, depending upon the sign and magnitude of v_1 and v_2, five types of behavior for the isobaric univariant curves can be distinguished. For the sake of completeness, a sixth type has been added to Greenwood's list, and shown schematically in Fig. 6.1.

Type 1: $v_1 = 0$, $v_2 > 0$. This is tantamount to saying that F_2 is actively participating in the reaction, F_1 is merely the inert diluent of the former. The isobaric univariant curve traverses the T - X_2 section with a positive slope all the way to $X_2 = 1$. As $X_2 \to 0$, $(\partial T/\partial X_2)_P \to +\infty$.

Type 2: $v_1 > 0$, $v_2 = 0$. F_1 is now actively participating in the reaction, F_2 is the inert diluent. The isobaric univariant curve commences at $X_2 = 0$ and crosses the T - X_2 plane with a negative slope. This time, $(\partial T/\partial X_2)_P \rightarrow -\infty$ as $X_2 \rightarrow 1$.

Type 3: $v_1 > 0$, $v_2 > 0$. That is, both F_1 and F_2 are liberated by the reaction. The isobaric univariant curve passes through a maximum on a T - X_2 isobar. The fluid composition at T_{max} is given by the stoichiometry of the reaction; $X_2(T_{max}) = v_2/(v_1 + v_2)$.

Type 4: $v_1 > 0$, $v_2 < 0$. Both F_1 and F_2 are active participants; while F_2 is consumed by the reaction, F_1 is set free. The isobaric univariant curve passes through an inflection point (rather than a maximum) and has a negative slope on the T - X_2 section.

Type 5: $v_1 < 0$, $v_2 > 0$. F_1 will be consumed now and F_2 released by the reaction. The isobaric univariant curve will show a positive slope and again pass through an inflection point.

Type 6: $v_1 = 0$, $v_2 = 0$. Neither fluid component is now actively involved in the reaction. In other words, it is a solid-solid reaction taking place in the presence of the binary fluid mixture. Its equilibrium temperature will depend on $P_{tot}(= P_{F1} + P_{F2})$ only.

It is evident from the above that for the equilibria of the types 1, 2, 4, and 5, there is a unique solution for X_2. By contrast, at any T, a double solution for X_2 is obtained for a type 3 reaction.

Following this digression, we take up the thermodynamic formalism for calculating mixed-volatile equilibria. Choosing the standard state for solids as the pure substances at 1 bar and any T of interest, and that for the fluids as the hypothetical ideal gases at 1 bar and specified T, the condition of equilibrium of (6.1) is

$$\Delta\mu = 0 = \Delta G_T^\circ + \int_1^{P_{tot}} \Delta V_{T,s}(P)dP + RT\ln K .$$

(5.7)

The equilibrium constant for the reaction (6.1) is

$$K(6.1) = \frac{a_C^c a_D^d (f_1^\nabla)^{\nu_1} (f_2^\nabla)^{\nu_2}}{a_B^b} ,$$

(6.3)

where f^∇_i, an abbreviation for f^∇_{Fi}, denotes the fugacity of the fluid species i in the fluid mixture at P, T, and X_i. Ignoring $\alpha(T)$ and $\beta(P)$ of the solids, and recalling that the activities of the end-member solids are unity at T and P, substitution of (6.3) into (5.7) yields

$$\Delta\mu = 0 = \Delta G_T^\circ + \Delta V_{298,s}^\circ(P_{tot} - 1) + \nu_1 RT\ln f_1^\nabla + \nu_2 RT\ln f_2^\nabla$$

$$= \Delta H_{298}^\circ - T\Delta S_{298}^\circ + \int_{298}^T \Delta C_P^\circ(T)dT - T\int_{298}^T \left[\frac{\Delta C_P^\circ(T)}{T}\right]dT$$

$$+ \Delta V_{298,s}^\circ(P_{tot} - 1) + \nu_1 RT\ln f_1^\nabla + \nu_2 RT\ln f_2^\nabla .$$

(6.4)

The basic data required to perform such computations are $H°_{f,298}$, $S°_{298}$ and $C°_P(T)$ of every phase, solid or fluid, participating in the reaction, $V°_{298}$ of the solids, and the fugacities of the fluid components 1 and 2 in the fluid mixture, f^V_1 and f^V_2, at T, P, and X_2. Given the fugacity of the pure fluid i at P and T, along with its activity in the fluid mixture at T, P, and X_i, $f^V_i(T,P,X_i)$ is obtained from

$$f_i^V = f_i^* a_i.$$ (3.10b)

For the H_2O-CO_2 mixtures, these data will be taken from Jacobs and Kerrick (1981a).

It is evident from the above that a_i-X_i data for the pertinent fluid mixture are required for computing mixed-volatile reactions. If these data are lacking, an ideal mixing of nonideal fluids is sometimes tacitly assumed. Ideal mixing implies that the activity coefficient, γ_i, is unity across the entire composition range (see Sect.1.2.2.3), and a_i of Eq.(3.10b) may be replaced by X_i. Thus, to obtain $f^{V,id}_i$, fugacity of the component i in an ideal mixture of fluids, (3.10b) is recast to[1]

$$f_1^{V,id} = f_1^*(1-X_2) \text{ and}$$ (6.5a)

$$f_2^{V,id} = f_2^* X_2.$$ (6.5b)

---

[1] Eqs.(6.5a) and (6.5b) are known as the Lewis and Randall rule.

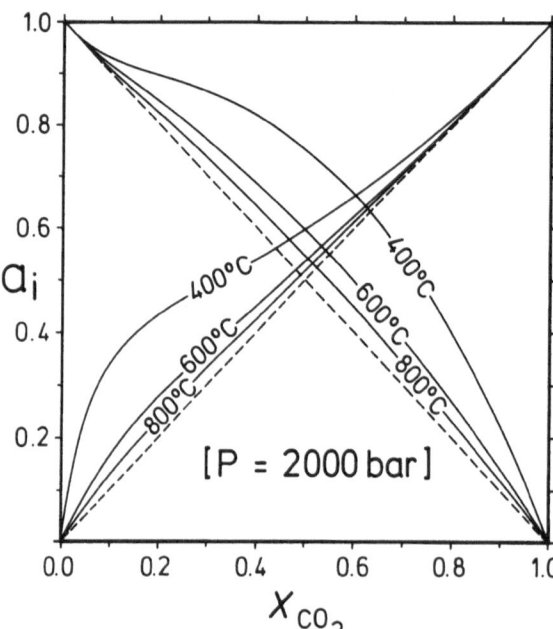

Fig. 6.2 Activity-composition relation of the H_2O-CO_2 binary at 2000 bar P_{tot} and three temperatures (Jacobs and Kerrick 1981a). The *dashed diagonals* apply to ideal mixing of the two species

Substituting (6.5a) and (6.5b) into (6.4), the condition for the equilibrium coexistence of end-member solids and a binary ideally mixing fluid may be recovered,

$$\Delta\mu = 0 = \Delta G_T^\circ + \Delta V_{298,s}^\circ(P_{tot}-1) + v_1 RT \ln f_1^{\nabla,id} + v_2 RT \ln f_2^{\nabla,id}. \tag{6.6}$$

In reality, fluid mixtures seldom behave ideally under geological conditions. Fig. 6.2 illustrates, for example, the a_i-X_i relation for the system H$_2$O-CO$_2$ following Jacobs and Kerrick (1981a). It is apparent that at T less than 600°C, a strong deviation from ideal mixing behavior sets in. Only at temperatures above 800°C does the H$_2$O-CO$_2$ mixture behave nearly ideally. Given a_i-X_i data, the ideal mixing approximation is not recommended for computing any mixed-volatile equilibrium.

6.3 Sample Calculation of Mixed-Volatile (H$_2$O-CO$_2$) Equilibria

The objective of the following exercises is to familiarize the reader with some of the practical aspects of computing mixed-volatile equilibria. The worked examples in this section are based on calorimetric data of the participating phases, as summarized in Table 6.1. The thermodynamic extrapolation and linear summation of the experimental phase equilibria data will not be addressed until the next section.

As a prelude to computing H$_2$O-CO$_2$ mixed-volatile reactions, let us first consider the decarbonation reaction

$$1 \text{ magnesite (M)} = 1 \text{ periclase (P)} + 1 \text{ CO}_2. \tag{6.7}$$

If it were to take place in the presence of pure CO$_2$, the relevant univariant equilibrium curve could have been computed in a manner outlined in Chap. 5, and displayed on a P$_{CO2}$-T diagram. But the question we wish to explore now is the equilibrium involving the end-member solids, magnesite and periclase, with a fluid mixture of CO$_2$ and a second, as yet unspecified, fluid species. For now, we assume that the second fluid species acts as an inert diluent of CO$_2$, which is actively participating in the reaction. This corresponds to Greenwood's (1967) type 1 reaction ($v_1 = 0$, $v_2 > 0$). Diluting CO$_2$ by a second fluid species is equivalent to reducing its a_{CO2}. Consequently, we may display the equilibrium in terms of the P$_{tot}$-T-a_{CO2} space. The bivariant equilibrium of the reaction (6.7) will span a curved surface in a P$_{tot}$-T-a_{CO2} space, separating the trivariant stability fields of the reactant and the products. Rather than using an unwieldy three dimensional phase diagram, we choose the option of projecting that surface along the a_{CO2} axis onto the P$_{tot}$-T plane. The lines for a constant a_{CO2}, thus generated, are called the a_{CO2} isopleths.

Table 6.1. Thermodynamic data for computing H_2O-CO_2 mixed-volatile equilibria in the problem sets 6.1 and 6.2

Phases	Abbreviations	$H^{\circ}_{f,298}$ (J mol^{-1})	S°_{298} (J K^{-1} mol^{-1})	V°_{298} (J bar^{-1} mol^{-1})
Periclase	(P)	-601490.0(290)	26.94(0.17)	1.1248(0.0004)
Talc	(Tc)	-5915900.0(4330)	260.83(0.63)	13.6250(0.0260)
Forsterite	(Fo)	-2170400.0(1400)[a]	94.11(0.10)[a]	4.3790(0.0030)
Magnesite	(M)	-1113280.0(1339)	65.09(0.14)	2.8018(0.0013)
H_2O		-241814.0(42)	188.83(0.04)	-
CO_2		-393510.0(130)	213.79(0.04)	-

$$C^{\circ}_p(T) \ (J \ K^{-1} \ mol^{-1}) = a + b{\cdot}T + c{\cdot}T^{-2} + d{\cdot}T^{-0.5} + e{\cdot}T^2 + f{\cdot}T^{-3}$$

Phase Abbrev.	a	b	c	d	e	f
(P)	65.211	-1.2699E-3	-4.6185E5	-3.8724E2	0.0	0.0
(Tc)[b]	664.11	0.0	-2.1472E6	-5.1872E3	0.0	-3.2737E8
(Fo)[a]	87.36	0.08717	-3.699E6	8.436E2	-2.237E-5	0.0
(M)[b]	194.08	0.0	-1.533E5	-2.1210E3	0.0	1.7491E8
(H_2O)	7.3680	2.7468E-2	-2.2316E5	3.6174E2	-4.8117E-6	0.0
(CO_2)	87.820	-2.6442E-3	7.0641E5	-9.9886E2	0.0	0.0

[a] Robie et al. (1982).
[b] The molar heat capacities of Tc and M have been measured to only 639 and 743 K, respectively. To use them in our problem sets, they must be extrapolated over a large range of temperature. To ensure proper extrapolations, the Berman-Brown type polynomials have been used (Berman and Brown 1985, Table 3). Note that the $C^{\circ}_p(T)$ data for P, H_2O and CO_2 are valid between 298.15 - 1800 K, and require no extrapolation. All other data are from Robie et al. (1979).

Problem Set 6.1

1. Compute the equilibrium pressures of the reaction

$$1 \ \text{magnesite (M)} = 1 \ \text{periclase (P)} + 1 \ CO_2 \qquad (6.7)$$

at specified values of a_{CO2} (1.0, 0.7, 0.4, 0.1, and 0.02) and T, proceeding in 100° steps from 500 to 800°C.
2. Display the results as a_{CO2} isopleths on a P_{tot}-T projection. How do the results compare with the experimental data obtained by Harker and Tuttle (1955) and Johannes and Metz (1968), on the one hand, and by Harker (1958), on the other? What alternative options are available to depict these results?

Solutions to the Problem Set 6.1

Problem 1. In order to solve for the equilibrium pressure of the reaction (6.7) at desired T and a_{CO2}, we utilize Eq.(6.4). Noting that for (6.7), $v_1 = 0$ and v_2 ($\equiv v_{CO2}$) = 1, and inserting (3.10b), (6.4) reduces to

$$\Delta\mu = 0 = \Delta G_T^\circ + \Delta V_{298,s}^\circ (P_{tot} - 1) + RT \ln f_{CO_2}^* + RT \ln a_{CO_2}. \qquad (6.8)$$

At any specified T and a_{CO2}, P is then be solved for by iteration. For this purpose, a suitable initial pressure is chosen, followed by computation of $\Delta\mu$, the sum of the terms on the RHS of (6.8). If $\Delta\mu > 0$, the reactant (magnesite) is stable; by contrast, $\Delta\mu < 0$ implies that the products (periclase + CO$_2$) are stable. Examples of iteration cycles for one temperature (700°C) and the specified set of a_{CO2} are reproduced in Table 6.2. Note that the iteration cycles have been cut off as soon as the equilibrium pressure was established to within $\pm$ 10 bar. Proceeding in an analogous manner, any degree of approach to equilibrium P is possible.

Problem 2. The examples of calculation of reaction (6.7), given in Table 6.2, are for the 700°C isotherm only. Not reproduced are the results for the 500, 600, and 800°C isotherms. Fig. 6.3 assembles the entire set of results in a P_{tot}-T plot of the a_{CO2} isopleths. It demonstrates how the decreasing a_{CO2} in the coexisting fluid dramatically reduces the stability of magnesite.

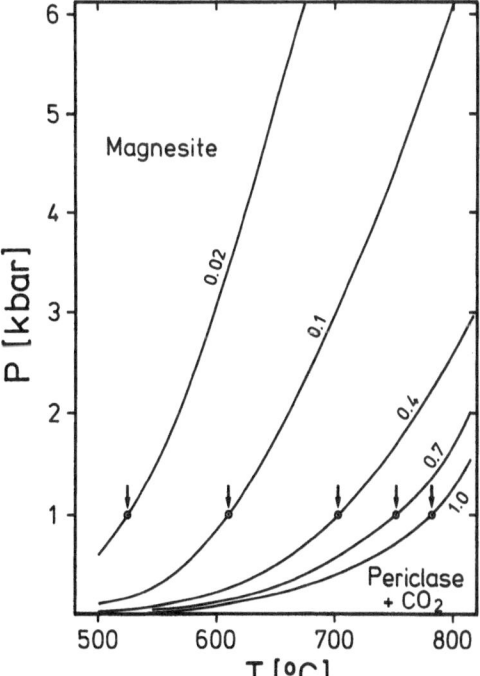

Fig. 6.3 A P-T projection of calculated a_{CO_2} isopleths for the reaction magnesite = periclase + CO$_2$. The equilibrium temperatures at 1000 bar pressure, highlighted by the *arrows*, were utilized to construct the T-a_{CO_2} isobar of Fig. 6.4

Table 6.2. Iterative solutions for the equilibrium pressure of the reaction magnesite = periclase + CO_2 at 700 °C and specified a_{CO_2}

P (bar)	ΔG_T° (J)	$\Delta V_{298,s}^\circ (P_{tot}-1)$ (J)	$f_{CO_2}^*$	$RTlnf_{CO_2}^*$ (J)	$RTlna_{CO_2}$ (J)	$\Delta\mu$ (J)
		$a_{CO_2} = 1.0$				
340	-48579.2	-568.5	362.50	47680.8	0.0	-1467
380	-48579.2	-635.6	409.26	48662.5	0.0	-552
420	-48579.2	-702.7	457.13	49557.5	0.0	+276
400	-48579.2	-669.1	433.04	49119.5	0.0	-129
		$a_{CO_2} = 0.7$				
500	-48579.2	-836.8	556.25	51145.4	-2885.9	-1157
540	-48579.2	-903.9	607.61	51859.9	-2885.9	-509
580	-48579.2	-971.0	660.27	52532.4	-2885.9	+96
560	-48579.2	-937.4	633.81	52201.5	-2885.9	-201
		$a_{CO_2} = 0.4$				
800	-48579.2	-1339.9	974.24	55679.9	-7413.8	-1653
900	-48579.2	-1507.6	1131.75	56892.5	-7413.8	-608
1000	-48579.2	-1675.3	1299.50	58010.8	-7413.8	+342
980	-48579.2	-1641.8	1265.08	57793.6	-7413.8	+159
960	-48579.2	-1608.2	1231.20	57573.9	-7413.2	-27
		$a_{CO_2} = 0.1$				
2000	-48579.2	-3352.3	3683.6	66440.9	-18630.4	-4121
3000	-48579.2	-5029.3	7945.8	72661.0	-18630.4	+422
2900	-48579.2	-4861.6	7405.2	72090.9	-18630.4	+20
2880	-48579.2	-4828.1	7300.5	71975.6	-18630.4	-62
		$a_{CO_2} = 0.02$				
5000	-48579.2	-8383.3	27550.5	82721.2	-31652.4	-5894
6000	-48579.2	-10060.3	47588.4	87143.5	-31652.4	-3148
7000	-48579.2	-11733.3	79720.2	91318.0	-31652.4	-647
7200	-48579.2	-12072.7	88123.0	92128.8	-31652.4	-176
7300	-48579.2	-12240.4	92619.5	92531.4	-31652.4	+59
7280	-48579.2	-12206.9	91704.0	92451.1	-31652.4	+13
7260	-48579.2	-12173.3	90796.5	92370.6	-31652.4	-34

The results of the equilibria calculations may now be compared to their experimental counterparts. The decomposition of magnesite to periclase + pure CO_2 has been experimentally explored by Harker and Tuttle (1955), and later on by Johannes and Metz (1968). At a P_{tot} (= P_{CO2}) of 500 bar, both obtained reaction reversals between 700-710°C, approximately 10°C below the curve calculated from the thermodynamic data (cf. Fig. 6.3, a_{CO2} = 1.0

isopleth). At 1000 bar pressure, the high temperature reversal bracket indicated by the former authors very nearly agrees with our calculated curve, whereas the latter authors' bracket approaches the curve to within 12°C. This picture is similar to that encountered on previous occasions. The reversal brackets do not generally straddle the calculated curve, although they do overlap within the range of uncertainty of the latter, a fact that could have been shown also in the present case by estimating $2\sigma_T$ of the computed curve.

Of immediate relevance in this context is the experimental study of the magnesite-periclase equilibrium in the presence of a mixed-volatile comprising CO_2 and the inert gas Ar (Harker 1958, Fig. 5). These experiments demonstrated that increasing dilution of CO_2 by Ar reduces the dissociation temperature of magnesite to periclase at any P_{tot}, in unison with the above calculations (cf. Fig. 6.3).

There are, of course, other options available to delineate the stability relations of magnesite and periclase in the presence of mixed fluids. One could, for instance, display the equilibrium of reaction (6.7) on an isobaric ($P_{tot} = P_{fluid}$) T-a_{CO2} section (Fig. 6.4). At any P_{tot}, the T-a_{CO2} section can be easily constructed by replotting the equilibrium depicted in Fig. 6.3. To do that at 1000 bar P_{tot}, we read off the equilibrium T from the individual a_{CO2} isopleths (highlighted in Fig. 6.3 by arrowheads) and plot them on the T-a_{CO2} diagram (Fig. 6.4). Yet another possibility is to display the phase relations in terms of the isobaric T-X_{CO2} sections, as advocated and popularized in geochemistry by Greenwood (1967). For an *ideal mixture* of fluids, the T-a_{CO2} diagram becomes identical to its T-X_{CO2} diagram, simply because a_{CO2} equals X_{CO2} for those mixtures. Even when the fluid mixture is nonideal, it is possible to translate a T-a_{CO2} diagram to its T-X_{CO2} counterpart, given the

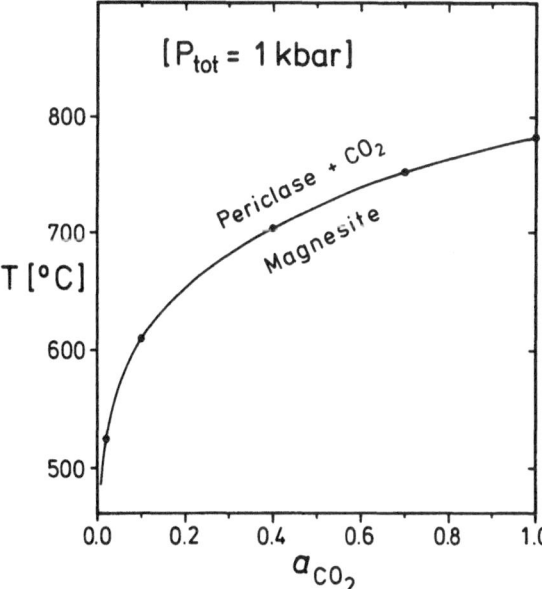

Fig. 6.4 A calculated T-a_{CO_2} isobaric (P_{tot} = 1000 bar) section for the reaction magnesite = periclase + CO_2

appropriate a_i-X_i relations at T and P of interest. Fortunately, there is a simpler, and more direct, technique for computing the T-X_{CO2} section at any desired P_{tot}. Let us now address this problem.

Problem Set 6.2

1. Calculate the isobaric T-X_{CO2} curve of the reaction

$$1 \text{ magnesite (M)} = 1 \text{ periclase (P)} + 1 \text{ CO}_2 \qquad (6.7)$$

at 1000 bar P_{tot} (= P_{H2O} + P_{CO2}), employing the thermodynamic data summarized in Table 6.1. Compare the results of computations with those obtained experimentally by Johannes and Metz (1968).
2. Employing the thermodynamic data from Table 6.1, calculate the isobaric T-X_{CO2} curve for the reaction

$$1 \text{ talc} + 5 \text{ magnesite} = 4 \text{ forsterite} + 1 \text{ H}_2\text{O} + 5 \text{ CO}_2 \qquad (6.9)$$
$$\text{(Tc)} \qquad \text{(M)} \qquad\qquad \text{(Fo)}$$

at a total pressure (= P_{H2O} + P_{CO2}) of 1000 bar.
3. Repeat these calculations, assuming an ideal mixing of H_2O and CO_2. Do the reversal brackets by Johannes (1969) agree within $\pm 2\sigma_T$ of the T-X_{CO2} curve computed for nonideal H_2O-CO_2 mixing?

Solutions to the Problem Set 6.2

Problem 1. The isobaric T-X_{CO2} section of reaction (6.7) was computed from

$$\Delta\mu = 0 = \Delta G_T^\circ + \Delta V_{298,s}^\circ(P_{tot}-1) + RT\ln f_{CO_2}^* + RT\ln a_{CO_2}. \qquad (6.8)$$

This time, (6.8) was solved for X_{CO2} at desired T and P_{tot} (Table 6.3), using f_{CO2}^* and a_{CO2}[= f(X_{CO2})] from the MRK for the H_2O-CO_2 fluid mixture

Table 6.3. Solutions for X_{CO_2} at 700°C and 1000 bar P_{tot} for the reaction magnesite = periclase + CO_2

X_{CO_2}	ΔG_T° (J)	$\Delta V_{298,s}^\circ(P_{tot}-1)$ (J)	$RT\ln f_{CO_2}^*$ (J)	a_{CO_2}	$RT\ln a_{CO_2}$ (J)	$\Delta\mu$ (J)
0.30	-48579.2	-1675.3	58010.8	0.3182	-9264.9	-1509
0.32	-48579.2	-1675.3	58010.8	0.3377	-8783.6	-1027
0.34	-48579.2	-1675.3	58010.8	0.3571	-8331.7	-575
0.36	-48579.2	-1675.3	58010.8	0.3765	-7903.6	-147
0.38	-48579.2	-1675.3	58010.8	0.3959	-7497.1	+259
0.37	-48579.2	-1675.3	58010.8	0.3862	-7697.8	+58

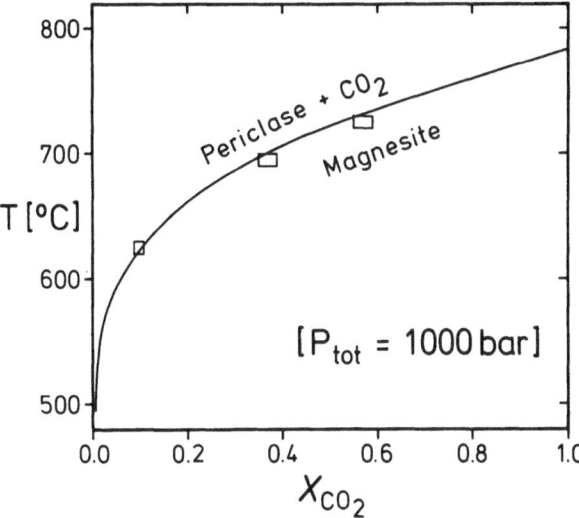

Fig. 6.5 An isobaric (P_{tot} = 1000 bar) section displaying the calculated T-X_{CO_2} curve for the reaction magnesite = periclase + CO_2. The experimental data (*box symbols*), reported by Johannes and Metz (1968), are indicated for comparison

(Jacobs and Kerrick 1981a). As before, $\Delta\mu$, the sum of the terms on the RHS of (6.8) was used to monitor the fields of stability of the reactant ($\Delta\mu > 0$) and the products ($\Delta\mu < 0$). Note that at 700°C and 1 kbar, the equilibrium X_{CO2} has been approached to within ± 0.005. The complete T-X_{CO2} curve has been obtained in this fashion, and displayed in Fig. 6.5. A comparison of the T-X_{CO2} section with the T-a_{CO2} diagram of Fig. 6.4 reveals a near-quantitative agreement at T > 700°C. This is due to the near-ideal mixing of the fluids at these conditions. Also shown in Fig. 6.5 (box symbols) are the experimental reversals for this equilibrium (Johannes and Metz 1968). The mutual agreement between the theory and the experiment is reassuring.

Problem 2. To calculate the T-X_{CO2} isobar of reaction

$$1 \text{ talc} + 5 \text{ magnesite} = 4 \text{ forsterite} + 1 \text{ H}_2\text{O} + 5 \text{ CO}_2, \tag{6.9}$$

its condition of equilibrium was first deduced following Eq.(6.4):

$$0 = \Delta G_T^\circ + \Delta V_{298,s}^\circ(P_{tot}-1) + RT\ln f_{H_2O}^* + 5 \, RT\ln f_{CO_2}^*$$
$$+ RT\ln a_{H_2O} + 5 \, RT\ln a_{CO_2}. \tag{6.10}$$

From (6.10) it is apparent that, in addition to the standard state thermodynamic properties of the solids and fluids (Table 6.1), we need to know f_i^* and a_i for both CO_2 and H_2O. In order to preserve the internal consistency of data on the thermodynamic properties of the fluids, it is *essential* that f_i^* and a_i for *both* fluids be taken from the *same* equation of

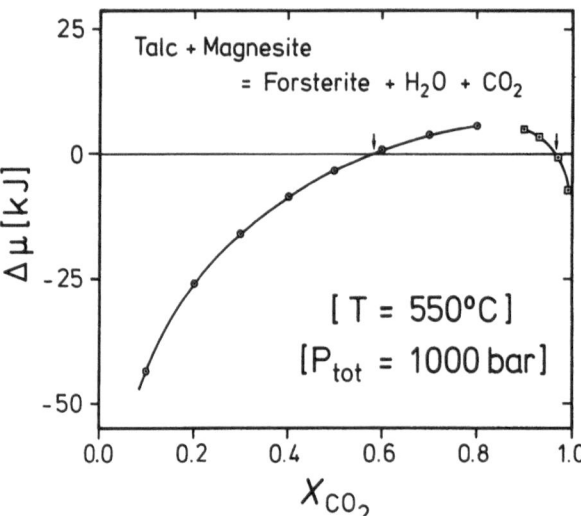

Fig. 6.6 A $\Delta\mu$ vs X_{CO_2} plot of the reaction 1 talc + 5 magnesite = 4 forsterite + 1 H_2O + 5 CO_2 at 550°C and 1 kbar P_{tot}. The dual solutions for X_{CO_2} are indicated by arrows

state. We shall utilize Jacobs and Kerrick's (1981a) MRK for f^*_i and a_i of both CO_2 and H_2O.

Before initiating the calculations, we note that (6.9) belongs to Greenwood's (1967) type 3 reaction, in which both H_2O and CO_2 are liberated. The isobaric univariant curve of such a reaction passes through a maximum and yields dual solutions for X_{CO2} at any T below T_{max}. Using the equation indicated in Section 6.2, X_2(at T_{max}) = $v_2/(v_1 + v_2)$, and noting that $v_1 = 1$ and $v_2 = 5$, we have $X_{CO2}(T_{max})$ = 0.833. We shall use this information in the following calculations. Because the equilibrium has dual solutions for X_{CO2} at any P_{tot} and T, $\Delta\mu$ will change its sign twice as one goes from low to successively higher X_{CO2}. Both solutions may be sought by iteration, and the complete T-X_{CO2} curve obtained by repeating the process at several T. To locate the maximum on the T-X_{CO2} curve, iteration in T is executed at a fixed X_{CO2} = 0.833. The resulting T-X_{CO2} curve for reaction (6.9) is delineated as a solid line in Fig. 6.7 (see later). Aside from this, a graphical technique may also be devised to solve for X_{CO2} at any set of T and P_{tot}. To achieve this, $\Delta\mu$ is calculated at arbitrarily fixed steps of X_{CO2}, and plotted against X_{CO2}. Fig. 6.6 is such a plot for 550°C and 1 kbar P_{tot}. Because $X_{CO2}(T_{max})$ = 0.833, at T < T_{max} two solutions are anticipated, one below and the other above 0.833 X_{CO2}. Figure 6.6 (small circles) delineates $\Delta\mu$ for 0.1 to 0.8 X_{CO2} in steps of 0.1 X_{CO2}. A smooth line drawn through them crosses the $\Delta\mu = 0$ line at 0.58 X_{CO2}; this is one solution. The second one was obtained by plotting $\Delta\mu$ from 0.90 to 0.99 X_{CO2} in steps of 0.03 (box symbols), and was found to be approximately 0.965 X_{CO2}. Again, the iterative method is recommended for more precise results.

Problem 3. The isobaric T-X_{CO2} curve of reaction (6.9) is to be recalculated, assuming ideally mixing fluids. Replacing the a_i terms in Eq.(6.10) by X_i, its condition of equilibrium becomes

$$\Delta\mu = 0 = \Delta G_T^\circ + \Delta V_{298,s}^\circ(P_{tot}-1) + RT\ln f_{H_2O}^* + 5\,RT\ln f_{CO_2}^*$$
$$+ RT\ln(1-X_{CO_2}) + 5\,RT\ln X_{CO_2}. \tag{6.11}$$

The T-X_{CO2} curve shown by a dashed line (Fig. 6.7) was generated by solving (6.11) for X_{CO2} at 1 kbar and a set of T. Also shown in that diagram is the T-X_{CO2} curve obtained by considering the nonideality of the fluid mixture (solid curve). Attention is drawn to the fact that the two calculated curves diverge progressively with decreasing T and X_{CO2}, reflecting enhanced nonideality of mixing of the fluids at those conditions, in consonance with Fig. 6.2. The box symbols in Fig. 6.7 depict the experimental T-X_{CO2} data reported by Johannes (1969). Again, they do not straddle the curve computed from the calorimetric data. The question we must explore then is whether or not the experimental data points lie within the $\pm 2\sigma_T$ range of the computed curve.

The error propagation formalism is similar to that outlined in Chap. 5, except that for this reaction between end-member solids and a H₂O-CO₂ fluid mixture, we need to consider $\sigma_{\Delta\mu}$, rather than $\sigma_{\Delta G^*}$. It is obtained in a manner analogous to Eq.(5.13):

$$\sigma_{\Delta\mu}^2 = [\sigma_{\Delta H_{298}^\circ}^2 + T^2\sigma_{\Delta S_{298}^\circ}^2 + (P_{tot}-1)^2\sigma_{\Delta V_{298,s}^\circ}^2$$
$$+ (v_{H_2O})^2\sigma_{RT\ln f_{H_2O}}^2 + (v_{CO_2})^2\sigma_{RT\ln f_{CO_2}}^2]^{0.5}. \tag{6.12}$$

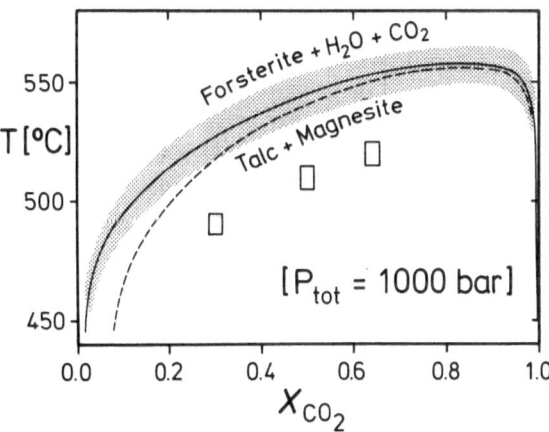

Fig. 6.7 A calculated isobaric (P_{tot} = 1 kbar) T-X_{CO_2} curve for the reaction 1 talc + 5 magnesite = 4 forsterite + 1 H₂O + 5 CO₂, compared with its reversals (*box symbols*) by Johannes (1969). The *solid curve* with $\pm 2\sigma_T$ envelope (*stippled*) is for nonideal mixing, the *dashed curve* is for ideal mixing of H₂O and CO₂

Note that the $\sigma_{RTlnf^\nabla_i}$ comprises the sigmas of $RTlnf^*_i$ *and* $RTlna_i$. Because σ_{RTlna_i} is unknown, and $\sigma_{RTlnf^*_i}$ was earlier assumed to be 0.5% of $RTlnf^*_i$, $\sigma_{RTlnf^\nabla_i}$ is estimated to be 1% of $RTlnf^\nabla_i$. Having derived $\sigma_{\Delta\mu}$, we only require an expression for σ_T. Proceeding in a manner analogous to that outlined in Section 5.3, we may write

$$\sigma_T = \frac{\sigma_{\Delta\mu}}{\left| -\Delta S^\circ_T + \nu_{H_2O}\left[Rlnf^\nabla_{H_2O} + \frac{RT}{f^\nabla_{H_2O}}\left(\frac{\partial f^\nabla_{H_2O}}{\partial T}\right)\right] + \nu_{CO_2}\left[Rlnf^\nabla_{CO_2} + \frac{RT}{f^\nabla_{CO_2}}\left(\frac{\partial f^\nabla_{CO_2}}{\partial T}\right)\right]\right|}$$

(6.13)

The stippled envelope in Fig. 6.7 shows the $2\sigma_T$ of reaction (6.9). It is clear that the experimental reversals of reaction (6.9) can *not* be reconciled with the phase diagram derived from calorimetric data. It is interesting to note that this apparent discrepancy has been traced back by Berman et al. (1986) to a hefty 16 kJ mol^{-1} error in the calorimetric $H^\circ_{f,298}$ of talc, an error far beyond its reported 2σ uncertainty of ± 4.33 kJ mol^{-1} (Robie et al. 1979).

6.4 Interpolation, Extrapolation and Linear Summation of Reaction Reversal Data

In Section 5.5 it was demonstrated that the thermodynamic extrapolation of the reaction reversal data often suffices to gain insight into some geochemical problems. The theoretical formalisms developed there will now be extended to treat the H_2O-CO_2 mixed-volatile reactions.

The technique of handling reaction reversal data, outlined in Chap. 5, and extended below to H_2O-CO_2 mixed-volatile reactions, invariably requires that *a number of* tight reversals (Case 1) be available for each independent reaction. In actual practice, this requirement is not always fulfilled. As an alternative, we may use *one* tight reversal (Case 2) from which to start our extrapolation, provided accurate S°_{298}, $C^\circ_p(T)$, and V°_{298} data for all the phases participating in the reaction are available.

Case 1. A set of tight reversal brackets is available

This case was treated earlier, while dealing with dehydration reactions. Two situations arise, depending upon whether or not the $C^\circ_p(T)$ data are available. If $C^\circ_p(T)$ data are lacking, $\Delta H^\circ_{<T>}$ and $\Delta S^\circ_{<T>}$, obtained from the RHS of

$$- \Delta H^\circ_{<T>} + T\Delta S^\circ_{<T>} = \Delta V^\circ_{298,s}(P_{tot}-1) + RTlnK,$$

(5.22)

are utilized for interpolation (a significant extrapolation beyond the range of the reversals is not possible). The $\Delta H^{\circ}_{<T>}$ and $\Delta S^{\circ}_{<T>}$ data are reported in two alternative fashions:

$$\Delta G^{\circ}_{T} = \Delta H^{\circ}_{<T>} - T\Delta S^{\circ}_{<T>}, \tag{5.23}$$

or, together with $\Delta V^{\circ}_{298,s}$, recast to the form

$$\log K = -\frac{A}{T} + B + C \cdot \frac{(P_{tot} - 1)}{T}. \tag{5.26}$$

An example of application of Eq. (5.22) to a number of dehydration equilibria has been provided in Section 5.5.2. The H_2O-CO_2 mixed-volatile reactions are treated in an analogous manner, noting that its K will now comprise f^{V}_{H2O} and f^{V}_{CO2} in the mixed fluid, rather than f^{*}_{H2O}, the fugacity of pure H_2O. Skippen's (1971) pioneering study of the phase relations in metamorphosed siliceous marbles is an instructive example in this regard. The system comprises eight end-member solids plus a CO_2-H_2O fluid mixture, related to each other through 46 equilibria, stable or metastable. Of these, only five are linearly independent. Choosing a reference pressure of 2000 bar, Skippen (1971, Table 10) calculated $\log K$ ($\log K_1$ through $\log K_5$) for the five *independent* equilibria. Their linear summation gave $\log K_N$ ($\log K_6$ through $\log K_{46}$) of all the 41 linearly *dependent* reactions, following the general scheme

$$\log K_N = k_1 \log K_1 + k_2 \log K_2 + k_3 \log K_3 + k_4 \log K_4 + k_5 \log K_5. \tag{6.14}$$

He listed the linear summation coefficients k_1 to k_5 and $\log K_N$ for the linearly dependent reactions, expressing the latter as

$$\log K_N = -\frac{A_N}{T} + B_N + C_N \cdot \frac{(P_{tot} - 2000)}{T}. \tag{6.15}$$

The phase relations for the metamorphosed siliceous marbles were thus elucidated from experimental data on merely five equilibria. Simple and elegant though this approach is, the results obtained are somewhat marred by the huge uncertainties generally associated with them (see below, Problem Set 6.3).

If the $C^{\circ}_P(T)$ data of all the phases are available, the basic equation employed to handle reaction reversal data is

$$-\Delta H^{\circ}_{298} + T\Delta S^{\circ}_{298} = \int_{298}^{T} \Delta C^{\circ}_P(T) dT - T \int_{298}^{T} \left[\frac{\Delta C^{\circ}_P(T)}{T}\right] dT$$

$$+ \Delta V^{\circ}_{298,s}(P_{tot} - 1) + RT\ln K . \tag{5.27}$$

Again, the RHS of (5.27) is evaluated for the linearly independent reactions, followed by the derivation of their $\Delta H°_{298}$ and $\Delta S°_{298}$, a technique dubbed as the "second-law method" in thermochemistry. The linearly dependent equilibria are then handled by the linear summation of these $\Delta H°_{298}$ and $\Delta S°_{298}$ data. An example of this type of calculation for dehydration reactions has been documented in Section 5.5.2. To extend this to the mixed-volatile reactions, K, the equilibrium constant, needs to be modified only, inserting the appropriate f^{v}_i terms in lieu of f^{*}_i.

Having elucidated the thermodynamic basis for the manipulation of experimental phase equilibria data, let us analyze a simple set of equilibria among the end-member solids anorthite ($Ca[Al_2Si_2O_8]$, An), wollastonite ($CaSiO_3$, Woll), grossular ($Ca_3Al_2[SiO_4]_3$, Gr), calcite ($CaCO_3$, Cc), quartz (SiO_2, Q), and a fluid mixture of H_2O and CO_2. Four reactions may be written between them:

$$1 \text{ calcite} + 1 \text{ quartz} = 1 \text{ wollastonite} + 1 \text{ } CO_2, \tag{6.16}$$

$$1 \text{ grossular} + 1 \text{ quartz} = 1 \text{ anorthite} + 2 \text{ wollastonite}, \tag{6.17}$$

$$1 \text{ calcite} + 1 \text{ anorthite} + 1 \text{ wollastonite} = 1 \text{ grossular} + 1 \text{ } CO_2, \tag{6.18}$$

$$2 \text{ calcite} + 1 \text{ anorthite} + 1 \text{ quartz} = 1 \text{ grossular} + 2 \text{ } CO_2. \tag{6.19}$$

The isobaric univariant $T\text{-}X_{CO2}$ curves of these reactions intersect at an isobaric invariant point. The worked problem set given below is concerned with the phase relations around that invariant point. Note that only two of the four reactions are linearly independent. In the following, reactions (6.16) and (6.17) are considered as independent, because a number of tight reversals are available for both. The reactions (6.18) and (6.19) may then be computed by linear summation:

$$(6.18) = 1 \cdot (6.16) - 1 \cdot (6.17), \text{ and} \tag{6.20}$$

$$(6.19) = 2 \cdot (6.16) - 1 \cdot (6.17). \tag{6.21}$$

Problem Set 6.3

1. Employing $V°_{298}$ and $C°_p(T)$ data listed in Table 6.4, along with the experimental reversal brackets of (6.16) and (6.17) summarized in Tables 6.5 and 6.6 (see later), respectively, derive $\Delta H°_{298}$ and $\Delta S°_{298}$ for those two reactions.

2. For a P_{tot} = 2 kbar, calculate the isobaric $T\text{-}X_{CO2}$ equilibrium curves of reactions (6.16) through (6.19). Compare the results with the experimental reversals reported by Gordon and Greenwood (1971). Do they agree within $\pm 2\sigma_T$ of the calculated curve?

Solutions to the Problem Set 6.3

Problem 1. Numerous reversal brackets for the reaction

$$1 \text{ calcite} + 1 \text{ quartz} = 1 \text{ wollastonite} + 1 \text{ CO}_2 \tag{6.16}$$

were reported by Ziegenbein and Johannes (1974). Of these, a batch of ten were selected such that they cover the entire P_{tot}, T, and X_{CO2} space explored by them. These were combined with the single bracket obtained by Jacobs and Kerrick (1981c, Table 2). Table 6.5 summarizes these experimental data. To obtain $\Delta H°_{298}$ and $\Delta S°_{298}$ of (6.16), its condition of equilibrium was first written as

$$-\Delta H°_{298} + T\Delta S°_{298} = \int_{298}^{T} \Delta C°_P(T)dT - T \int_{298}^{T} \left[\frac{\Delta C°_P(T)}{T}\right]dT$$

$$+ \Delta V°_{298,s}(P_{tot} - 1) + RT\ln f^{\triangledown}_{CO_2}. \tag{6.22}$$

Its RHS was then evaluated for each bracketing point, using $V°_{298}$ and $C°_P(T)$ data from Table 6.4, and taking $f^{\triangledown}_{CO2}$ from Kerrick and Jacobs' MRK for the $H_2O\text{-}CO_2$ mixture. The results are reproduced in the last column of Table 6.5 and plotted in Fig. 6.8. Each bracket appears as a box or a diamond symbol indicating the uncertainty of the experiment. The bold line and the pair of thinner straight lines denote the linear least squares fit to the data and its $\pm 1\sigma$, respectively. From that fit, we obtain

$\Delta H°_{298}(6.16) = 96372 \text{ J},$
$\Delta S°_{298}(6.16) = 167.758 \text{ J K}^{-1}$, and
$\rho_{\Delta H \Delta S(6.16)} = -0.9878.$ $\tag{6.23}$

The $\Delta H°_{298}$ and $\Delta S°_{298}$ of the solid-solid reaction

$$1 \text{ grossular} + 1 \text{ quartz} = 1 \text{ anorthite} + 2 \text{ wollastonite} \tag{6.17}$$

was derived likewise from the appropriate condition of equilibrium

$$-\Delta H°_{298} + T\Delta S°_{298} = \int_{298}^{T} \Delta C°_P(T)dT - T \int_{298}^{T} \left[\frac{\Delta C°_P(T)}{T}\right]dT$$

$$+ \Delta V°_{298}(P_{tot} - 1). \tag{6.24}$$

The RHS of (6.24) for all four reversals by Huckenholz (1975) are reproduced in the last column of Table 6.6. A linear least squares fit to the data yields

$\Delta H°_{298}(6.17) = 55818 \text{ J},$
$\Delta S°_{298}(6.17) = 73.970 \text{ J K}^{-1}$, and
$\rho_{\Delta H \Delta S(6.17)} = -0.9982.$ $\tag{6.25}$

Table 6.4. V°_{298}, S°_{298}, and $C^\circ_P(T)$ data for the phases pertinent to problem sets 6.3 and 6.4

Phases	Composition	V°_{298}(J bar^{-1} mol^{-1})	S°_{298}(J K^{-1} mol^{-1})
Calcite	$CaCO_3$	3.6934	91.71
Anorthite	$Ca[Al_2Si_2O_8]$	10.0790	199.30
Quartz	SiO_2	2.2688	41.46
Grossular	$Ca_3Al_2[SiO_4]_3$	12.5300	255.50
Wollastonite	$CaSiO_3$	3.9930	82.01
Carbon dioxide	CO_2	-	213.79

$C^\circ_P(T)$ (in J K^{-1} mol^{-1}) = a + b·T + C·T^{-2} + d·T$^{0.5}$ + e·T^2 + f·T^{-3}

Phases	a	b	c	d	e	f
Calcite[a]	193.24	0.0	0.0	-2.0409E3	0.0	1.9946E8
Anorthite (validity 298.15-1800 K)	516.83	-9.2492E-2	-1.4085E6	-4.5885E3	4.1883E-5	0.0
Low quartz (validity 298.15-844 K)	44.603	3.7754E-2	-1.0018E6	0.0	0.0	0.0

T_{tr} = 844 K, ΔH°_{tr} = 476 J, ΔS°_{tr} = 0.55 J K^{-1}

Phases	a	b	c	d	e	f
High Quartz (validity 844-1800 K)	58.928	1.0031E-2	0.0	0.0	0.0	0.0
Grossular[b]	519.40	0.0	-2.80063E7	-0.631E2	0.0	3.51073E9
Wollastonite (validity 298.15-1400 K)	111.25	1.4373E-2	-2.7779E6	1.6936E1	0.0	0.0
Carbon dioxide (validity 298.15-1800 K)	87.82	-2.6442E-3	7.0641E5	-9.9886E2	0.0	0.0

[a] The high-temperature $C^\circ_P(T)$ measurements of calcite extend to 780 K only. To ensure its proper extrapolation to higher temperatures, the Berman-Brown polynomial has been used.
[b] The high-temperature $C^\circ_P(T)$ function of grossular, tabulated by Robie et al. (1979), is based on C°_P measurements to 987 K, although it is believed to be extrapolable to 1200 K. To avoid error due to unconstrained extrapolation (see Sect. 2.2.1) to higher temperature, its $C^\circ_P(T)$ polynomial, indicated by Berman and Brown (1985), has been preferred. It is based on the same set of high-temperature $C^\circ_P(T)$ measurements, but supplemented by drop calorimetric H°_T-H°_{298} data to 1273 K. All other data are from Robie et al. (1979).

Table 6.5. Summary of experimental reversal data for the reaction 1 calcite + 1 quartz = 1 wollastonite + 1 CO_2 (Ziegenbein and Johannes 1974, and Jacobs and Kerrick 1981c). The last column lists the RHS of Eq. (6.22) for each experimental data point

P_{tot} (kbar)	T (°C)	T (K)	X_{CO_2}	Stable assemblage	RHS of Eq. (6.22) (J)
2	705	978.15	0.97	Woll + CO_2	67415.0
2	695	968.15	0.97	Cc + Q	66588.5
2	680	953.15	0.74	Woll + CO_2	63289.5
2	670	943.15	0.77	Cc + Q	62778.8
2	630	903.15	0.34	Woll + CO_2	54474.4
2	630	903.15	0.36	Cc + Q	54800.9
2	580	853.15	0.15	Woll + CO_2	47093.7
2	570	843.15	0.16	Cc + Q	46839.9
2	555	828.15	0.04	Woll + CO_2	39054.4
2	555	828.15	0.07	Cc + Q	42092.4
4	790	1063.15	0.92	Woll + CO_2	82270.2
4	790	1063.15	0.95	Cc + Q	82547.9
4	765	1038.15	0.72	Woll + CO_2	78026.2
4	765	1038.15	0.78	Cc + Q	78654.5
4	720	993.15	0.41	Woll + CO_2	70328.1
4	720	993.15	0.44	Cc + Q	70761.1
6[a]	880	1153.15	0.83	Woll + CO_2	95586.5
6[a]	880	1153.15	0.89	Cc + Q	96221.9
6	765	1038.15	0.25	Woll + CO_2	76989.5
6	765	1038.15	0.29	Cc + Q	77811.9
6	685	958.15	0.08	Woll + CO_2	64733.5
6	685	958.15	0.11	Cc + Q	66476.3

[a] From Jacobs and Kerrick (1981c). All other reversal brackets are from Ziegenbein and Johannes (1974).

Table 6.6. Summary of reversal brackets for the reaction 1 quartz + 1 grossular = 1 anorthite + 2 wollastonite (Huckenholz 1975). The last column lists the RHS of Eq. (6.24) for each bracketing point

P (kbar)	T (°C)	T (K)	Stable assemblage	RHS of Eq. (6.24) (J)
1	530	803.15	Gr + Q	3752.8
1	545	818.15	An + Woll	3816.0
2	570	843.15	Gr + Q	7195.2
2	580	853.15	An + Woll	7234.7
3	615	888.15	Gr + Q	10685.2
3	635	908.15	An + Woll	10790.6
4	670	943.15	Gr + Q	14237.6
4	680	953.15	An + Woll	14291.1

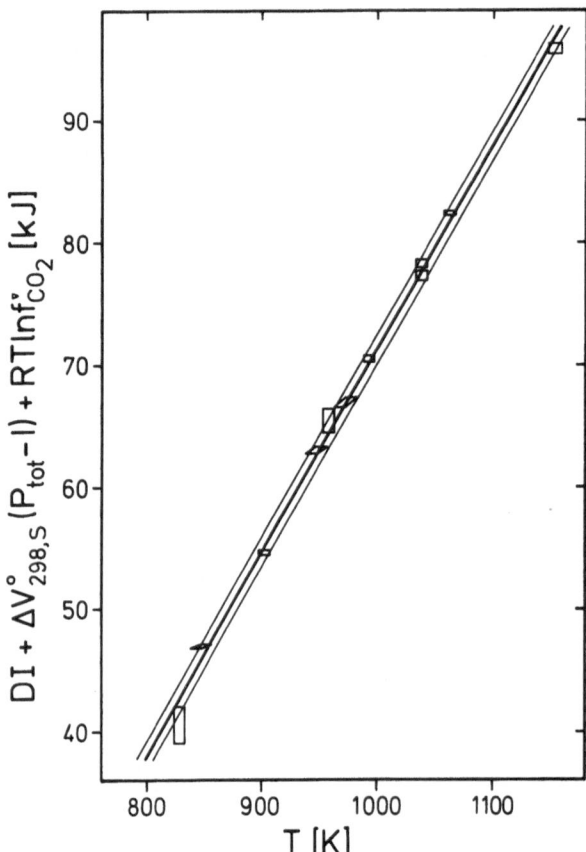

Fig. 6.8 The RHS of Eq.(6.22) plotted against T for the reversals of the reaction calcite + quartz = wollastonite + CO_2 (Ziegenbein and Johannes 1974; Jacobs and Kerrick 1981c). The sum of the terms $_{298}\int^T \Delta C_P^\circ dT - T_{298}\int^T (\Delta C_P^\circ /T) dT$ has been abbreviated DI. The linear regression fit to the data, and its standard deviation, are indicated by *bold* and *thin lines*, respectively

Problem 2. The above data will be utilized now to compute the isobaric univariant curves of reactions (6.16) through (6.19). The first step toward that goal is deriving ΔH°_{298} and ΔS°_{298} of reactions (6.18) and (6.19). Given ΔH°_{298} and ΔS°_{298} of (6.16) and (6.17), their linear addition [according to Eqs.(6.20) and (6.21)] yields:

$$\Delta H^\circ_{298}(6.18) = 40554 \text{ J}, \quad \Delta S^\circ_{298}(6.18) = 93.788 \text{ J K}^{-1}, \text{ and}$$

$$\Delta V^\circ_{298,s}(6.18) = -5.2354 \text{ J bar}^{-1}; \tag{6.26}$$

$$\Delta H^\circ_{298}(6.19) = 136926 \text{ J}, \quad \Delta S^\circ_{298}(6.19) = 261.546 \text{ J K}^{-1}, \text{ and}$$

$$\Delta V^\circ_{298,s}(6.19) = -7.2046 \text{ J bar}^{-1}. \tag{6.27}$$

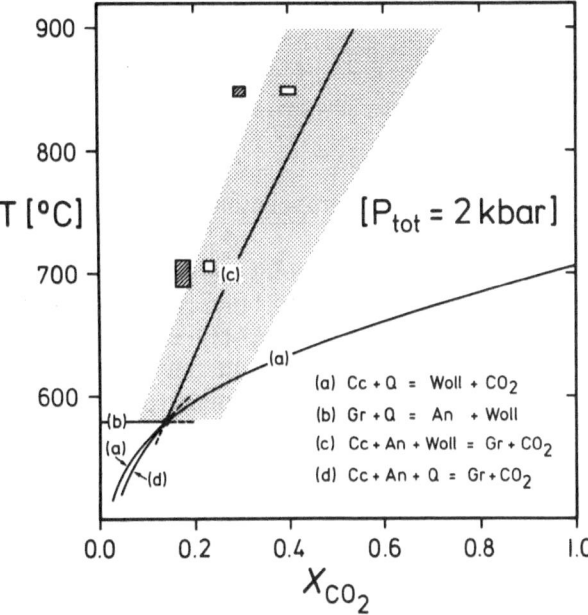

Fig. 6.9 A T-X_{CO_2} isobar (2000 bar P_{tot}) showing the computed equilibria involving calcite, wollastonite, grossular, anorthite, quartz, and the H_2O-CO_2 fluid. The experimental reversal brackets for Eq. (6.18), reported by Gordon and Greenwood (1971), are *hatched* (grossular stable) and *open* (anorthite + wollastonite + calcite stable) *box symbols*. The reversals agree with computation within its one σ_T range (stippled envelope)

Employing the standard state data, the isobaric T-X_{CO2} curves were computed, and displayed in Fig. 6.9. Also plotted are two brackets for reaction (6.18), reported by Gordon and Greenwood (1971). Note that the computed curve of that reaction, labeled (c) in Fig. 6.9, misses the reversal brackets by a clear margin. In order to check if the reversal brackets are compatible with the computed T-X_{CO2} curve within its $\pm 2\sigma_T$ range, the error propagation formalisms outlined in Section 5.5.2 were used. Following the arguments given there, $\sigma_{\Delta\mu(6.18)}$ was obtained from

$$\sigma_{\Delta\mu(6.18)} = \left[\left(\frac{\partial\Delta\mu(6.18)}{\partial\Delta\mu(6.16)} \right)^2 \sigma^2_{\Delta\mu(6.16)} + \left(\frac{\partial\Delta\mu(6.18)}{\partial\Delta\mu(6.17)} \right)^2 \sigma^2_{\Delta\mu(6.17)} \right]^{0.5} .(6.28a)$$

Taking the derivatives with respect to (6.20), (6.28a) reduces to

$$\sigma_{\Delta\mu(6.18)} = [\sigma^2_{\Delta\mu(6.16)} + \sigma^2_{\Delta\mu(6.17)}]^{0.5}. \tag{6.28}$$

Using $\sigma_{\Delta\mu(6.18)}$, and $\Delta S^\circ_{298}(6.18)$ given above, σ_T for reaction (6.18) may be computed from

$$\sigma_T = \frac{\sigma_{\Delta\mu}}{| -\Delta S^\circ_{298}|} . \tag{6.29}$$

Figure 6.9 shows σ_T as a stippled envelope. The reversal brackets clearly overlap with the computed curve even within its $\pm\sigma_T$ range. As emphasized earlier, σ_T of equilibria obtained by the method of linear summation is generally very large; the errors accumulate despite use of $C°_P(T)$ of the participating phases.

Case 2. A selected bracket is to be utilized for extrapolation

To apply this method, it is necessary to know the point within the bracket at which $\Delta\mu$ equals zero. Usually, the midpoint of the bracket is *assumed* to satisfy that condition. Given data on $S°_{298}$, $V°_{298}$, and $C°_P(T)$, this assumption allows the derivation of $\Delta H°_{298}$ of the reaction. This is the traditional "third-law" method (Lewis and Randall 1961, p.178) of thermochemistry. Knowing $\Delta H°_{298}$, and using $\Delta S°_{298}$, $\Delta C°_P(T)$, etc., the phase relations are computed in a straightforward manner. Thus, comprehensive phase diagrams may be derived starting from the selected reversal brackets for a desired number of reactions.[2] This simple approach will be illustrated in the next problem set.

Problem Set 6.4

1. Employing the standard state thermodynamic data from Table 6.4 and the equation of state for H_2O-CO_2 fluids (Kerrick and Jacobs 1981), calculate $\Delta H°_{298}$ of the reaction

$$1 \text{ calcite} + 1 \text{ quartz} = 1 \text{ wollastonite} + 1 \text{ CO}_2, \qquad (6.16)$$

assuming equilibrium at 6000 bar P_{tot}, 880°C, and 0.86 X_{CO2} (cf. Jacobs and Kerrick 1981c, Table 2).
2. Employing $\Delta H°_{298}$ derived above, calculate the isobaric T-X_{CO2} curve for Eq.(6.16) at 2000 and 6000 bar P_{tot}. Compare the results with the reversal brackets established independently by Ziegenbein and Johannes (1974), and listed in Table 6.5.

Solutions to the Problem Set 6.4

Problem 1. To compute $\Delta H°_{298}$ of reaction (6.16), its condition of equilibrium, given earlier as Eq.(6.22), is rewritten as

$$-\Delta H_{298}° = -T\Delta S_{298}° + \int_{298}^{T} \Delta C_P°(T)dT - T\int_{298}^{T}\left[\frac{\Delta C_P°(T)}{T}\right]dT$$

$$+ \Delta V_{298,s}°(P_{tot} - 1) + RT\ln f_{CO_2}^{\triangledown}. \qquad (6.22a)$$

---

[2] For an alternative method of computing phase relations starting from the midpoints of selected reversal brackets, see Slaughter et al. (1976) and Perkins et al. (1987).

Given the data on $\Delta S°_{298}$, $\Delta C°_p(T)$, $\Delta V°_{298,s}$ (Table 6.4) and f^V_{CO2}, Eq.(6.22a) involves only one unknown, $\Delta H°_{298}$. We may then use the specified point at 6000 bar P_{tot}, 880°C, and 0.86 X_{CO2}, to solve (6.22a) for $\Delta H°_{298}$:

$$- \Delta H^{\circ}_{298}(6.16) = -1153.15(162.63) - 14319.66 + 22216.59 \qquad (6.30a)$$
$$+ (-1.9692(5999)) + 99823.30 \text{ (J), or}$$

$$\Delta H^{\circ}_{298}(6.16) = 91630 \text{ J.} \qquad (6.30)$$

Before proceeding to Problem 2, a comment on $\Delta H°_{298}$ derived above by the third-law and the second-law methods seems warranted. It is emphasized that $\Delta H°_{298}$, obtained by third-law treatment, is highly dependent on the *input* value of $\Delta S°_{298}$. Consequently, $\Delta H°_{298}$, thus derived, may not be consonant with that obtained by the second-law method, where *both* $\Delta H°_{298}$ *and* $\Delta S°_{298}$ are simultaneously extracted from the reversal brackets. Reaction (6.16) is a case in point. Its $\Delta H°_{298}$, derived from the second-law and the third-law formats, differs substantially from each other. Therefore, to compute the T-X_{CO2} curve, we must rely either on the $\Delta H°_{298}$ *and* $\Delta S°_{298}$ given in Eq.(6.23), or stick with the $\Delta H°_{298}$ given in (6.30), *along with* the calorimetric $\Delta S°_{298}$ from Table 6.4. It is also pointed out that in the general practice of thermochemistry, the third-law technique is applied to a *set* of reversal

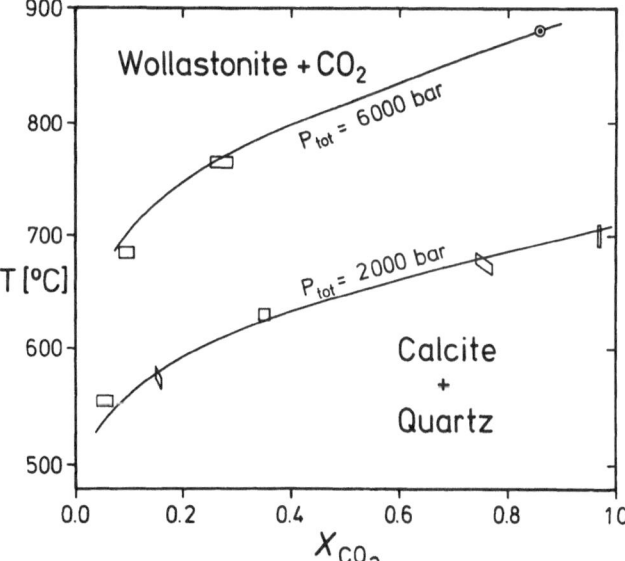

Fig. 6.10 Two calculated T-X_{CO_2} isobars of the reaction calcite + quartz = wollastonite + CO_2 based on thermodynamic data obtained from the midpoint (*encircled dot*) of the reversal bracket given by Jacobs and Kerrick (1981c). Also displayed are a set of reversal brackets obtained by Ziegenbein and Johannes (1974). The latter are clearly compatible with the calculated curves

brackets, rather than to a single equilibrium point. The average of the ΔH°_{298} is then quoted, along with its standard deviation. A good example of such evaluation is given by Robie and Hemingway (1984b, Fig. 5; see also Lewis and Randall 1961, Fig. 15-2). Only if ΔS°_{298}, $\Delta C^\circ_p(T)$, and $\Delta V^\circ_{298,s}$ are consistent with the reversal brackets, does a unique ΔH°_{298} result from each bracket.

Problem 2. Using ΔH°_{298} from Eq.(6.30), along with the ancillary data reproduced in Table 6.4, the isobaric T-X_{CO2} curve for (6.16) was computed at 2 and 6 kbar P_{tot}. The results have been displayed in Fig. 6.10. The starting point of the extrapolation (6000 bar P_{tot}, 880°C and 0.86 X_{CO2}) appears in this figure as an encircled dot. Also indicated are the reversal brackets by Ziegenbein and Johannes (1974). Note that the two curves are consistent with the majority of the reversals, missing two of them only marginally.

6.5 Concluding Remarks

At the very beginning of this chapter it was pointed out that geochemically relevant fluid mixtures usually comprise a multitude of species. Nonetheless, calculation of the solid-fluid equilibria are limited in this book to binary fluids only. The computational procedure used here could be extended in a straightforward manner to fluid mixtures comprising more than two species, given activity data of the fluids of interest in the requisite multispecies fluid mixture.[3] Jacobs and Kerrick (1981b, pp.611-613, Fig.5) extended the computation of mixed-volatile reactions from the H_2O-CO_2 binary to the H_2O-CO_2-CH_4 ternary following this principle.

---

[3] Such data are rather rare. Jacobs and Kerrick (1981b) and Bowers and Helgeson (1983) have succeeded in developing equations of state of the ternary mixtures H_2O-CO_2-CH_4 and H_2O-CO_2-NaCl, respectively. Their success should spawn further research efforts.

Chapter 7

Derivation of an Internally Consistent Thermodynamic Dataset by Mathematical Programming

7.1 Introduction

A reliable prediction of phase relations in mineral systems is central to geochemical research. As demonstrated in the foregoing chapters, a thermodynamic dataset based *exclusively* on calorimetry seldom allows computation of phase diagrams sufficiently accurate for geochemical applications. Nevertheless, calorimetry is, and will remain, the fundamental source of thermodynamic data, even if such data need to be fine-tuned for further use. The preceding chapters have also taught us that reaction reversals are effective constraints for calorimetric data. Thus, a combination of reversal brackets and calorimetry must give us a thermodynamic dataset best suited to accomplish our goal of computing precise phase diagrams. A thermodynamic dataset, compatible with calorimetry and reaction reversals alike, is called an internally consistent thermodynamic dataset.

Two contrasting techniques have been employed so far to derive sets of internally consistent thermodynamic data:

1. Multiple least squares regression (e.g. Helgeson et al. 1978; Haas et al. 1981; Hemingway et al. 1982; Robinson et al. 1982; Holland and Powell 1985), and

2. Mathematical programming (e.g. Gordon 1973; Day and Kumin 1980; Halbach and Chatterjee 1982b, 1984; Berman et al. 1985, 1986; Berman 1988).

The method of least squares analysis has been reviewed elsewhere in considerable detail (Powell and Holland 1985, 1988). In this book, we focus on mathematical programming. The worked example of derivation of an internally consistent dataset concerns the system Al_2SiO_5. Readers seeking to go into the further ramifications of mathematical programming will find Berman et al. (1986) helpful.

Prior to continuing with the next section, it is pointed out that, following Iyanaga and Kawada (1977), we employ mathematical programming (MP) as a collective term comprising the techniques of linear programming (LP), quadratic programming (QP), and nonlinear programming (for definitions, see Sect. 7.4.2).

7.2 The Nature of the Problem

In order to comprehend the nature of the problems involved in deriving reliable thermodynamic data, let us start by calculating the phase diagram for the system Al_2SiO_5. The *calorimetric* and ancillary data, on the basis of which these calculations are to be done, are listed in Table 7.1 (in Sect. 7.5.1). The calculated phase diagram is displayed in Fig. 7.1. The heavy lines are the univariant curves for the three equilibria, kyanite = andalusite, kyanite = sillimanite, and andalusite = sillimanite. Both the sequence of the curves and the location of the invariant point are contrary to geological observations. The conclusion thus appears inescapable that the calorimetric data do not provide a meaningful phase diagram for the Al_2SiO_5 unary. This, however, is an outcome of the fact that the calorimetric data are a priori not *mutually* consistent, because they are obtained and evaluated on a phase by phase basis (see Chap. 2).

Recall that a tabulated calorimetric datum is the average of a number of measurements, and that its "correct" value lies with 95% confidence within a range of plus or minus two standard deviations ($\pm 2\sigma$) of the mean. The $\pm 2\sigma$ uncertainties for the Al_2SiO_5 phases translate to more or less large ranges of T or P in the calculated phase relation. These are also depicted in Fig. 7.1, and

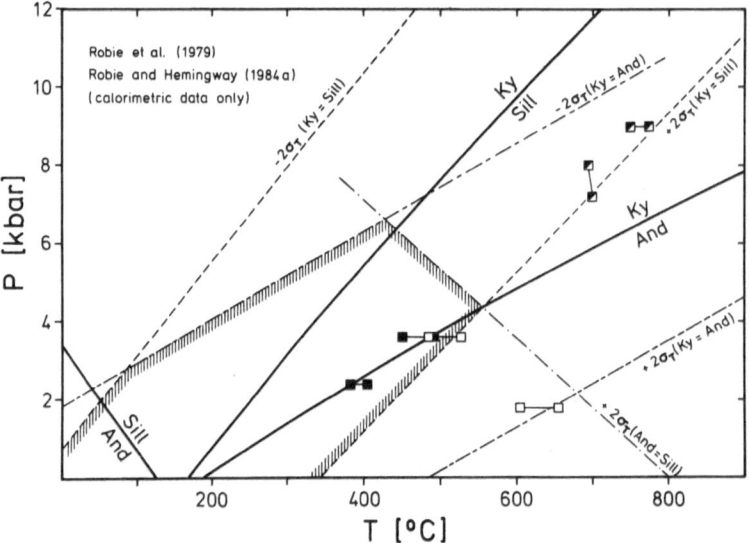

Fig. 7.1 The Al_2SiO_5 phase diagram calculated on the basis of calorimetric data. The univariant curves of the reactions And = Sill, Ky = And, and Ky = Sill are depicted by *bold lines*. The 2σ uncertainties of the computed curves are also shown and labeled in a self-evident manner. The hatched P-T field delimits the possible location of the invariant point. Superimposed on this diagram are two reversal brackets for each of the three reactions. Note that the size of the symbols reflects the P-T uncertainties of the reversal experiments

labeled accordingly. In other words, the "true" univariant curves need not necessarily coincide with those calculated from the tabulated mean calorimetric values, although they will be expected to lie within their respective uncertainty envelopes. It is interesting to note now that the correct topology of the Al_2SiO_5 phase diagram can be restored simply by sliding the univariant curves of the Ky = Sill and And = Sill reactions within their permissible ranges to higher temperatures. This displaces the invariant point to a geologically reasonable T and P. The shaded P-T field in Fig. 7.1, as deduced from the overlapping uncertainty ranges of the individual curves, depicts the range of permissible location for the invariant point.

What has been said above in connection with the Al_2SiO_5 unary, applies in general to other mineral systems. Due to this, earlier geochemists sometimes avoided using calorimetric data altogether, relying instead on reaction reversals. Given reversal brackets for some linearly independent equilibria, phase diagrams were obtained by linear summation (see Skippen 1971; Evans et al. 1976; Jacobs and Kerrick 1981c). Although this may be a viable alternative in a few cases, it has two drawbacks. First, the number of equilibria accessible to computation by linear summation is limited, owing to the lack of reversal data on the relevant independent equilibria. And secondly, errors pile up prohibitively in most cases of linear summation, as demonstrated in the foregoing chapters. Viewed that way, there is no substitute for a reliable thermodynamic database. Given such a database, the prediction of the phase diagram can eventually get off the ground.

The obvious way to refine existing calorimetric data would be to include into the input database other independent measurements that exert additional constraints on the results of calorimetry, and solve them simultaneously for the thermodynamic parameters of the individual phases. That reaction reversals are very useful for this purpose is easily appreciated from Fig. 7.1. Two reversal brackets each for the three reactions have been plotted there. The brackets of the Ky = And reaction (solid boxes) cut the calculated curve, implying a very good agreement between calorimetry and the phase equilibria experiments. This is no longer true for the other two equilibria. For example, the two And = Sill brackets, depicted in Fig. 7.1 as open boxes; do not straddle the calculated curve. However, they do lie within the 2σ range of uncertainty, implying that the results still agree within $\pm 2\sigma$ of the calorimetric data. Likewise, there is also an agreement between the calorimetric data and the reversal brackets for the Ky = Sill reaction (half-shaded boxes), albeit only marginal. An agreement between the reversal brackets and the uncertainty envelopes of the curves computed from calorimetric data suggests that the calorimetric data are correct within $\pm 2\sigma$. Therefore, it must be possible to find a set of values for $H°_{f,298}$, $S°_{298}$ etc. for the three Al_2SiO_5 phases which, though not identical to the calorimetric *mean* values indicated in tables, still lie within their $\pm 2\sigma$ range, and at the same time satisfy the reversal brackets. They will evidently be better suited to compute phase diagrams. Collectively, they represent a set of internally consistent thermodynamic data.

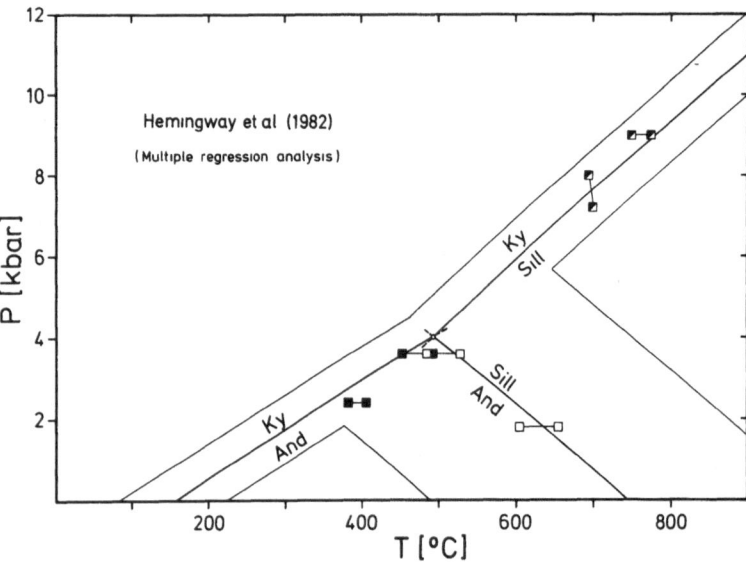

Fig. 7.2 The Al$_2$SiO$_5$ phase diagram generated from thermodynamic data derived by Hemingway et al. (1982) by the method of multiple regression analysis of calorimetric and phase equilibria data. The uncertainty envelopes (shown by *thinner lines*) correspond to ±2σ of the thermodynamic data. The reversal brackets are from Fig. 7.1

As stated above, two contrasting methods have been utilized to derive an internally consistent thermodynamic dataset. The merits and demerits of the two have been critically appraised by Berman et al. (1986, Table 1). The most notable advantage of mathematical programming is that it automatically ensures consistency with all primary input data, which is not always achieved by the regression method. This is best seen by calculating the Al$_2$SiO$_5$ phase diagram using alternative datasets obtained by the two methods. Figure 7.2 displays the Al$_2$SiO$_5$ phase diagram computed from Hemingway et al. (1982) data, derived by multiple regression analysis. Compared to the earlier results based on calorimetry alone (cf. Fig. 7.1), a dramatic improvement is evident. Also, the uncertainty envelopes have shrunk considerably. Nevertheless, the calculated Ky = And univariant curve fails to cut across all the brackets, a situation often encountered with some other databases obtained by multiple regression analysis. This may be contrasted with Fig. 7.3, which depicts the Al$_2$SiO$_5$ phase diagram computed by the program "PTX-system" (Perkins et al. 1986), employing the Berman et al. (1985) database derived by mathematical programming. The computed curves are now consistent with the brackets, if only marginally. Note, however, that no uncertainty envelopes appear in this P-T diagram, because the uncertainties are undefined. Mathematical programming has been rightly criticized on this ground (see Powell and Holland 1988, p.182). Although the uncertainties remain to be defined, it should be borne in mind that the uncertainty envelopes of the P-T diagram can be no larger than the reversal brackets themselves.

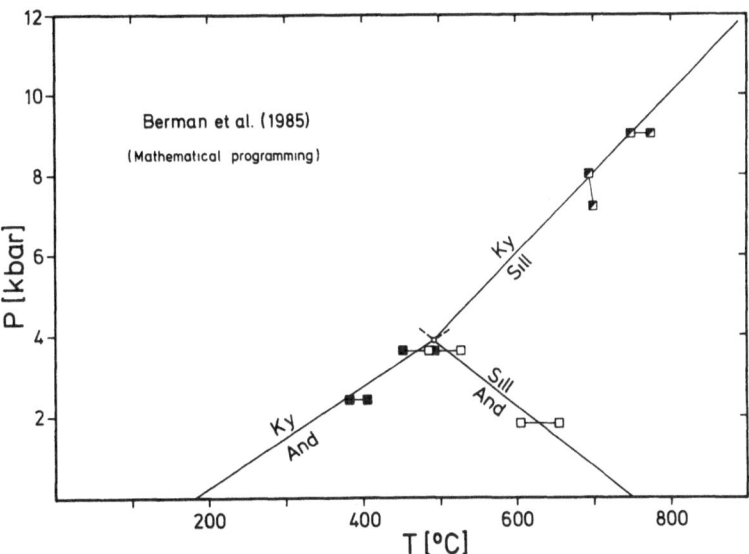

Fig. 7.3 Phase diagram for the Al_2SiO_5 system calculated from a set of internally consistent thermodynamic data obtained by Berman et al. (1985) by processing calorimetric and reversal data by MP. The reversal brackets are the same as in Fig. 7.1

7.3 A Thermodynamic Model for Mathematical Programming

Derivation of an internally consistent thermodynamic dataset by mathematical programming requires us to outline a thermodynamic model, which is appropriate for handling the job. Let us build the model in the context of the Al_2SiO_5 system. Considering that this system involves solid-solid reactions only, we generalize them as

$$bB = cC + dD, \tag{4.1}$$

that is, the reactant, b mole of the solid B, decomposes to yield the products, c and d moles of the solids C and D, respectively. In the specific example of the system Al_2SiO_5, both reactants and products are end-member phases.[1] Choosing the pure substances at 1 bar and T as the standard state, the condition of equilibrium of (4.1) is written as,

$$\Delta G^* = 0 = \Delta H^{\circ}_{298} - T\Delta S^{\circ}_{298} + \int_{298}^{T} \Delta C^{\circ}_{P}(T)dT - T\int_{298}^{T} \left[\frac{\Delta C^{\circ}_{P}(T)}{T}\right]dT$$

$$+ \left[\Delta V^{\circ}_{298} + \Delta(\alpha_{\circ}V^{\circ}_{298})(T-298) + \frac{1}{2}\Delta(\alpha_1 V^{\circ}_{298})(T^2-298^2)\right.$$

$$\left. - \left\{\frac{1}{2}\Delta(\beta_0 V^{\circ}_{298}) + \frac{1}{6}\Delta(\beta_1 V^{\circ}_{298})(P+2)\right\}(P-1)\right](P-1). \tag{4.15}$$

---

[1] This formalism can be extended in a straightforward fashion to include crystalline solutions, pure fluids and fluid mixtures (see Berman et al. 1986).

Next, the quantities on the RHS of (4.15) are subdivided into two categories according to whether or not they are to be *adjusted* within their $\pm 2\sigma$ uncertainty limits. It is essential to keep in mind that model building is always an arbitrary process. A simple, yet successful practice, is to permit the model to adjust $H°_{f,298}$ and $S°_{298}$ of the phases, taking the input values of the remaining quantities for granted. Nonadjusted quantities would then include $C°_p(T)$, $V°_{298}$, α_o, α_1, β_o, and β_1. Such a model is justified when comparatively large uncertainties in $H°_{f,298}$ and $S°_{298}$ are likely, relative to those of the other quantities. As indicated in Chap. 2, $H°_{f,298}$ for solids is derived by linear summation of a number of $\Delta H°_{soln}$ terms; the uncertainty of each $\Delta H°_{soln}$ thus adds up to cause a large uncertainty of $H°_{f,298}$. Although the $S°_{298}$ of solids is usually measured fairly accurately, large uncertainties might yet arise owing to improper appraisal of their configurational entropy. These considerations justify this model.

The above model may not be adequate in all cases, however. To give just one example, consider a system in which high-temperature $C°_p(T)$ data are lacking for some of the phases. In this case, our model will have to be extended to incorporate all the coefficients of the estimated $C°_p(T)$ functions of the relevant phases into the category of quantities to be adjusted by the model (e.g. Halbach and Chatterjee 1982b, p.478). In this way, we could go on adding more and more variables on the RHS of Eq.(4.15) into the category of adjustable quantities, ending up more or less by adjusting all the variables within their uncertainty limits (Berman et al. 1986; Berman 1988).

7.4 Outlines of Mathematical Programming

7.4.1 Input Constraints

Having elucidated the thermodynamic model, let us address the input constraints applied to derive a set of internally consistent thermodynamic data. Basically, there are two types of experimental measurements whose results comprise the input database. The first variety of experiments measures the phase properties. It includes calorimetric measurements of $H°_{f,298}$, $S°_{298}$, and $C°_p(T)$ on the one hand, and $V°_{298}$, $\alpha(T)$, and $\beta(P)$ of solids on the other. The second category of experiments is the reaction reversal, which measures the property difference, $\Delta G^*(T,P)$, due to a reaction. It should be noted, however, that the results of these two types of independent experiments are mutually related. They may be reconciled with each other, because linear summation of the phase properties yields the property differences due to a reaction. The challenge, then, is to process the two contrasting types of input data simultaneously and rigorously.

Another aspect of the two varieties of input constraints has to be examined. Each measurement of a phase property like $H°_{f,298}$, $S°_{298}$, $C°_p(T)$ or $V°_{298}$ is executed a number of times, and the mean value of the quantity, Y_i', is reported along with $2\sigma(Y_i')$. If our thermodynamic model treats the quantity

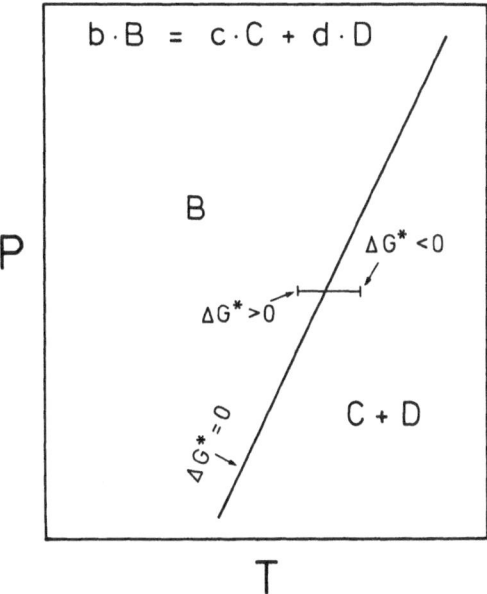

Fig. 7.4 A schematic P-T diagram showing a reversal bracket for the reaction (4.1). At the left bracketing point, the reactant B is stable ($\Delta G^* > 0$), while the products C and D are stable (i.e. $\Delta G^* < 0$) at the right bracketing point. The equilibrium curve, on which $\Delta G^* = 0$, is unknown; it is depicted only schematically

as an adjustable variable, to be designated henceforth as Y_i, it is immediately obvious that Y_i must lie within the range

$$Y_i' - 2\sigma(Y_i') \le Y_i \le Y_i' + 2\sigma(Y_i').$$ (7.1)

The second type of input constraints employed in mathematical programming is derived from phase equilibria reversal experiments. The type of information supplied by reaction reversal is indicated in Fig. 7.4. In terms of reaction (4.1), the low-temperature bracket denotes the T-P condition at which the relative stability of the reactant (in our case, phase B) has been experimentally demonstrated; at that T and P, $\Delta G^* > 0$. By contrast, the higher-temperature bracket shows the relative stability of the products, C and D; thus, $\Delta G^* < 0$ at this T and P. Note, in particular, that the location of the P-T curve, on which B, C, and D are mutually at equilibrium, and $\Delta G^* = 0$, although shown schematically in this figure, remains unknown. The constraints derived from the reaction reversals may then be formalized as

$\Delta G^* > 0$ (reactants stable), and (7.2a)

$\Delta G^* < 0$ (products stable). (7.2b)

It is quite clear by now that both categories of constraints, utilized to derive an internally consistent thermodynamic dataset, have to be handled as *linear inequalities*. As pointed out earlier, dealing simultaneously with several equilibria in a given chemical system amounts to handling a series of *linear equalities*, because the linear summation of the phase properties provides the property difference due to the reaction. The mathematical technique used to process a bunch of simultaneous linear inequalities *and* equalities is known as *mathematical programming*.

7.4.2 Basic Formalisms and Definitions

Basically, a mathematical programming problem is one in which a function, known as objective function, F, is optimized (that is, minimized or maximized), subject to a set of constraints involving both inequalities and equalities. Mathematical programming, MP, is a collective term comprising a variety of techniques like linear, quadratic, and nonlinear programming. In the problem to be handled below, the constraints invariably comprise linear inequalities and equalities; the objective function, nevertheless, may be linear or quadratic. A *linear programming* problem, LP, optimizes a *linear* objective function subject to a set of *linear* constraints, while a *quadratic programming* problem, QP, optimizes a *quadratic* objective function subject to *linear* constraints. We may note that nonlinear constraints also arise in mineralogical thermodynamics, and may be handled by the technique of nonlinear programming. This, however, is beyond the scope of this book.

In the simplest case, that of linear programming, we need to solve for the n variables Y_1, Y_2, ... Y_n, such that the linear objective function

$$F_1 = c_1 Y_1 + c_2 Y_2 + ... + c_n Y_n \qquad (7.3)$$

assumes a minimum value subject to *two* types of constraints (Gill et al. 1983):

$$l_i \leq Y_i \leq u_i, \text{ with } i = 1 \text{ to n}, \qquad (7.4a)$$

l_i and u_i denoting the lower and the upper bounds for the variable Y_i, *and* m general constraints of the form

$$
\begin{aligned}
l_1 &\leq a_{11} Y_1 + \text{...............} &+ a_{1n} Y_n &\leq u_1 \\
l_2 &\leq a_{21} Y_1 + \text{...............} &+ a_{2n} Y_n &\leq u_2 \\
&\quad . &\quad . \\
&\quad . &\quad . \\
&\quad . &\quad . \\
l_m &\leq a_{m1} Y_1 + \text{...............} &+ a_{mn} Y_n &\leq u_m.
\end{aligned}
\qquad (7.4b)
$$

Note that c_i and a_{ij} of Eqs.(7.3) and (7.4b) are constants. Note also that l_i must be less than or equal to u_i; when the two bounds become identical, we have an equality. Moreover, the upper and the lower bounds of these constraints are such that either can be ±∞.

In many cases, it is found to be more desirable to optimize a nonlinear (e.g. a quadratic) objective function subject to a set of linear constraints. For this type of application, Gill et al. (1983) have devised an algorithm which permits optimization of a nonlinear objective function, F_n, of the general form

$$F_n = \sum_i \sum_j d_{ij} Y_i Y_j + \sum_i c_i Y_i. \tag{7.5}$$

It should be noted that this nonlinear objective function reduces to F_1, the linear objective function of (7.3), if $d_{ij} = 0$. On the other hand, for $c_i = 0$ and d_{ij} $(i \neq j) = 0$, F_n is identical to F_q, the quadratic objective function of Eq.(7.15), because d_{ij} $(i = j)$ is still nonzero. Thus, the algorithm by Gill et al. (1983) can be used to solve both linear and quadratic programming problems.

7.5 Worked Example: The System Al₂SiO₅

Let us apply mathematical programming to derive an internally consistent thermodynamic dataset for the three phases, sillimanite (Sill), kyanite (Ky), and andalusite (And), of the system Al₂SiO₅. The primary objective of this exercise is to document the approach explicitly, and in a step by step manner.

7.5.1 Input Database

The first step is to collect all relevant experimental data on the phases of interest. Tables 7.1 and 7.2 list the phase property and the phase equilibria reversal data, respectively. They warrant two comments. First, the reversals reported for the reaction And = Sill by Richardson et al. (1969) have been excluded from the input database. Their exclusion is justified on the ground that they are incompatible with the naturally occurring assemblages chloritoid-sillimanite-quartz (Holdaway 1978; Milton 1986) and paragonite-sillimanite-quartz (Grambling 1984). And secondly, every reversal bracket has been expanded to take into account the uncertainties of temperature and pressure in the reversal experiments. This has been done in such a manner that the adjusted data form the widest permissible brackets for each reaction. Each half-bracket has been labeled $\Delta G^* > 0$ or $\Delta G^* < 0$, depending on the relative stability of the reactant or the product in that experiment (cf. Fig. 7.4).

Table 7.1. Summary of phase property data ($\pm 2\sigma$ in parentheses) for the system Al_2SiO_5. Unless otherwise stated, the calorimetric and V°_{298} data are from Robie et al. (1979), whereas α_0, α_1, β_0 and β_1 are from Clark (1966)

Phases	Abbreviations	$H^\circ_{f,298}$ (J mol^{-1})	S°_{298} (J K^{-1} mol^{-1})	V°_{298} (J bar^{-1} mol^{-1})
Kyanite	(Ky)	-2591730.0(1900)	82.30(0.13)[a]	4.415(0.0050)[a]
Andalusite	(And)	-2587525.0(2100)	91.39(0.14)[a]	5.152(0.0050)[a]
Sillimanite	(Sill)	-2585760.0(1740)	95.79(0.14)[a]	4.986(0.0050)[a]

Phases	α_0(K^{-1})	α_1(K^{-2})	β_0(bar^{-1})	β_1(bar^{-2})
Ky	6.56E-6	2.42E-8	0.70E-6[a]	0.00[a]
And	5.80E-6	3.46E-8	0.67E-6[a]	0.00[a]
Sill	4.06E-6	2.05E-8	0.65E-6[a]	0.00[a]

	$C^\circ_P(T)$(J K^{-1} mol^{-1}) $= a + b \cdot T + c \cdot T^{-2} + d \cdot T^{-0.5} + e \cdot T^2$				
Phases	a	b	c	d	e
Ky[a]	3.0390E2	-1.3390E-2	-8.9520E5	-2.9043E3	0.00
And[a]	2.9040E2	-1.0520E-2	-1.1090E6	-2.6278E3	0.00
Sill[a]	2.2610E2	1.4070E-2	-2.4400E6	-1.3760E3	0.00

[a] Robie and Hemingway (1984a).

7.5.2 Thermodynamic Analysis

7.5.2.1 Introduction

The thermodynamic model developed above requires us to handle two adjustable parameters, $H^\circ_{f,298}$ and S°_{298}, for each phase, that is, a total of six variables for the three phases And, Sill, and Ky. However, dealing with six variables simultaneously precludes a diagrammatic representation of the results, as may be desirable to make every step of mathematical programming understood. For this reason, we shall begin by simplifying the problem, and developing the concepts of MP in terms of ΔH°_{298} and ΔS°_{298} of one reaction, for example, Ky = Sill. This will enable us to show the adjustable parameters of the model in two-dimensional diagrams. Once that is done, we shall proceed by splitting ΔH°_{298} and ΔS°_{298} of the Ky = Sill reaction into $H^\circ_{f,298}$ and S°_{298} of the phases Ky and Sill. The final step will comprise a

Table 7.2. Reaction reversal data to be used as input for deriving an internally consistent thermodynamic dataset for the phases of the system Al$_2$SiO$_5$. Note that the nominal reversal brackets have been expanded according to the reported uncertainties of T and P. The relative stability of the reactants is indicated by $\Delta G^* > 0$, that of the products by $\Delta G^* < 0$

Reaction 1: 1 Ky = 1 And (Newton 1966a; Richardson et al. 1969; Holdaway 1971)

T(°C)	P(bar)	ΔG^*
377	2500	> 0
409	2300	< 0
447	3700	> 0
499	3500	< 0
557	4900	> 0
590	4700	< 0
690	6700	> 0
710	5900	< 0
740	7200	> 0
760	5900	< 0
790	7600	> 0
810	6800	< 0
820	7900	> 0
830	7250	< 0

Reaction 2: 1 Ky = 1 Sill (Newton 1966b; Richardson et al. 1968; Newton, in Holdaway 1971, p.120). The restriction numbers are referred to in the text

T(°C)	P(bar)	ΔG^*	Restriction No.
740	8900	> 0	1
760	7300	< 0	2
690	8100	> 0	3
705	7100	< 0	4
745	9100	> 0	5
780	8900	< 0	6
795	10100	> 0	7
830	9900	< 0	8
890	12400	> 0	9
910	10600	< 0	10

Table 7.2. continued

Reaction 3: 1 And = 1 Sill (Holdaway 1971; Weill, in Holdaway 1971)

T(°C)	P(bar)	ΔG^*
600	1700	> 0
660	1900	< 0
486	3500	> 0
532	3700	< 0
750	1	> 0
800	1	< 0

simultaneous treatment of all three reactions involving the three phases And, Ky, and Sill.

7.5.2.2 Phase Equilibria Constraints

Phase equilibria constraints for a reaction may be derived in a straightforward manner from the reversal data listed in Table 7.2. Inserting $C^\circ_p(T)$, V°_{298}, α_o, α_1, β_o, and β_1 data from Table 7.1 into (4.15), the inequalities (7.2a) and (7.2b) may be written in the general form

$$\Delta H^\circ_{298} - T_i \Delta S^\circ_{298} + r_i > 0 \text{ (or} < 0), \tag{7.6}$$

with r_i denoting the sum of the terms on the RHS of (4.15), except $\Delta H^\circ_{298} - T_i \Delta S^\circ_{298}$. To be more specific, let us consider the first half-bracket at 1013.15 K and 8900 bar for the reaction Ky = Sill (restriction No. 1 of reaction 2 in Table 7.2), for which $\Delta G^* > 0$. Inserting the appropriate data on $C^\circ_p(T)$, V°_{298}, α_o, α_1, β_o and β_1 of Ky and Sill from Table 7.1 into Eq.(4.15), and recalling that $\Delta G^* > 0$, this restriction may be written as

$$\Delta H^\circ_{298} - 1013.15 \cdot \Delta S^\circ_{298} + 5614.6 \text{ (J)} > 0. \tag{7.6a}$$

An evaluation of the next restriction (No. 2 in Table 7.2) at 1033.15 K and 7300 bar yields

$$\Delta H^\circ_{298} - 1033.15 \cdot \Delta S^\circ_{298} + 4764.7 \text{ (J)} < 0, \tag{7.6b}$$

and so on. Thus, the reversal brackets of the reaction Ky = Sill may be readily translated into a system of linear inequalities in the $\Delta H^\circ_{298} - \Delta S^\circ_{298}$ space.

A system of linear inequalities has no unique solution; thus, no unique values can be assigned to ΔH°_{298} and ΔS°_{298}. However, it often has *ranges* of

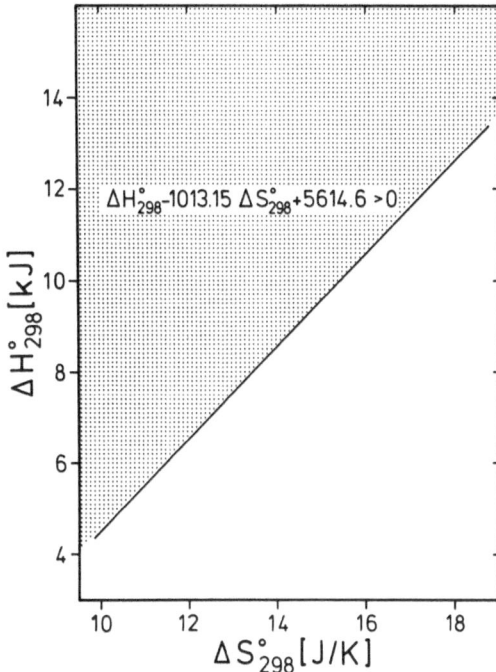

Fig. 7.5 A ΔH°_{298} vs ΔS°_{298} plot showing the feasible solutions set (*stippled area*) for the inequality (7.6a) corresponding to the first half-bracket (Table 7.2) of the Ky = Sill reaction

values, to be designated henceforth as a set of feasible solutions. Because the concept of a feasible solutions set is central to mathematical programming, we must examine it carefully. Let us do that by deriving the feasible solutions set for the reaction Ky = Sill in a ΔH°_{298} - ΔS°_{298} space. To do this, we start with the first restriction. Figure 7.5 depicts a straight line plot of the *equality*

$$\Delta H^\circ_{298} - 1013.15 \cdot \Delta S^\circ_{298} + 5614.6 \ (J) = 0 \qquad (7.6a')$$

on the ΔH°_{298} - ΔS°_{298} graph. From this diagram, it is immediately apparent that the set of feasible solutions for the *inequality* (7.6a) comprises all elements of the set above the straight line (the stippled region in Fig. 7.5). Proceeding in this way, the set of feasible solutions for the second restriction of the Ky = Sill reaction, written as a linear inequality in (7.6b), is obtained. It is evident that the set of feasible solutions now extends below the straight line, because the sign of the inequality is reversed. Therefore, each restriction generates a set of feasible solutions corresponding to a half-plane either above or below the straight lines on a ΔH°_{298} - ΔS°_{298} graph, depending upon the sign of ΔG^*. Simultaneous consideration of these half-planes for any particular reaction may lead to either of the two cases. The half-planes are non-overlapping; the overall solutions set is empty. This implies that the reversal brackets are mutually inconsistent. The second possibility is that the half-planes do overlap, generating a non-empty feasible solutions set; this time, the reversal brackets are mutually compatible. Figure 7.6 illustrates the

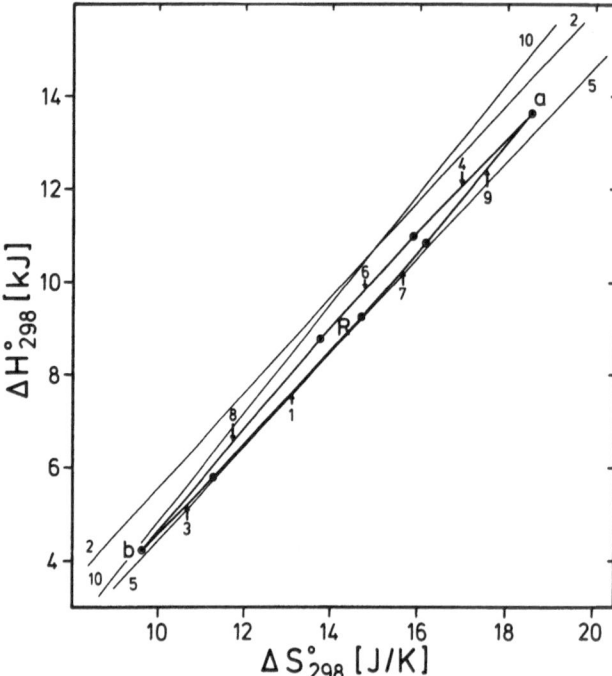

Fig. 7.6 A ΔH°_{298} vs ΔS°_{298} diagram showing the set of feasible solutions, R, for the equilibrium Ky = Sill. The restrictions 3, 1, 7, 9, 4, 6, and 8 (of Table 7.2) define the set. The redundant restrictions 2, 5, and 10 are shown by *thinner lines*

resulting feasible solutions set, R, for the reaction Ky = Sill; it is bounded on all sides by straight lines. The restrictions (numbered as in Table 7.2) defining the feasible solutions set are shown by heavier lines, the redundant ones by lighter ones. Note, however, that a feasible solutions set need not necessarily be bounded on all sides; it may also extend to infinity on one side.

Before proceeding further, let us examine some properties of the feasible solutions set R. First of all, *every* element of the set R (corresponding to a pair of values for ΔH°_{298} and ΔS°_{298}), including the extreme ones, is invariably consistent with *all* the reversal brackets. In other words, a P-T curve calculated from *any* of them must pass through all reversal brackets; the slope of the curve will, however, depend on the pair of the values of ΔH°_{298} and ΔS°_{298} chosen. Thus, choosing the upper right extremal pair of values for ΔH°_{298} and ΔS°_{298} (indicated by the letter a in Fig. 7.6) leads to the maximum permissible slope of the computed curve, because the slope of the curve is dictated by the relation

$$\frac{dP}{dT} = \frac{\Delta S}{\Delta V} = \frac{\Delta H}{T \Delta V} . \tag{7.7}$$

Conversely, the curve computed from the pair of ΔH°_{298} and ΔS°_{298}, corresponding to the lower left corner of R (point b of Fig. 7.6), gives the lowest possible slope. It is important to realize, however, that these extremal values of $\Delta H°_{298}$ and $\Delta S°_{298}$ are *not independent* of each other. That is, if we choose to use the $\Delta H°_{298}$ corresponding to the point a for computing the curve, we *must* also use the single corresponding value of $\Delta S°_{298}$. For this reason, it is *not permissible* to use the extremal values of $\Delta H°_{298}$ or $\Delta S°_{298}$ to estimate *an uncertainty*; the statistical method of deriving the standard error does not apply to a feasible solutions set.

7.5.2.3 Application of Linear Programming

It has been emphasized above that each element of the set R is consistent with all reversal brackets. The question then is how to select one element out of the entire set? If we are dealing with only two variables, like $\Delta H°_{298}$ and $\Delta S°_{298}$, we may easily read off the permissible pair of values from Fig. 7.6. This is no longer possible though as soon as we split $\Delta H°_{298}$ and $\Delta S°_{298}$ into $H°_{f,298}$ and $S°_{298}$ of Ky and Sill; we must now deal with a four-dimensional space. Picking one element from such a multidimensional feasible solution set is done by MP, which involves an appropriately chosen objective function (linear or nonlinear). However, to clarify our concept of LP, we shall continue with the two-dimensional diagram (Fig. 7.6), but nevertheless apply LP to choose a pair of values for $\Delta H°_{298}$ and $\Delta S°_{298}$ of the reaction Ky = Sill.

As emphasized earlier, a linear programming problem is one in which a linear objective function is optimized subject to a set of constraints involving both linear inequalities and equalities. For the Ky = Sill reaction, only the constraints 1, 3, 4, 6, 7, 8, and 9 (see Fig. 7.6) of Table 7.2 define the feasible solutions set R. More specifically, these are

$$\Delta H°_{298} - 1013.15 \cdot \Delta S°_{298} + 5614.6 \ (J) > 0 \ (\text{constraint } 1), \tag{7.6a}$$
$$\Delta H°_{298} - \ 963.15 \cdot \Delta S°_{298} + 5051.3 \ (J) > 0 \ (\text{constraint } 3), \tag{7.6c}$$
$$\Delta H°_{298} - \ 978.15 \cdot \Delta S°_{298} + 4522.8 \ (J) < 0 \ (\text{constraint } 4), \tag{7.6d}$$
$$\Delta H°_{298} - 1053.15 \cdot \Delta S°_{298} + 5711.6 \ (J) < 0 \ (\text{constraint } 6), \tag{7.6f}$$
$$\Delta H°_{298} - 1068.15 \cdot \Delta S°_{298} + 6421.7 \ (J) > 0 \ (\text{constraint } 7), \tag{7.6g}$$
$$\Delta H°_{298} - 1103.15 \cdot \Delta S°_{298} + 6399.4 \ (J) < 0 \ (\text{constraint } 8), \text{and} \tag{7.6h}$$
$$\Delta H°_{298} - 1163.15 \cdot \Delta S°_{298} + 7957.1 \ (J) > 0 \ (\text{constraint } 9). \tag{7.6i}$$

Figure 7.7 displays the resulting feasible solutions set R, which is a convex polygon, defined by seven straight lines (8, 6, 4, 9, 7, 1 and 3), having the seven vertices c, d, a, e, f, g, and b. In order to pick one element out of this set R, we must first specify a suitable objective function. The linear objective function given in (7.3) will be recast in an appropriate manner to deal with this case. To apply to the variables $\Delta H°_{298}$ and $\Delta S°_{298}$ of the Ky = Sill reaction (Fig. 7.7), we may, for example, specify the function

$$F_1 = \left(\frac{1}{1000 \ J}\right)\Delta H°_{298} + \left(\frac{1}{J K^{-1}}\right)\Delta S°_{298} . \tag{7.8}$$

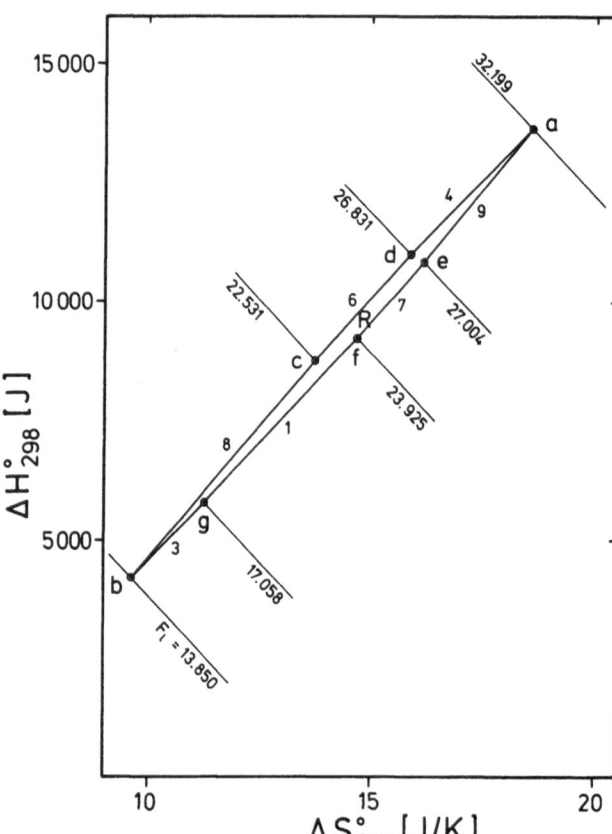

Fig. 7.7 A ΔH°_{298} vs ΔS°_{298} plot showing the feasible solutions set R for the equilibrium Ky = Sill. The boundary constraints and vertices of R are labeled as in the text. The objective function F_1, and its numerical values at the vertices, are also displayed

The reason for choosing the coefficients of ΔH°_{298} and ΔS°_{298} in this form is simply to meet two requirements: first, the nature of the function F_1 must be clear from Fig. 7.7, and secondly, ΔH°_{298} and ΔS°_{298} must be normalized such that F_1 becomes dimensionless.

If F_1 [of (7.8)] is optimized subject to the constraints (7.6a, 7.6c, 7.6d, 7.6f-i), either of the two following routes will be taken by the optimization routine. At first, any two constraints, for instance 6 and 8, may be chosen and solved simultaneously (by treating them as equalities) to compute the ΔH°_{298} and ΔS°_{298} values corresponding to the vertex c. Substituting these data into (7.8), the value of F_1 (= 22.531) is obtained. Figure 7.7 depicts a plot of F_1, indicating its value at vertex c. After this, the optimization routine replaces one of the two above constraints by the next one, and repeats the same process of calculations. If the constraint 8 is replaced by 4, it obtains the values of ΔH°_{298} and ΔS°_{298} at the vertex d. As shown in Fig. 7.7, the value of F_1 is higher at d compared to that at c. Proceeding in this fashion, the optimization algorithm consecutively determines the values of F_1 at each of the vertices. If F_1 is to be maximized, it passes the vertices c-d-a and stops only at e, because F_1 decreases from a to e and, therefore, F_1 hits the maximum at a ($F_1 = 32.199$). If F_1 is to be minimized instead, it will reverse its course at d (because F_1 increases from c to d) and backtrack the vertices d-c-b-g, stopping at g and establishing the vertex b as the minimum ($F_1 = 13.850$). Figure 7.7 shows the function F_1 at each vertex of the set R, along with the corresponding value of F_1.

The features of the feasible solutions set described above for the two-dimensional case can also be recast to a multidimensional optimization problem involving n variables and m restrictions. In such real-life situations, a more generalized algorithm is used to optimize the objective function. This step will be deferred until we have examined the constraints due to the phase property data.

7.5.2.4 Phase Property Constraints

Phase property constraints follow from the data listed in Table 7.1. They are rewritten according to the general relation,

$$Y_i' - u(Y_i') \leq Y_i \leq Y_i' + u(Y_i'), \tag{7.1a}$$

with Y_i and Y_i', indicating the adjustable i-th parameter and its directly measured value, respectively. Note also that, $u(Y_i')$, the uncertainty of Y_i', is a short-hand notation for 2σ of Y_i', quoted in Table 7.1. Recalling that our thermodynamic model requires two parameters, ΔH°_{298} and ΔS°_{298}, to be adjusted, (7.1a) is recast as

$$\Delta H^\circ_{298}' - u(\Delta H^\circ_{298}') \leq \Delta H^\circ_{298} \leq \Delta H^\circ_{298}' + u(\Delta H^\circ_{298}'), \text{ and} \tag{7.9}$$

$$\Delta S^\circ_{298}' - u(\Delta S^\circ_{298}') \leq \Delta S^\circ_{298} \leq \Delta S^\circ_{298}' + u(\Delta S^\circ_{298}'). \tag{7.10}$$

Processing the calorimetric data on $H^\circ_{f,298}$ and S°_{298} for Sill and Ky from Table 7.1 accordingly, we have

$$3394 \text{ J} \leq \Delta H^\circ_{298} \leq 8546 \text{ J}, \text{ and} \tag{7.11}$$

$$13.299 \text{ J K}^{-1} \leq \Delta S^\circ_{298} \leq 13.681 \text{ J K}^{-1}. \tag{7.12}$$

Figure 7.8 displays the phase property constraints as the set C in the ΔH°_{298} - ΔS°_{298} space; the dot at the center of the rectangle indicating the calorimetric mean values, $\Delta H^\circ_{298}'$ and $\Delta S^\circ_{298}'$, for the reaction Ky = Sill. Superimposed onto this diagram is the set R obtained earlier by evaluating the phase equilibria constraints. Recall that an internally consistent set of thermodynamic data comprises only the values of the adjustable parameters, which are consistent with both phase property (set C) and phase equilibria (set

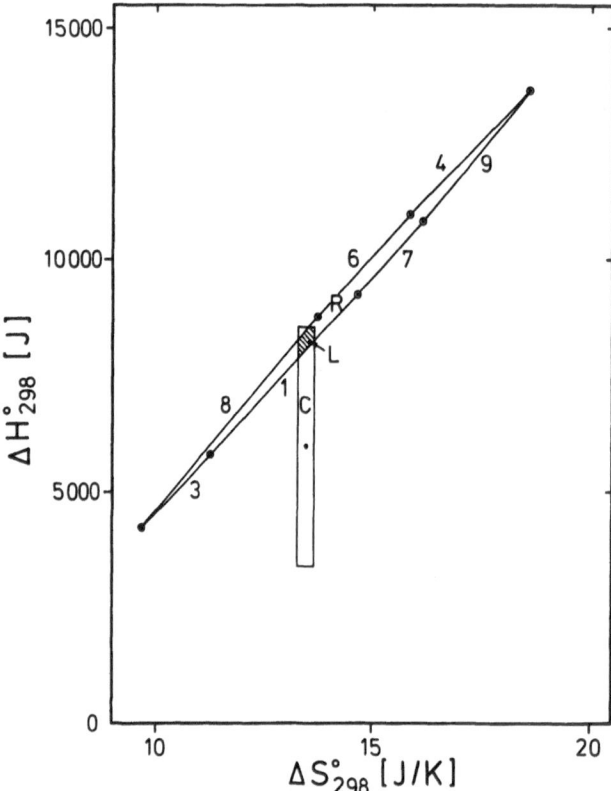

Fig. 7.8 A ΔH°_{298} vs ΔS°_{298} plot showing the feasible solutions sets C and R for the equilibrium Ky = Sill. Each element of the intersection set L (*hatched*) corresponds to one pair of ΔH°_{298} and ΔS°_{298}, which are internally consistent

R) constraints. Thus, in terms of ΔH°_{298} and ΔS°_{298}, only the elements of the set L (hatched region in Fig. 7.8) delimit the internally consistent thermodynamic dataset for the reaction Ky = Sill, with L given by

$$L = C \cap R. \tag{7.13}$$

From now onward, therefore, we shall be exclusively concerned with the set L. Note that L is defined by as few as four restrictions. Two of these are phase equilibria restrictions [numbers 1 and 8, corresponding to the inequalities (7.6a) and (7.6h)], whereas the others are the phase property restrictions given earlier as (7.11) and (7.12).

7.5.2.5 Application of Quadratic Programming

Having established the feasible solutions set, L, which is valid for both phase equilibria and phase property constraints, we once again need to select one pair of values for $\Delta H°_{298}$ and $\Delta S°_{298}$ for the reaction Ky = Sill. Although we could have done the job by mere inspection, we shall again utilize MP to do this. The problem this time is not just to select *any* one pair of values for $\Delta H°_{298}$ and $\Delta S°_{298}$, but to choose the "*best*" one. But what criterion shall we apply to define the "best" value? Day and Kumin (1980, p.272) suggested a criterion based on the minimization of the sum of the absolute values of the differences between the adjustable variable (Y_i) inside the feasible solutions set and the corresponding phase property (Y_i') measured directly. To obtain the desired solution, they used the nonlinear[2] objective function

$$F_n = \sum_i \left| \frac{Y_i - Y_i'}{u(Y_i')} \right| ,$$ (7.14)

u denoting the uncertainty of the measured property Y_i'. However, such a function has one disadvantage. As pointed out by Day and Kumin (1980, p.282), F_n must be transformed appropriately prior to optimization, such that the new function can be handled by linear programming. To do this, an additional variable must be introduced for every adjustable parameter (see Wagner 1959, p.207-208). This is one disadvantage of using (7.14) to optimize the thermodynamic data for a very large number of phases of any chemical system.

As an alternative criterion to derive the thermodynamic "best" values, Berman et al. (1985, 1986, p.1342-1343) recommended using a quadratic objective function, F_q, which is far better suited for obtaining optimized thermodynamic data for any number of phases in a chemical system. They suggested using the same function which is minimized in least squares regression analysis,

$$F_q = \sum_i \frac{(Y_i - Y_i')^2}{u^2(Y_i')} .$$ (7.15)

Application of (7.15) leads to a close match between the directly measured property Y_i' and the variable Y_i within the framework of a feasible solutions set. Thus, the results obtained by (7.15) are somewhat similar to those generated by the objective function (7.14). The real advantage of (7.15) is that this function can be readily applied without having to introduce additional variables. Thus, the number of variables depends solely on the thermodynamic model and the number of phases to be optimized.

In the following, we shall employ F_q, the quadratic objective function, substituting the 2σ uncertainties for u in (7.15). This leads us to quadratic programming, QP, involving the optimization of a quadratic objective function subject to linear constraints.

---

[2] The objective function (7.14) is nonlinear due to the absolute value terms. A function is said to be linear only if it fulfills the criterion f(x+y) = f(x) + f(y).

7.5.2.6 An Internally Consistent Thermodynamic Dataset for the Three Al$_2$SiO$_5$ Phases by Quadratic Programming

In this section, we shall derive an internally consistent set of thermodynamic data for the three Al$_2$SiO$_5$ polymorphs. This will be achieved in two steps. Let us begin by applying QP to optimize $\Delta H°_{298}$ and $\Delta S°_{298}$ of the reaction Ky = Sill. This will allow us to display the quadratic objective function on a $\Delta H°_{298}$ - $\Delta S°_{298}$ plot like Fig. 7.8, which may be helpful in appreciating the optimizing procedure. After that, we will simultaneously process the reversal brackets of all three reactions (Ky = Sill, Ky = And, and And = Sill), by splitting $\Delta H°_{298}$ and $\Delta S°_{298}$ into $H°_{f,298}$ and $S°_{298}$ of the phases.

Optimization of the quadratic objective function with the Gill et al. (1983) algorithm demands that F_q of Eq.(7.15) be appropriately recast to comply with the general form of (7.5). To achieve this, F_q may be rewritten as

$$F_q = \sum_i \frac{Y_i^2 - 2Y_iY_i' + Y_i'^2}{u^2(Y_i')}$$

$$= \sum_i \frac{Y_i^2}{u^2(Y_i')} - \sum_i \frac{2Y_iY_i'}{u^2(Y_i')} + \sum_i \frac{Y_i'^2}{u^2(Y_i')}$$

$$= \sum_i \left[\frac{1}{u^2(Y_i')}\right]Y_i^2 - \sum_i \left[\frac{2Y_i'}{u^2(Y_i')}\right]Y_i + \sum_i \left[\frac{Y_i'^2}{u^2(Y_i')}\right]. \qquad (7.16)$$

Note that the square-bracketed terms are constants, since Y_i' and $u(Y_i')$ have specified values (Table 7.1) for every phase. From a comparison of the last line of Eq.(7.16) with the function (7.5), used in the optimization routine of Gill et al. (1983), it is apparent that the first two terms of (7.16) comply with (7.5), but the last term of the former does not show up in the latter. This calls for the manipulations described below.

To achieve our goal of making Eq.(7.15) consonant with (7.5), we introduce the new variable δY_i, a difference function, defined by

$$\delta Y_i \equiv (Y_i - Y_i'). \qquad (7.17)$$

Substituting (7.17) into (7.15), we have

$$F_q = \sum_i \left[\frac{(\delta Y_i)^2}{u^2(Y_i')}\right]. \qquad (7.18)$$

The coefficient $[1/u^2(Y_i')]$ of the variable $(\delta Y_i)^2$ in (7.18) then corresponds to d_{ij} (with i = j) of (7.5), whereas d_{ij} (with i ≠ j) and c_i equal zero. As a result of this, (7.18) becomes amenable to a straightforward optimization by the Gill et al. (1983) algorithm. It will be demonstrated later that the introduction of the new variable δY_i is tantamount to a coordination transformation in the adjustable variables space.

Let us next address the derivation of a pair of "best" values for $\Delta H°_{298}$ and $\Delta S°_{298}$ of the Ky = Sill equilibrium. The immediate consequence of operating with the variable δY_i is that both phase property (7.1) and phase equilibria (7.6) restrictions have to be properly transformed. With the phase property constraints, this is achieved simply by subtracting Y_i' from each term of (7.1a). This leads to

$$- u(Y_i') \le (Y_i - Y_i') \le + u(Y_i').$$ (7.19a)

Inserting (7.17) into (7.19a), the appropriately transformed phase property constraint is recovered,

$$- u(Y_i') \le \delta Y_i \le + u(Y_i').$$ (7.19)

To apply to the $\Delta H°_{298}$ - $\Delta S°_{298}$ space of the reaction Ky = Sill, shown in Fig. 7.8, (7.19) is evaluated from Table 7.1,

$$-2576 \ (J) \le \delta \Delta H°_{298} \le + 2576 \ (J), \text{ and}$$ (7.20a)

$$-0.191 \ (J \ K^{-1}) \le \delta \Delta S°_{298} \le + 0.191 \ (J \ K^{-1}).$$ (7.20b)

The phase property restrictions given in (7.20a) and (7.20b) agree with the format specified in (7.4a) for use with the optimization routine of Gill et al. (1983), the variables $\delta \Delta H°_{298}$ and $\delta \Delta S°_{298}$ of (7.20a) and (7.20b) corresponding to Y_i in (7.4a).

Next, the phase equilibria constraints need to be recast to the format of (7.4b), also prescribed by that optimization routine. This time, the starting point is

$$\Delta H°_{298} - T_i \Delta S°_{298} + r_i > 0 \ (\text{or} < 0),$$ (7.6)

r_i being the sum of all the terms on the RHS of (4.15), except the first two. Inserting $\Delta H°_{298}'$ and $\Delta S°_{298}'$ into (7.6) such that the inequality is kept in tact, it may be rewritten as

$$(\Delta H°_{298} - \Delta H°_{298}') - T_i(\Delta S°_{298} - \Delta S°_{298}') + [\Delta H°_{298}' - T_i \Delta S°_{298}' + r_i] > 0 \ (\text{or} < 0). \ \ (7.21)$$

Substituting the transposed variables $\delta \Delta H°_{298}$ and $\delta \Delta S°_{298}$ into (7.21), we have

$$\delta \Delta H°_{298} - T_i \delta \Delta S°_{298} + [\Delta H°_{298}' - T_i \Delta S°_{298}' + r_i] > 0 \ (\text{or} < 0).$$ (7.22)

The numerical values of r_i have been evaluated earlier [see Eqs.(7.6a-f)] for each half-bracket of the reaction Ky = Sill. Using the calorimetric values of $H°_{f,298}$ and $S°_{298}$ for Ky and Sill listed in Table 7.1, $\Delta H°_{298}'$ and $\Delta S°_{298}'$ of the Ky = Sill reaction amount to 5970 J and 13.490 J K^{-1}, respectively.

Substituting them into (7.22), the restriction 1 of reaction 2 of Table 7.2 is found to be

$$\delta\Delta H^\circ_{298} - 1013.15\cdot\delta\Delta S^\circ_{298} + [5970 - 1013.15\cdot 13.49 + 5614.6]\ (J) > 0,\ or \quad (7.23a')$$

$$\delta\Delta H^\circ_{298} - 1013.15\cdot\delta\Delta S^\circ_{298} - 2082.8\ (J) > 0. \quad\quad (7.23a)$$

Proceeding in a similar manner, the next inequality (corresponding to the restriction 2 of reaction 2 in Table 7.2) will be

$$\delta\Delta H^\circ_{298} - 1033.15\cdot\delta\Delta S^\circ_{298} - 3202.5\ (J) < 0, \quad\quad (7.23b)$$

and so on.

To comply to the format required by the optimization routine [see (7.4b)], we now recast (7.23a), (7.23b), and the rest of the phase equilibria restrictions for the reaction Ky = Sill to

$$
\begin{aligned}
2082.8\ (J) &\le \delta\Delta H^\circ_{298} - 1013.15\cdot\delta\Delta S^\circ_{298} \le +\infty\ (J), && (7.24a)\\
-\infty\ (J)\ &\le \delta\Delta H^\circ_{298} - 1033.15\cdot\delta\Delta S^\circ_{298} \le 3202.5\ (J), && (7.24b)\\
1917.4\ (J) &\le \delta\Delta H^\circ_{298} - 963.15\cdot\delta\Delta S^\circ_{298} \le +\infty\ (J), && (7.24c)\\
-\infty\ (J)\ &\le \delta\Delta H^\circ_{298} - 978.15\cdot\delta\Delta S^\circ_{298} \le 2702.4\ (J), && (7.24d)\\
2026.2\ (J) &\le \delta\Delta H^\circ_{298} - 1018.15\cdot\delta\Delta S^\circ_{298} \le +\infty\ (J), && (7.24e)\\
-\infty\ (J)\ &\le \delta\Delta H^\circ_{298} - 1053.15\cdot\delta\Delta S^\circ_{298} \le 2525.4\ (J) && (7.24f)\\
2017.6\ (J) &\le \delta\Delta H^\circ_{298} - 1068.15\cdot\delta\Delta S^\circ_{298} \le +\infty\ (J), && (7.24g)\\
-\infty\ (J)\ &\le \delta\Delta H^\circ_{298} - 1103.15\cdot\delta\Delta S^\circ_{298} \le 2512.1\ (J), && (7.24h)\\
1763.8\ (J) &\le \delta\Delta H^\circ_{298} - 1163.15\cdot\delta\Delta S^\circ_{298} \le +\infty\ (J),\ and && (7.24i)\\
-\infty\ (J)\ &\le \delta\Delta H^\circ_{298} - 1183.15\cdot\delta\Delta S^\circ_{298} \le 2984.5\ (J) && (7.24j)
\end{aligned}
$$

Note that the variables $\delta\Delta H^\circ_{298}$ and $\delta\Delta S^\circ_{298}$ in these inequalities correspond to Y_i in (7.4b).

Having computed the lower and the upper bounds for both phase property and phase equilibria constraints, we only need to express the objective function (7.18) specifically for the reaction Ky = Sill. Using data from Table 7.1, we have

$$F_q = \left[\frac{(\delta\Delta H^\circ_{298})^2}{(2576\ J)^2} + \frac{(\delta\Delta S^\circ_{298})^2}{(0.191\ J\,K^{-1})^2}\right] \quad\quad (7.25)$$

Because the numerators and denominators in (7.25) have identical dimensions, the objective function is dimensionless.

The data listed in (7.20a,b) and (7.24a-j) are then entered into the optimization routine of Gill et al. (1983) and run on a digital computer, minimizing the objective function (7.25). As illustrated in Fig. 7.8, the feasible solutions space, L, is bounded only by the four restrictions (7.20a,b) and (7.24a, h); the latter two correspond to the half-bracket Nos. 1 and 8. This implies that eight out of the ten phase equilibria restrictions are redundant.

Consequently, they could have been deleted from the input prior to minimizing F_q. The optimization routine automatically takes care of such preparatory data sorting. Minimizing F_q, we obtain:

$$\delta \Delta H^\circ_{298} = 2071 \text{ J and} \tag{7.26a}$$

$$\delta \Delta S^\circ_{298} = -0.012 \text{ J K}^{-1}, \tag{7.26b}$$

the value of the objective function being

$$Fq = 0.6501. \tag{7.27}$$

The "best" values of ΔH°_{298} and ΔS°_{298} for the reaction Ky = Sill are then obtained from Eq.(7.17),

$$\Delta H^\circ_{298} = \Delta H^\circ_{298}{}' + \delta \Delta H^\circ_{298}, \text{ and} \tag{7.28a}$$

$$\Delta S^\circ_{298} = \Delta S^\circ_{298}{}' + \delta \Delta S^\circ_{298}. \tag{7.28b}$$

Using the calorimetric $H^\circ_{f,298}$ and S°_{298} data of Ky and Sill given in Table 7.1, we have $\Delta H^\circ_{298}{}' = 5970$ J and $\Delta S^\circ_{298}{}' = 13.490$ J K^{-1}. Combining these values with $\delta \Delta H^\circ_{298}$ and $\delta \Delta S^\circ_{298}$ of (7.26a,b), the "best" values are recovered:

$$\Delta H^\circ_{298} = (5970 + 2071) \text{ J} = 8041 \text{ J, and} \tag{7.29a}$$

$$\Delta S^\circ_{298} = (13.490 - 0.012) \text{ J K}^{-1} = 13.478 \text{ J K}^{-1}. \tag{7.29b}$$

Having achieved our goal of deriving the "best" pair of values for ΔH°_{298} and ΔS°_{298} of the Ky = Sill reaction, let us focus on a depiction of the results in terms of the ΔH°_{298}-ΔS°_{298} space. This is done in Fig. 7.9. Plotted on it is the feasible solutions set L, shown earlier in Fig. 7.8. Note that both plots are identical, except that the scales of ΔH°_{298} and ΔS°_{298} are highly magnified in Fig. 7.9. The calorimetric values, $\Delta H^\circ_{298}{}'$ and $\Delta S^\circ_{298}{}'$, for this reaction are also shown on this diagram by the circumscribed dot. This point serves as the origin of the transformed coordinate system for the new variables $\delta \Delta H^\circ_{298}$ and $\delta \Delta S^\circ_{298}$. The search for the solution by the optimization routine commences at the origin of the $\delta \Delta H^\circ_{298}$-$\delta \Delta S^\circ_{298}$ coordinate system (at which the value of F_q is zero), and continues for successively increasing values of F_q, until the final solution is reached. It is clear from (7.25) that the function F_q corresponds to an ellipse on the $\delta \Delta H^\circ_{298}$-$\delta \Delta S^\circ_{298}$ space. Three such ellipses have been drawn in Fig. 7.9, with F_q rising from inside outward. The outermost ellipse ($F_q = 0.6501$) grazes the set L. Therefore, it corresponds to the *lowest* value of F_q (*the minimum!*) at which an element common to both F_q and L is detected. This element, expressed in terms of the $\delta \Delta H^\circ_{298}$-$\delta \Delta S^\circ_{298}$ coordinates of (7.26a, b), yields the optimal solution, and is indicated by an arrow in Fig. 7.9. It is clear from the foregoing discussions that if ΔH°_{298} and

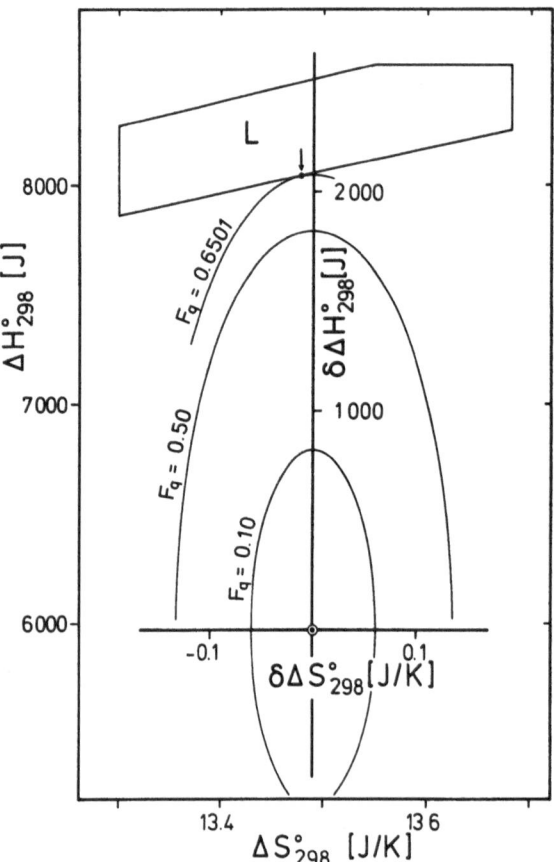

Fig. 7.9 A ΔH°_{298} vs ΔS°_{298} plot of the feasible solutions set L for the reaction Ky = Sill. The calorimetric data from Table 7.1 of the same reaction, $\Delta H^{\circ}_{298}{}' = 5970$ J and $\Delta S^{\circ}_{298}{}' = 13.490$ J K^{-1}, are indicated by the *encircled dot*; it serves as the origin of the transformed coordinate system in the $\delta\Delta H^{\circ}_{298}$-$\delta\Delta S^{\circ}_{298}$ space. Plots of F_q (= 0.1, 0.5, and 0.6501) appear as ellipses on the $\delta\Delta H^{\circ}_{298}$-$\delta\Delta S^{\circ}_{298}$ plane. The optimized (minimized) solution for $\delta\Delta H^{\circ}_{298}$ and $\delta\Delta S^{\circ}_{298}$ is indicated by the *arrow*

ΔS°_{298} were to be split into $H^{\circ}_{f,298}$ and S°_{298} for Ky and Sill, the solution will again be an element common to both the sets L and F_q. The only difference between the two cases is that the bounding lines of L as well as the function F_q become hypersurfaces in a four-dimensional space of $H^{\circ}_{f,298}$ and S°_{298} for Ky and Sill and, therefore, can no longer be depicted. It should also be apparent by now that if the calorimetric $\Delta H^{\circ}_{298}{}'$ and $\Delta S^{\circ}_{298}{}'$ values of any reaction happen to fall *within* the feasible solution set L, the optimal solutions are $\Delta H^{\circ}_{298} = \Delta H^{\circ}_{298}{}'$ and $\Delta S^{\circ}_{298} = \Delta S^{\circ}_{298}{}'$, F_q equaling zero. Our Ky = And reaction very nearly approaches that case, with $F_q = 0.0004$.

The next item on our agenda is to consider the reactions Ky = Sill, Ky = And, and And = Sill *simultaneously* and to solve for the "best" values of $H^{\circ}_{f,298}$ and S°_{298} of the phases And, Ky and Sill. This involves only a straightforward extension of (7.20), (7.24), and (7.25), taking into account $\delta H°_{f,298}$ and $\delta S°_{298}$ of the phases, rather than $\delta \Delta H°_{298}$ and $\delta \Delta S°_{298}$ of the reactions. Let us begin by considering the phase property restrictions. They must be written in a manner analogous to (7.20a,b). Using the data from Table 7.1, they may be formulated as follows:

$$- 1900 \text{ J mol}^{-1} \leq \delta H°_{f,298,Ky} \leq 1900 \text{ J mol}^{-1} \tag{7.30a}$$
$$- 2100 \text{ J mol}^{-1} \leq \delta H°_{f,298,And} \leq 2100 \text{ J mol}^{-1} \tag{7.30b}$$
$$- 1740 \text{ J mol}^{-1} \leq \delta H°_{f,298,Sill} \leq 1740 \text{ J mol}^{-1} \tag{7.30c}$$
$$- 0.13 \text{ (J K}^{-1} \text{ mol}^{-1}) \leq \delta S°_{298,Ky} \leq 0.13 \text{ (J K}^{-1} \text{ mol}^{-1}) \tag{7.30d}$$
$$- 0.14 \text{ (J K}^{-1} \text{ mol}^{-1}) \leq \delta S°_{298,And} \leq 0.14 \text{ (J K}^{-1} \text{ mol}^{-1}) \tag{7.30e}$$
$$- 0.14 \text{ (J K}^{-1} \text{ mol}^{-1}) \leq \delta S°_{298,Sill} \leq 0.14 \text{ (J K}^{-1} \text{ mol}^{-1}). \tag{7.30f}$$

Next, the phase equilibria restrictions are rewritten in terms of $\delta H°_{f,298}$ and $\delta S°_{298}$, with no accompanying change in the numerical values of the upper and lower bounds. Thus, the half-brackets of the Ky = Sill reaction [cf.(7.24a-j)] now appear as

$$2082.8 \text{ (J mol}^{-1}) \leq -1 \cdot \delta H°_{f,298,Ky} + 1 \cdot \delta H°_{f,298,Sill}$$
$$+ 1013.15 \cdot \delta S°_{298,Ky} - 1013.15 \cdot \delta S°_{298,Sill} \leq +\infty \text{ (J mol}^{-1}), \tag{7.31a}$$

$$-\infty \text{ (J mol}^{-1}) \leq -1 \cdot \delta H°_{f,298,Ky} + 1 \cdot \delta H°_{f,298,Sill}$$
$$+ 1033.15 \cdot \delta S°_{298,Ky} - 1033.15 \cdot \delta S°_{298,Sill} \leq 3202.5 \text{ (J mol}^{-1}), \tag{7.31b}$$

and so on. In this manner, the phase equilibria restrictions for all three reactions are reformulated. Finally, using the data listed in Table 7.1, the relevant objective function is seen to be

$$F_q = \left(\frac{\delta H°_{f,298,Ky}}{1900 \text{ J mol}^{-1}} \right)^2 + \left(\frac{\delta H°_{f,298,And}}{2100 \text{ J mol}^{-1}} \right)^2 + \left(\frac{\delta H°_{f,298,Sill}}{1740 \text{ J mol}^{-1}} \right)^2$$
$$+ \left(\frac{\delta S°_{298,Ky}}{0.13 \text{ J K}^{-1} \text{ mol}^{-1}} \right)^2 + \left(\frac{\delta S°_{298,And}}{0.14 \text{ J K}^{-1} \text{ mol}^{-1}} \right)^2$$
$$+ \left(\frac{\delta S°_{298,Sill}}{0.14 \text{ J K}^{-1} \text{ mol}^{-1}} \right)^2. \tag{7.32}$$

The minimization of (7.32), subject to the restrictions (7.30) and (7.31), yields the "best" values for Ky, And, and Sill:

$$H°_{f,298,Ky} = -2592530 \text{ J mol}^{-1}, \tag{7.33a}$$
$$H°_{f,298,And} = -2588462 \text{ J mol}^{-1}, \tag{7.33b}$$
$$H°_{f,298,Sill} = -2584446 \text{ J mol}^{-1}, \tag{7.33c}$$
$$S°_{298,Ky} = 82.302 \text{ J K}^{-1} \text{ mol}^{-1}, \tag{7.33d}$$
$$S°_{298,And} = 91.396 \text{ J K}^{-1} \text{ mol}^{-1}, \text{ and} \tag{7.33e}$$
$$S°_{298,Sill} = 95.781 \text{ J K}^{-1} \text{ mol}^{-1}; \text{ with} \tag{7.33f}$$

$$F_q = 0.9532. \tag{7.34}$$

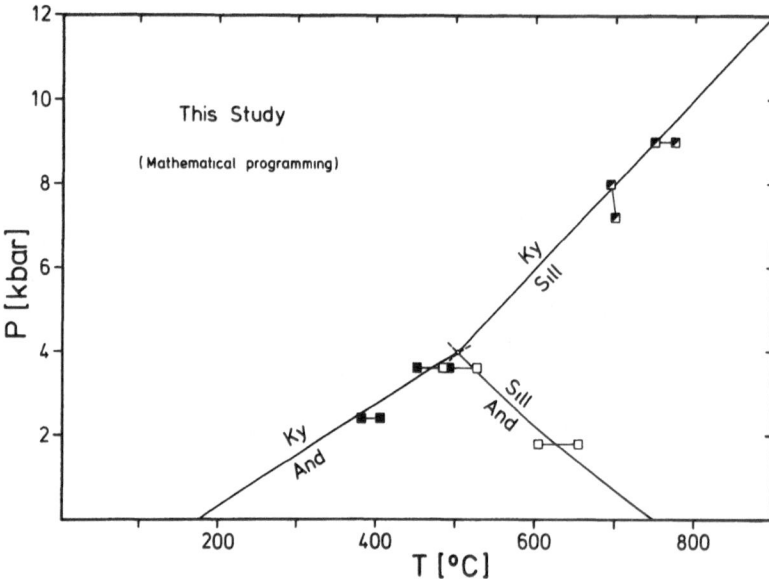

Fig. 7.10 The Al_2SiO_5 phase diagram calculated from the internally consistent thermodynamic data, Eqs.(7.33a-f), in conjunction with the $C_P^\circ(T)$, α_o, α_1, β_o, and β_1 of Table 7.1

The extent to which MP has been succeessful in deriving useful thermodynamic data for the three Al_2SiO_5 phases may be judged now by computing the phase diagram for the Al_2SiO_5 unary. To calculate the phase diagram, we only need to combine the $H^\circ_{f,298}$ and S°_{298} data from Eqs.(7.33a-f) with the $C^\circ_P(T)$, V°_{298}, α_o, α_1, β_o, and β_1 values from Table 7.1, and solve Eq.(4.15) for P at any given T. The results are shown in Fig. 7.10. Also shown are two reversal brackets for each of the three reactions, Ky = Sill, And = Sill, and Ky = And. A very good mutual agreement is apparent, which is reassuring for the computation of geological phase diagrams in general.

7.6 Future Perspectives and Concluding Remarks

The capability of mathematical programming to derive a set of internally consistent thermodynamic data has been demonstrated. If the calorimetric data are accurate, and many reactions involving a phase are well reversed, its thermodynamic data, derived by MP, are likely to be reliable. The internally consistent thermodynamic dataset derived by Berman (1988) may be regarded as belonging to this category, and is already widely used. A task for the future is to identify all phases for which there is a need for further refinement, from the viewpoints of both calorimetry and reversal experiments. Another important task is to define the uncertainties of the thermodynamic data derived by MP. A lack of knowledge of the uncertainties is indeed a great drawback of the thermodynamic datasets derived by mathematical programming.

For the users of internally consistent thermodynamic datasets, irrespective of whether they were obtained by regression (such as Holland and Powell 1985) or by mathematical programming (Berman 1988), a word of caution is necessary prior to wrapping up this chapter. No matter how tempting, *thermodynamic data from unrelated sources should not be combined* to compute phase relations. This will destroy their internal consistency and lead to a meaningless phase diagram.

Chapter 8

Thermodynamics of Crystalline Solutions

8.1 Introduction, Scope, Definitions

Minerals are *crystalline solutions* showing a very wide range of chemical and structural variations. As opposed to *molecular solutions*, in which the *molecules mix* with each other, *crystalline solutions* are characterized by the *mixing of ions* (or atoms) on a variety of crystal structure sites. A comprehensive treatment of the thermodynamics of crystalline solutions, a rather daunting task, is beyond the scope of this book. In order to do justice to this subject within a single chapter, we shall concentrate on a few selected topics. Our treatment focuses on the thermodynamics of simple crystalline solutions (defined below). Its extension to complex crystalline solutions will be somewhat cursory, and order-disorder phenomena, observed in numerous minerals, will be barely touched. Readers interested in such topics may find the monograph by Ganguly and Saxena (1987) helpful.

Minerals may be subdivided into simple and complex crystalline solutions depending up the number of independent components and end-members involved (Brown 1977). Simple crystalline solutions[1] are those in which the number of independent components equals the number of end-members. By contrast, *complex solution*s have more end-members than there are independent components. By that count, the olivine, $(Mg,Fe,Mn)_2SiO_4$, is a simple crystalline solution, with three independent components, Mg_2SiO_4, Fe_2SiO_4, and Mn_2SiO_4, as well as three end-members, forsterite, fayalite, and tephroite. Note that the three ions Mg, Fe, and Mn mix on two six-coordinated M-sites. In other words, the site mixing occurs solely in the M-sublattice. Similarly, the garnet, $(Ca,Mg)_3Al_2[SiO_4]_3$, is also a simple solution; it has two independent components, $Ca_3Al_2[SiO_4]_3$ and $Mg_3Al_2[SiO_4]_3$, and two end-members, grossular and pyrope. This garnet solution is the result of mixing of Ca and Mg on three eight-coordinated C-sites, while there is no site mixing on the M- and the four-fold T-sites, these being solely occupied by Al and Si, respectively. Again, the site mixing is restricted to one sublattice. If we now permit mixing of Al and Cr on the two M-sites, in addition to Ca and Mg mixing in the C-sites, retaining the T-sites occupied exclusively by Si, we generate a garnet with four end-members: grossular, pyrope, uvarovite, and knorringite. However, the number of independent components is now three;

---

[1] Not to be confused with a simple solution mentioned in Section 1.2.2.6.

we may choose any three of $Ca_3Al_2[SiO_4]_3$, $Mg_3Al_2[SiO_4]_3$, $Ca_3Cr_2[SiO_4]_3$, and $Mg_3Cr_2[SiO_4]_3$. Therefore, $(Ca,Mg)_3(Al,Cr)_2[SiO_4]_3$ garnet, with *independent* mixing on two sublattices, is no longer a simple, but a complex (also known as a reciprocal[2]) solution. Note, however, that a *coupled* substitution on two sublattices, as in the pyroxene $(Ca,Na)(Mg,Al)[Si_2O_6]$, may be described as a simple crystalline solution, because we again end up with two independent components, $CaMg[Si_2O_6]$ and $NaAl[Si_2O_6]$, and two end-members alike, diopside and jadeite. Although numerous rock-forming minerals are complex solutions, they will not be treated in this book in any detail. It will suffice to emphasize that a treatment of the thermodynamics of complex solutions requires that the conditions of internal equilibrium also be applied. This takes care of the energy arising due to the reciprocal ("cross-site") interaction, i.e. interaction between ions mixing on energetically distinct sublattices. Readers seeking to pursue the thermodynamics of complex solutions may turn to Wood and Nicholls (1978), Ganguly and Saxena (1987) and others.

8.2 Extension of the Thermodynamic Theory of Molecular Solutions to Crystalline Solutions

8.2.1 Introduction

The thermodynamic theory of solutions, developed in Chap. 1, is applicable primarily to molecular solutions. In order to apply it to crystalline solutions, it must be modified and extended. This is the main objective of this section. The first step (Sect.8.2.2) is a derivation of equations dealing with simple solutions (with site mixing on one sublattice or those with coupled substitutions involving two sublattices). The second step (Sect.8.2.3) tackles the complex crystalline solutions characterized by independent site mixing of ions on more than one sublattice. In accordance with the restricted scope of this chapter, the thermodynamics of complex solutions is presented in bare outline, referring to appropriate literature for rigorous derivations.

---

[2] The arbitrary choice of a set containing three of these four species as components is possible, because we can express the composition of the fourth species in terms of the three chosen as independent components. For example, we may write

$Mg_3Cr_2[SiO_4]_3 = Ca_3Cr_2[SiO_4]_3 + Mg_3Al_2[SiO_4]_3 - Ca_3Al_2[SiO_4]_3$.

Writing such an identity is permissible; it merely requires us to handle a negative concentration variable for the $Ca_3Al_2[SiO_4]_3$ component. Systems having such a property are known as reciprocal systems, and such solutions are reciprocal solutions.

8.2.2 Derivation of Equations for Simple Crystalline Solutions

8.2.2.1 Solutions with Site Mixing on One Sublattice

The molar Gibbs energy, G, of a molecular solution was written earlier as

$$G = \sum_i X_i \mu_i^* + RT \sum_i X_i \ln X_i + RT \sum_i X_i \ln \gamma_i. \qquad (1.73)$$

Recall that the first term refers to the energy change arising due to mechanical mixing, G^{mech}, whereas the subsequent terms are due to ideal and excess chemical mixing, G^{id}_m and G^{ex}_m, respectively.

In order to be extended to crystalline solutions,
1. The standard chemical potential, μ_i^*, must consider the relevant structure of i at system T and P;
2. The ideal chemical mixing term, G^{id}_m, must be modified to take into account the site mixing in crystalline solutions; and
3. If so desired, the excess chemical mixing term, G^{ex}_m, may also be suitably adjusted to match the expression for G^{id}_m.

Let us examine the standard chemical potential term. According to the notations introduced in Chap. 1 (Sect. 1.2.2.3), it is evident that μ_i^* denotes the standard chemical potential of the pure component i at system T and P. It also signifies the choice of a Raoultian standard state, demanding that $\gamma_i \rightarrow 1$ (and $a_i \rightarrow 1$), as $X_i \rightarrow 1$. This specification, however, is not necessarily sufficient for a crystalline solution, since the component i may possess different structures, such that the numerical value of its standard chemical potential depends not only on T and P, but also on its structure. Thus, to make μ_i^* unambiguous, we must also specify the structure of pure i at T and P. To appreciate this point, consider a binary crystalline solution (A,B)R, where A and B are the ions mixing on one site, R denotes the remainder of the structure. AR and BR are then the components of this binary; for simplicity and brevity, in the following they will be denoted as A and B, rather than as AR and BR, and the solution (A,B)R will be abbreviated as A-B. Figure 8.1 schematically depicts G-X_B plots of this binary at specified T and P. Two different cases must be distinguished. In Fig. 8.1a, both components A and B have an identical structure, designated as 1. Such *isostructural* solutions are represented by *one* equation of state at any set of T and P; that is, *one* G-X_B curve extends from the uniquely fixed μ^*_{A1} to μ^*_{B1}. In our example, the G-X_B curve is concave upward; therefore, the crystalline solution 1 is stable and continuous throughout the composition range. However, in cases where each of the components A and B have two structures, 1 and 2, a pair of values for μ_i^* exists at constant T and P, one for each structure. The one at the lower energy level represents the stable phase, and is shown in Fig. 8.1b by a dot, whereas the metastable one has been depicted by a circle. The energy differences, due to the polymorphic transitions, that is, the reactions 1 = 2 for each component A and B, are $\Delta G^*_A = \mu^*_{A2} - \mu^*_{A1}$ and $-\Delta G^*_B = \mu^*_{B2} - \mu^*_{B1}$,

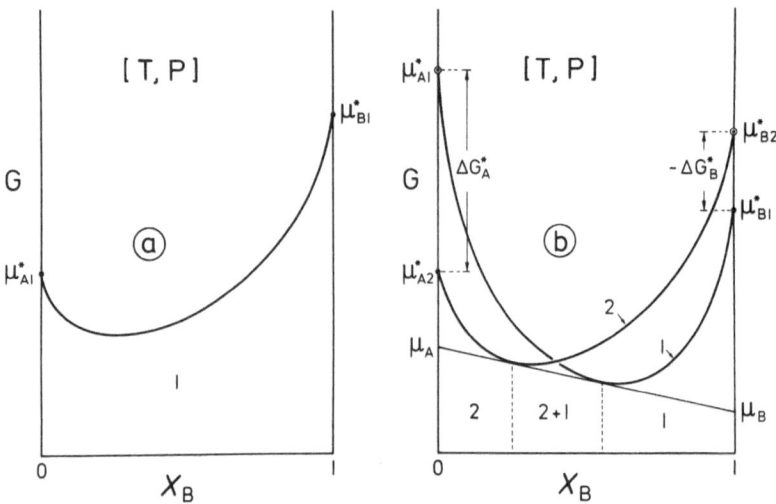

Fig. 8.1 G-X_B curves for isostructural (**a**) and non-isostructural (**b**) binary crystalline solutions A-B at a given T and P

respectively. Note that one G-X_B curve (labeled 2 in Fig. 8.1b) connects μ^*_{A2} of the stable structure A2 to μ^*_{B2} of the metastable structure B2. The other G-X_B curve (1 in Fig. 8.1b) is obtained likewise. Therefore, *two* equations of state emerge for this *non-isostructural* binary. Note that even if each G-X_B curve is concave upward, this time one solution is no longer stable across the entire composition range. Rather, two one-phase fields (for crystalline solutions 1 and 2) appear, interrupted by a two-phase field of the coexisting solutions 1 and 2. Such a binary is said to exhibit a miscibility gap.[3] The limiting compositions of both the coexisting phases are given by the common tangent (explained later) to the G-X_B curves, which implies equality of the chemical potential of each component in both structures (or phases).

Next, let us address the ideal chemical mixing term, G^{id}_m, and see how it is modified to meet the requirements of site mixing of ions in a crystalline solution. From Chap. 1, it is known that all solutions behave ideally at each end of the composition range; the excess terms vanish there. Consequently, ideal solutions may be treated in terms of the properties of solutions in the limits of extreme dilution. As emphasized by Thompson (1969, p. 350), the distribution of species on crystal structure sites may be regarded as a random distribution under the conditions of extreme dilution. Therefore, properties of ideal solutions may be modeled in terms of random mixing (RM) of ions on various crystal structure sites. This model is commonly accepted for solutions showing independent mixing of ions on any number of sublattices. But opinions differ on how best to model ideal crystalline solutions with

---

[3] A homogeneous isostructural solution may exsolve at some low T to yield two coexisting solutions. In such a case, we are handling a solvus (not a miscibility gap).

heterovalent coupled substitutions. We shall return to this issue later. The question right now is how to formulate equations expressing the properties of ideal crystalline solutions following the RM model.

Let us first split G^{id}_m into the enthalpy and entropy terms,

$$G^{id}_m = H^{id}_m - TS^{id}_m \ . \tag{8.1a}$$

Because $H^{id}_m = 0$ [see Eq.(1.78)], (8.1a) reduces to

$$G^{id}_m = - TS^{id}_m \ . \tag{8.1}$$

Clearly, calculation of G^{id}_m depends upon the correct formulation of S^{id}_m. Recall that S^{id}_m is the entropy difference due to mixing of pure end-members to give rise to a crystalline solution of some specified composition. Thus, to come up with S^{id}_m, we need to know the entropy of a solution in which there is random mixing of ions on certain sites, relative to those of pure end-members. Now, as indicated in Chap. 2, the entropy increment due to random mixing of ions on sites equals the ideal molar configurational entropy, $S^{\circ,id}_{cfg}$. Since there is no mixing of ions on the *relevant* sites in the pure end-members, they have no configurational entropy.[4] Consequently, S^{id}_m of any solution is identical to its $S^{\circ,id}_{cfg}$. Using Eq.(2.4a), for the general case of random mixing of ions on any number of sublattices, we may write

$$S^{id}_m = - R \sum_j q_j \sum_i X_{ij} \ln X_{ij}, \tag{8.2a}$$

where q_j denotes the number of sites in the sublattice j, X_{ij} the mole fraction of the ion i on the j-th sublattice, the summation extending over all ions i on the j-th sublattice, and over all the sublattices j. By implication, the mineral structure is envisioned as made up of a set of interpenetrating sublattices, each of which comprises a number of structural sites. The number of sites in any sublattice, q, is called the site multiplicity factor. For a given crystalline solution, where site mixing occurs on one sublattice only, (8.2a) simplifies to

$$S^{id}_m = - qR \sum_i X_i \ln X_i. \tag{8.2}$$

---

[4] This is not to be confused with the statement that the pure end-members necessarily have zero configurational entropy under all circumstances. To appreciate this, consider an alkali feldspar comprising the end-members high sanidine, $K(Al_{0.25}Si_{0.75})_4O_8$, and analbite, $Na(Al_{0.25}Si_{0.75})_4O_8$, both of which show complete Al/Si disorder on T-sites. Both have a nonzero configurational entropy (shown to be 18.7 J K^{-1} mol^{-1} in Chap. 2). Now, the formation of alkali feldspar solution, $(K,Na)[(Al_{0.25}Si_{0.75})_4O_8]$, involves K-Na mixing on the A-site. This is the *relevant* site for computing S^{id}_m of the Al/Si disordered alkali feldspar; we only need to consider the configurational entropy due to K-Na mixing on the A-site in order to compute S^{id}_m of the alkali feldspar solution. Alkali feldspar end-members indeed have no configurational entropy due to mixing on the A-site, because it is occupied entirely by K or Na.

Equations (8.2) [and (8.2a)] are central to the thermodynamics of crystalline solutions. Because S^{id}_m is independent of T and P, only q and X_i must be known to apply Eq.(8.2). Furthermore, since $0 \le X_i \le 1$, S^{id}_m is a positive quantity.

Having elucidated S^{id}_m of crystalline solutions, let us turn to their G^{id}_m. Substitution of (8.2a) into (8.1) helps recover the general expression:

$$G^{id}_m = RT \sum_j q_j \sum_i X_{ij} \ln X_{ij}. \tag{8.3a}$$

For a simple crystalline solution, with site mixing of ions on one sublattice only, (8.3a) reduces to

$$G^{id}_m = qRT \sum_i X_i \ln X_i. \tag{8.3}$$

Thus, at constant T, G^{id}_m of a crystalline solution is a function of X_i *and* q. *The q-dependences of* G^{id}_m *and* S^{id}_m *are characteristic for crystalline solutions*; they set them apart from all molecular solutions. Note, however, that unlike S^{id}_m, which is invariably a positive quantity, G^{id}_m is always negative, because $0 \le X_i \le 1$.

Before proceeding any further, it might be advantageous to explore the implicit significance of the site multiplicity factor, q, in the G^{id}_m term (8.3). For this purpose, and also for later considerations, let us switch to a binary solution $(A,B)_q R$. Again, $A_q R$ and $B_q R$ are the components, and R denotes the remainder of the structure. Denoting $A_q R$ and $B_q R$ as A and B, the solution $(A,B)_q R$ as A-B, G^{id}_m of A-B is obtained by rewriting (8.3) as

$$G^{id}_m = qRT(X_A \ln X_A + X_B \ln X_B). \tag{8.4}$$

To grasp the significance of q, consider the G^{id}_m-X_B plot for the solution A-B at a constant T = 1000 K (Fig. 8.2a). The curves labeled q = 1, 2, and 3 refer to crystalline solutions in which mixing of ions involves 1, 2, or 3 sites, respectively. Thus, at given T and X_B, G^{id}_m of a two-site solution (q = 2) is twice that of a one-site solution (q = 1). However, if we normalize the composition of *every* solution such that there is *only one* site per sublattice, G^{id}_m of *all solutions will be identical* (at stated T) for any given composition. This happens because Gibbs energy is an extensive quantity. Indicating the q-normalized mixing units (i.e. chemical entities which mix to yield the solution) by a and b, and defining them as

$$a \equiv \frac{A}{q}, \quad b \equiv \frac{B}{q}, \quad \text{and} \quad r \equiv \frac{R}{q}, \tag{8.5}$$

the ideal molar Gibbs energy of mixing of the binary (a,b)r turns out to be

$$\frac{G^{id}_m}{q} = RT(X_a \ln X_a + X_b \ln X_b). \tag{8.6}$$

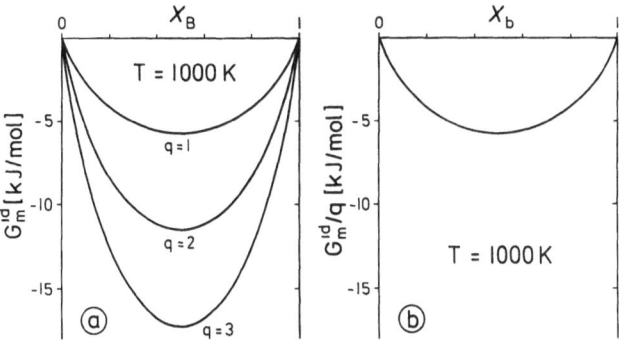

Fig. 8.2 An illustration of the composition dependence of G_m^{id} of a binary crystalline solution A-B at 1000 K. **a** A plot of G_m^{id} as a function of X_B. Note how G_m^{id} decreases as q increases. **b** A plot of G_m^{id}/q vs X_b for the same binary in terms of the q-normalized mixing units

Figure 8.2b depicts a G^{id}_m/q vs X_b plot at 1000 K for the binary a-b. Note that the variables X_B and X_b are numerically identical, as are X_A and X_a. Nonetheless, X_a and X_b are explicitly used in Eq.(8.6) to remind us that we are dealing with q-normalized mixing units. To clarify our concept of normalized mixing units, consider the olivine $(Mg,Fe)_2SiO_4$. Prior to normalization, Mg_2SiO_4 and Fe_2SiO_4 were the mixing units, that is, 2 Fe substituted for 2 Mg per mole of olivine. Following normalization, the mixing units are $MgSi_{1/2}O_2$ and $FeSi_{1/2}O_2$; now 1 Fe proxies for 1 Mg. Likewise, for a $(Ca,Mg)_3Al_2[SiO_4]_3$ garnet, the normalized mixing units would be $CaAl_{2/3}SiO_4$ and $MgAl_{2/3}SiO_4$, with 1 Mg substituting 1 Ca.

In developing the equations describing the other properties of crystalline solutions, we restrict ourselves to site mixing on one sublattice, keeping in mind that these equations will be amenable to an extension to any number of sublattices. For such a solution, the excess mixing term is given by

$$G_m^{ex} = G - G^{id} = G - (\sum_i X_i\mu_i^* + qRT \sum_i X_i\ln X_i). \tag{8.7}$$

For a specific composition, it indicates the nature and degree of deviation of a real solution from its hypothetical ideal analog. Rearrangment of (8.7) leads to the molar Gibbs energy, G, of that crystalline solution,

$$G = \sum_i X_i\mu_i^* + qRT \sum_i X_i\ln X_i + G_m^{ex}. \tag{8.8}$$

As shown in Chap. 1, G^{ex}_m may also be expressed in terms of the activity coefficient, γ_i. One way of relating the two quantites is to use (1.70), and express G^{ex}_m of the crystalline solution A-B as

$$G_m^{ex} = RT(X_A\ln\gamma_A + X_B\ln\gamma_B). \tag{8.9}$$

In other words, the expression derived for molecular solutions can also be applied to the *excess properties* of crystalline solutions, regardless of q. This notwithstanding, the mainstream practice is to write G^{ex}_m for a crystalline solution in a manner reminiscent of the expression for G^{id}_m in Eq.(8.6), that is, by including the multiplicity factor q. The connotation is clear; the mixing units (and the energy terms associated with them) are being normalized with regard to q,

$$\frac{G^{ex}_m}{q} \equiv X_a\left[\frac{RT\ln\gamma_A}{q}\right] + X_b\left[\frac{RT\ln\gamma_B}{q}\right]$$

$$= RT(X_a\ln\gamma_a + X_b\ln\gamma_b). \tag{8.10}$$

The reason for prefering (8.10) to (8.9) will become clear later, when activity-composition relations of crystalline solutions are treated.

As indicated in Chap. 1, the composition dependence of G^{ex}_m at any fixed T and P may be alternatively expressed by a suitable power series in mole fraction. Many different power functions exist. Employing the asymmetric Margules equation (1.86), popular among geochemists, we obtain

$$G^{ex}_m = X_A X_B[W_{G,BA} + (W_{G,AB} - W_{G,BA})X_B]. \tag{8.11}$$

Equation (8.11) may be rewritten for the normalized mixing units. This is also tantamount to normalizing $W_{G,ij}$. Note that $W_{G,AB}/q = W_{G,ab}$, and so on. On normalizing, we therefore have

$$\frac{G^{ex}_m}{q} = X_a X_b[W_{G,ba} + (W_{G,ab} - W_{G,ba})X_b]. \tag{8.12}$$

Equating (8.9) with (8.11),

$$RT(X_A\ln\gamma_A + X_B\ln\gamma_B) = X_A X_B[W_{G,BA} + (W_{G,AB} - W_{G,BA})X_B], \tag{8.13}$$

it may be demonstrated that

$$RT\ln\gamma_A = [W_{G,AB} + 2X_A(W_{G,BA} - W_{G,AB})](X_B)^2, \text{ and} \tag{8.14a}$$

$$RT\ln\gamma_B = [W_{G,BA} + 2X_B(W_{G,AB} - W_{G,BA})](X_A)^2. \tag{8.14b}$$

And likewise, equating (8.10) with (8.12), we obtain

$$RT\ln\gamma_a = [W_{G,ab} + 2X_a(W_{G,ba} - W_{G,ab})](X_b)^2, \tag{8.15a}$$

$$RT\ln\gamma_b = [W_{G,ba} + 2X_b(W_{G,ab} - W_{G,ba})](X_a)^2. \tag{8.15b}$$

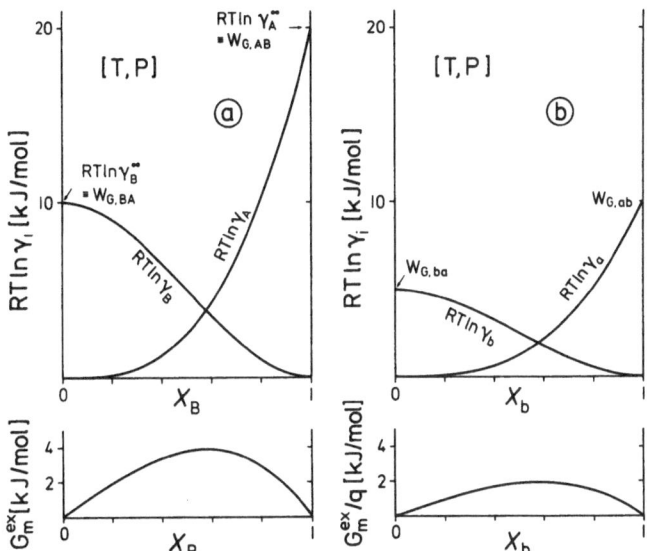

Fig. 8.3 $RT\ln\gamma_i$ and G^{ex}_m of a hypothetical crystalline solution $(A,B)_2R$ as functions of composition. **a** Displays these functions in terms of the whole mixing units A and B; **b** in terms of normalized mixing units a and b

Figure 8.3a shows $RT\ln\gamma_A$, $RT\ln\gamma_B$, and G^{ex}_m as functions of X_B, assigning $W_{G,AB} = 20000$ J and $W_{G,BA} = 10000$ J. Figure 8.3b depicts these quantities for the solution a-b as functions of X_b, assuming q = 2. Contrary to G^{id}_m, normalization of mixing units with regard to q does *not* lead to an identical composition dependence of G^{ex}_m for *all* solutions. This is because the nature and magnitude of the deviation from ideality at T and P are characteristic properties of each solution; it varies from solution to solution, regardless of the mixing unit used to express its properties.

Now, let us address the activity vs composition relations of a simple crystalline solution. They may be obtained by comparing the two sides of a general equation analogous to (1.71),

$$G_m = G^{id}_m + G^{ex}_m = RT(X_A\ln a_A + X_B\ln a_B). \tag{8.16}$$

Because G_m, appearing on the LHS, comprises G^{id}_m and G^{ex}_m, each of which can be expressed in two alternative ways, we may also expect two alternative expressions for a_A and a_B, appearing on the RHS of (8.16). Substituting Eqs.(8.6) and (8.10) for G^{id}_m and G^{ex}_m, G_m may be written as

$$\begin{aligned}G_m &= qRT(X_a\ln X_a + X_b\ln X_b) + qRT(X_a\ln\gamma_a + X_b\ln\gamma_b)\\ &= RT[qX_a\ln(X_a\gamma_a) + qX_b\ln(X_b\gamma_b)].\end{aligned} \tag{8.17}$$

Equating (8.16) with (8.17),

$$RT(X_A\ln a_A + X_B\ln a_B) = RT[qX_a\ln(X_a\gamma_a) + qX_b\ln(X_b\gamma_b)]. \tag{8.18}$$

Noting that X_A and X_a are numerically equal, as are X_B and X_b, we have

$$\ln a_A = q\ln(X_a\gamma_a), \text{ or } a_A = (X_a\gamma_a)^q = (a_a)^q, \text{ and} \tag{8.19a}$$

$$\ln a_B = q\ln(X_b\gamma_b), \text{ or } a_B = (X_b\gamma_b)^q = (a_b)^q. \tag{8.19b}$$

Alternatively, G_m of a crystalline solution is obtained using G^{id}_m and G^{ex}_m as in (8.4) and (8.9). Proceeding in the same fashion, we derive

$$\ln a_A = \ln(X_A)^q\gamma_A, \text{ or } a_A = (X_A)^q\gamma_A, \text{ and} \tag{8.20a}$$

$$\ln a_B = \ln(X_B)^q\gamma_B, \text{ or } a_B = (X_B)^q\gamma_B. \tag{8.20b}$$

It is important to realize that the two pairs of equations, (8.19) and (8.20), are merely alternative ways of expressing a_A and a_B; both are thermodynamically permissible. Therefore, the activities may be written alternatively in terms of both normalized or non-normalized components (mixing units). They may be translated back and forth, using the relations

$$a_a \equiv X_a\gamma_a = (a_A)^{\frac{1}{q}}, \text{ and} \tag{8.21a}$$

$$a_b \equiv X_b\gamma_b = (a_B)^{\frac{1}{q}}. \tag{8.21b}$$

Note also that in the event of ideal behavior, Eqs. (8.20a,b) reduce to the form

$$a^{id}_A = (X_A)^q, \text{ and} \tag{8.22a}$$

$$a^{id}_B = (X_B)^q. \tag{8.22b}$$

Therefore, the activity of B in an *ideal crystalline solution* is X_B raised to the power q. This is shown in Fig. 8.4a, in which a^{id}_B is plotted against X_B for crystalline solutions with q = 1, 2, and 3. Note that with increasing q, the a^{id}_B vs X_B relation deviates more and more from the familiar straight-line diagonal in a_i - X_i space. The only way to reconcile the a^{id}_i vs X_i relations to the traditional straight-line diagonal form is to express a^{id}_i in terms of normalized components (mixing units). This can be seen immediately from Eqs.(8.21a,b):

$$a^{id}_a = X_a, \text{ and} \tag{8.23a}$$

$$a^{id}_b = X_b. \tag{8.23b}$$

Figure 8.4b is such a plot. Like G^{id}_m, normalization of the mixing units leads to the identical form of the a^{id}_i-X_i relations for *all* crystalline solutions.

From the above discussions, it is now apparent that the total molar Gibbs energy of an A-B crystalline solution may also be written in two alternative

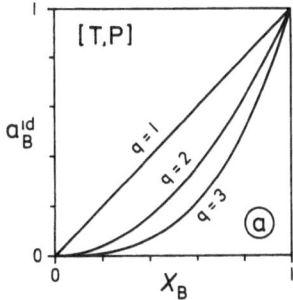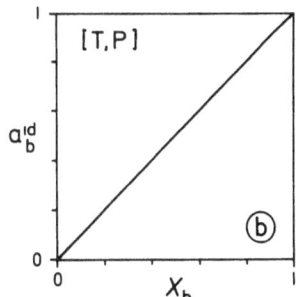

Fig. 8.4 Activity vs composition relations of *ideal* crystalline solutions $(A,B)qR$ at specified T and P. **a** Plots of a_B^{id} vs X_B for three solutions with q = 1, 2, and 3. The shapes of these curves depend on q; the familiar straight-line diagonal is fulfilled only for one-site solutions (q = 1). **b** Shows how normalizing the mixing units leads to a traditional straight-line diagonal a_b^{id} - X_b for all crystalline solutions

ways. To express G in terms of non-normalized components, all we need to do is to substitute Eq. (8.9) into (8.8),

$$G = X_A\mu_A^* + X_B\mu_B^* + qRT(X_A\ln X_A + X_B\ln X_B) + RT(X_A\ln\gamma_A + X_B\ln\gamma_B). \quad (8.24)$$

To express G (or rather G/q) in terms of normalized components, we must also normalize their standard potentials ($\mu_a^* \equiv \mu_A^*/q$, $\mu_b^* \equiv \mu_B^*/q$). Using these, and substituting (8.10) into (8.8), we have

$$\frac{G}{q} = X_a\mu_a^* + X_b\mu_b^* + RT(X_a\ln X_a + X_b\ln X_b) + RT(X_a\ln\gamma_a + X_b\ln\gamma_b). \quad (8.25)$$

Thus, normalizing the mixing units to *one* exchangeable ion reduces the total molar Gibbs energy of *every* crystalline solution to a form analogous to that of a molecular solution [Eq.(1.73)]. This helps explain why Eq.(1.73) also applies to one-site crystalline solutions, as stated in Chap. 1 (see Sect. 1.2.2.1.).

Finally, expressions for the chemical potentials of components of the solution may be derived from Eqs.(8.24) and (8.25). To achieve this, we simply relate G of a solution to the chemical potentials of its constituent components, using an identity like (1.50),

$$G = X_A\mu_A + X_B\mu_B. \quad (8.26)$$

Inserting (8.26) on the LHS of (8.24), we recover

$$\mu_A = \mu_A^* + qRT\ln X_A + RT\ln\gamma_A \text{ and} \quad (8.27a)$$

$$\mu_B = \mu_B^* + qRT\ln X_B + RT\ln\gamma_B. \quad (8.27b)$$

Likewise, from (8.25), we get

$$\mu_a = \mu_a^* + RT\ln X_a + RT\ln\gamma_a \text{ and} \qquad (8.28a)$$

$$\mu_b = \mu_b^* + RT\ln X_b + RT\ln\gamma_b. \qquad (8.28b)$$

The theoretical framework developed above will be employed in a subsequent section (see Sect. 8.5) to illustrate how thermodynamic modeling of simple A-B crystalline solutions is achieved. It will become clear that thermodynamic modeling of solutions always boils down to an exercise in finding appropriate analytical expressions for G^{ex}_m of A-B. For solutions requiring treatment in terms of two equations of state, the unknowns are ΔG^*_A ($\equiv \mu^*_{A2}-\mu^*_{A1}$) and $-\Delta G^*_B$ ($\equiv \mu^*_{B2}-\mu^*_{B1}$), in addition to G^{ex}_m (cf. Sect. 8.4.2).

8.2.2.2 Simple Crystalline Solutions with Charge-Coupled Site Mixing on Two Sublattices

Thermodynamic treatment of solutions with site mixing on more than one sublattice requires us to be conversant with the concepts of long- and short-range order. Briefly, a long-range order deals with exchange of ions occupying two sublattices (Thompson 1969, p. 352). It is expressed in terms of the site occupancies of the individual sublattices, e.g. mole fractions of Ca, Mg, and Fe in the eight- and six-fold sites of a clinopyroxene. By contrast, a short-range order is concerned with the immediate neighborhood of any given ion, and is thus a local phenomenon. We shall presently encounter it in a simple crystalline solution with charge-coupled substitution. It suffices to know that crystalline solutions often respond to changing T and P by adjusting their degrees of order. A rigorous treatment of heterogeneous equilibria demands that energy changes due to homogeneous reactions (e.g. an order-disorder process) also be taken into account.

   With that, let us turn to simple solutions involving charge-coupled substitution. The thermodynamic formalism required to deal with such solutions will be developed, taking the specific example of a clinopyroxene comprising two end-members, diopside, $CaMgSi_2O_6$, and jadeite, $NaAlSi_2O_6$. In the clinopyroxene structure, one Na^{1+} proxies for one Ca^{2+} on the eight-coordinated M2 site. The charge imbalance generated by this heterovalent substitution demands that the substitution 1 Al^{3+} for 1 Mg^{2+} be concurrently operative in the M1 site, such that bulk electrical neutrality is restored. To deal with such a solution thermodynamically, the prime question is how to formulate S^{id}_m. To this end, *two* contrasting approaches have been attempted; these will be discussed below.

   A straightforward way to handle the problem is to consider the clinopyroxene as a solution of the two mixing units $(CaMg)Si_2O_6$, Di for short, and $(NaAl)Si_2O_6$, Jd. According to our nomenclature, this is analogous to a one-site solution $(A,B)R$, where A $\equiv$ (CaMg) and B $\equiv$ (NaAl). By implication, such a model requires that in the structure of the clinopyroxene solution, every Ca is inevitably tied to an Mg, and each Na to an Al; that is, the ion pairs (CaMg) and (NaAl) appear in the pyroxene structure like dumbbells. Such a model has been alternatively dubbed as the molecular mixing (MM) (Cohen 1986a) or the charge balance (CB) model (Ganguly and Saxena 1987). Applying Eq.(8.2), and recalling that q = 1, Ganguly (1973) wrote the S^{id}_m of these clinopyroxenes as

$$S^{id}_m = -R(X_{Jd}\ln X_{Jd} + X_{Di}\ln X_{Di}). \tag{8.29}$$

Following the derivations in Section 8.2.2.1, the G of such a solution may then be written as

$$G = X_{Jd}\mu^*_{Jd} + X_{Di}\mu^*_{Di} + RT(X_{Jd}\ln X_{Jd} + X_{Di}\ln X_{Di}) + G^{ex}_m . \tag{8.30}$$

Thermodynamic modeling of this isostructural solution will thus be synonymous with analytically expressing its G^{ex}_m as a function of temperature, pressure, and composition. Writing G^{ex}_m in terms of γ_i, rather than $W_{G,i}$, the chemical potentials of Jd and Di become

$$\mu_{Jd} = \mu^*_{Jd} + RT\ln X_{Jd} + RT\ln\gamma_{Jd} \text{ and} \tag{8.31a}$$

$$\mu_{Di} = \mu^*_{Di} + RT\ln X_{Di} + RT\ln\gamma_{Di}. \tag{8.31b}$$

Recalling that the sum of the last two terms on the RHS of (8.31a) and (8.31b) may be equated to $RT\ln a_{Jd}$ and $RT\ln a_{Di}$ [see Eq.(1.62)], it follows that

$$a_{Jd} = X_{Jd}\gamma_{Jd} \text{ and} \tag{8.32a}$$

$$a_{Di} = X_{Di}\gamma_{Di}. \tag{8.32b}$$

A second, more sophisticated approach for formulating S^{id}_m of clinopyroxenes on the jadeite-diopside join has been suggested by Cohen (1986a). From statistical mechanical arguments, he concluded that the S^{id}_m of charge-coupled solutions in the range of extreme dilution is best derived by applying a random mixing (RM) model to all relevant sites. This is equivalent to postulating that in a matrix of diopside, a tiny amount of the valence-balanced NaAl pair of the Jd component dissociates completely, as does the CaMg pair of the Di component at extreme dilution in a jadeite matrix. In the more concentrated range, however, this model will no longer apply due to short-range order. To ensure appropriate consideration of short-range order, an analytical expression for S^{ex}_m is introduced into the model, and invariably retained as a nonzero term for subsequent solution modeling. In other words, this model does *not* assume random mixing for the concentrated range; RM is used merely as a base to ensure that the limiting behavior is correct.

A total association (TA) of the valence-balanced NaAl and CaMg pairs in clinopyroxene solutions is another limiting case explored by Cohen (1986a). TA of the ion pairs generates the dumbbell-like configurations described earlier in the context of the CB (or MM) model. While the TA model admits that these dumbbells might assume more than one orientation in the structure of the clinopyroxene, the CB model ignores this possibility. Thus, the TA model computes a higher S^{id}_m than the CB model. The real world of charge-coupled solutions is likely to show a degree of association of ion pairs intermediate between RM and TA models.

Employing the RM model *as a base*, S^{id}_m of these clinopyroxenes may be obtained from Eq.(8.2a) as follows:

$$S^{id}_m = -R[(X_{Na}\ln X_{Na}+X_{Ca}\ln X_{Ca})_{M2} + (X_{Al}\ln X_{Al}+X_{Mg}\ln X_{Mg})_{M1}]. \qquad (8.33a)$$

Now, in our clinopyroxenes, $X_{Na}(M2) = X_{Al}(M1) = X_{Jd}$ and $X_{Ca}(M2) = X_{Mg}(M1) = X_{Di}$. Thus, (8.33a) reduces to

$$S^{id}_m = -2R[X_{Jd}\ln X_{Jd} + X_{Di}\ln X_{Di}]. \qquad (8.33)$$

This leads to the following identities:

$$G = X_{Jd}\mu^*_{Jd} + X_{Di}\mu^*_{Di} + 2RT(X_{Jd}\ln X_{Jd} + X_{Di}\ln X_{Di}) + G^{ex}_m \; ; \qquad (8.34)$$

$$\mu_{Jd} = \mu^*_{Jd} + 2RT\ln X_{Jd} + RT\ln\gamma_{Jd} \text{ and} \qquad (8.35a)$$

$$\mu_{Di} = \mu^*_{Di} + 2RT\ln X_{Di} + RT\ln\gamma_{Di}. \qquad (8.35b)$$

From Eqs(8.35a,b), it follows that

$$a_{Jd} = (X_{Jd})^2\gamma_{Jd} \text{ and} \qquad (8.36a)$$

$$a_{Di} = (X_{Di})^2\gamma_{Di}. \qquad (8.36b)$$

In comparing the relative merits of the CB model with his own, Cohen (1986b) points out that short-range order, a phenomenon of fundamental importance in charge-coupled solutions, is accounted for only in the latter model; the CB model ignores it. Thus, G^{ex}_m obtained from Cohen's model will be superior; it should be applied for extrapolation beyond the range of the input experimental data. Because errors in G^{id}_m will generally be compensated by the G^{ex}_m term, *any model* may be utilized if data are fitted *exclusively for interpolation*. The charge-coupled MgSi vs AlAl substitution in orthopyroxenes, handled in terms of the CB model (Wood and Banno 1973; Perkins et al. 1981), may serve as an example of successful data smoothing for interpolation.

Before concluding this section, attention is drawn to the "Al-avoidance principle" invoked from time to time for writing S^{id}_m of some charge-coupled solutions like plagioclase (see Newton et al. 1980) and orthopyroxene (Ganguly and Ghose 1979). This principle, introduced by Loewenstein (1953), suggests that tetrahedral Al-Si mixing in some silicates cannot be completely random, because it is energetically unfavorable for two neighboring Al to share one bridging oxygen. Kerrick and Darken (1975) developed an expression for S^{id}_m for such cases. Employing the Kerrick-Darken expression, in conjunction with phase equilibria data from Orville (1972), and their own calorimetric data on H^{ex}_m, Newton et al. (1980) modeled the plagioclase solutions.

8.2.3 Thermodynamic Treatment of Complex Crystalline Solutions

Independent, as opposed to charge-coupled, mixing of ions on two or more sublattices yields complex crystalline solutions. They are characterized by a larger number of end-members than independent components. The $(Ca,Mg)_3(Al,Cr)_2[SiO_4]_3$ garnet, cited in Section 8.1, is a specific example of a complex solution; many other rock-forming minerals belong to this group. The objective of this section is to present the thermodynamic formalisms of complex solutions in bare outlines, following Wood and Nicholls (1978) in a general way. Readers seeking to delve deeper should refer to the groundbreaking works of Flood et al. (1954) and Blander (1964).

Consider the crystalline solution $(A,B)_{q_1}(X,Y)_{q_2}R$, in which site mixing involves two interpenetrating sublattices 1 and 2, q_1 and q_2 being the numbers of sites in those sublattices. Figure 8.5 shows the composition space. The end-members are $A_{q_1}X_{q_2}R$, $A_{q_1}Y_{q_2}R$, $B_{q_1}X_{q_2}R$, and $B_{q_1}Y_{q_2}R$, abbreviated below

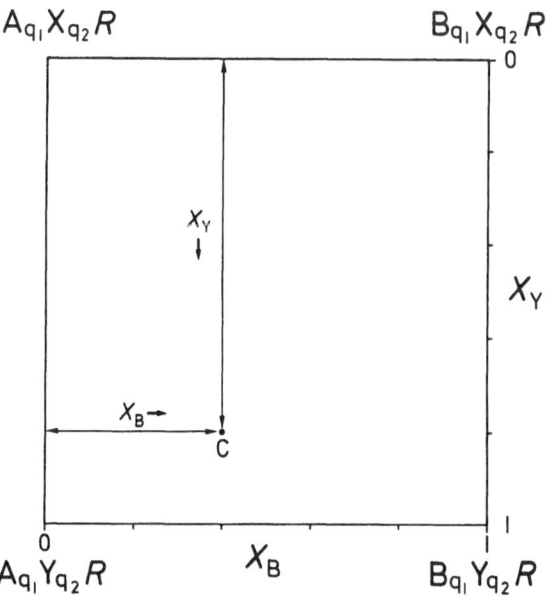

Fig. 8.5 The composition space of a $(A,B)_{q_1}(X,Y)_{q_2}R$ reciprocal solution, showing how the bulk composition C can be expressed in terms of the variables X_B and X_Y

as AX, AY, BX, and BY. Of these, any three can be chosen as independent components, because they are related to each other through the reciprocal reaction,

$$AY + BX = BY + AX. \tag{8.37}$$

For this reason, $(A,B)_{q_1}(X,Y)_{q_2}R$ type solutions are often referred to as reciprocal solutions. Because $(A,B)_{q_1}(X,Y)_{q_2}R$ is a ternary solution, its bulk composition can be unambiguously specified in terms of two variables. This has been demonstrated in Fig. 8.5, using X_B and X_Y to specify the composition C.

To derive the thermodynamic properties of $(A,B)_{q_1}(X,Y)_{q_2}R$, let us start with its S^{id}_m. Formulation of S^{id}_m is straightforward; it follows from Eq.(8.2a),

$$S^{id}_m = - q_1R(X_A\ln X_A + X_B\ln X_B) - q_2R(X_X\ln X_X + X_Y\ln X_Y). \tag{8.38}$$

To proceed further, let us assume for now that A-B mixing in sublattice 1 is ideal, as is X-Y mixing in 2. Thus, no G^{ex}_m terms appear due to these factors, and consequently, γ_i terms for mixing on sites in the sublattices 1 and 2 are unity. In analogy to the simple solutions treated above, the G of $(A,B)_{q_1}(X,Y)_{q_2}R$ could be written as

$$G = X_AX_X\mu^*_{AX} + X_AX_Y\mu^*_{AY} + X_BX_X\mu^*_{BX} + X_BX_Y\mu^*_{BY} - TS^{id}_m \ , \tag{8.39}$$

μ^*_{AX}, μ^*_{AY}, etc. being shorthand notations for the standard chemical potentials of Aq_1Xq_2R, Aq_1Yq_2R, and so on, with the pure components at system T and P representing the standard state. Nevertheless, the thermodynamic properties of the $(A,B)_{q_1}(X,Y)_{q_2}R$ solution is better handled in terms of its three independent components, which may be any of the following sets:

Set (a): AX, AY, and BX,
Set (b): AX, AY, and BY,
Set (c): BX, BY, and AX, and
Set (d): BX, BY, and AY.

Regardless of the set with which we choose to operate, G, and hence μ_i, of $(A,B)_{q_1}(X,Y)_{q_2}R$ will have to be unique at specified T, P, and composition. There is yet another requirement that must be fulfilled by the thermodynamic formalisms to be developed. It must satisfy the condition of internal equilibrium, which, from Eq.(8.37), is:

$$\mu_{AX} + \mu_{BY} = \mu_{AY} + \mu_{BX}. \tag{8.40}$$

As indicated by Wood and Nicholls (1978, p.390-391), treatment of the properties of a reciprocal solution in terms of independent components requires that a reciprocal term, G^{rec}, be introduced to write the molar Gibbs energy of the solution. That helps fulfill both the requirements indicated above. For instance, if we elect to formulate G of $(A,B)_{q_1}(X,Y)_{q_2}R$ in terms of AX, AY, and BX (set a), it will be written as (Wood and Nicholls 1978, p.390)

$$G = X_B\mu^*_{BX} + X_Y\mu^*_{AY} + (X_X\text{-}X_B)\mu^*_{AX} + G^{rec} - TS^{id}_m . \qquad (8.41)$$

From (8.41), the nature of the G^{rec} term is clear; it is analogous to G^{ex}_m. Contrary to the G^{ex}_m term handled thus far, it does not arise from nonideality of mixing in either sublattice; rather, it stems from the energetic interaction between the ions in the two sublattices 1 and 2. Equating (8.41) with (8.39), we may obtain an expression for G^{rec},

$$G^{rec} = X_A X_X \mu^*_{AX} + X_A X_Y \mu^*_{AY} + X_B X_X \mu^*_{BX} + X_B X_Y \mu^*_{BY}$$
$$- [X_B\mu^*_{BX} + X_Y\mu^*_{AY} + (X_X\text{-}X_B)\mu^*_{AX}]. \qquad (8.42a)$$

In order to solve for G^{rec}, note that $(A,B)_{q_1}(X,Y)_{q_2}R$ is a ternary solution with two independent composition variables. Referring to the composition space of $(A,B)_{q_1}(X,Y)_{q_2}R$, depicted in Fig. 8.5, and choosing X_B and X_Y as the independent variables, G^{rec} may be rewritten as

$$G^{rec} = (1\text{-}X_B)(1\text{-}X_Y)\mu^*_{AX} + (1\text{-}X_B)X_Y\mu^*_{AY} + X_B(1\text{-}X_Y)\mu^*_{BX}$$
$$+ X_B X_Y \mu^*_{BY} - X_B\mu^*_{BX} - X_Y\mu^*_{AY} - (1\text{-}X_Y\text{-}X_B)\mu^*_{AX}$$
$$= X_B X_Y (\mu^*_{AX} + \mu^*_{BY} - \mu^*_{AY} - \mu^*_{BX}) \qquad (8.42b)$$

The bracketed term on the RHS of the last line of (8.42b) is the difference in the standard chemical potentials of the end-members at T and P for the reciprocal reaction [see (8.37)]. Denoting it as ΔG^*_{12}, the energy difference arising from interactions between sublattices 1 and 2, and substituting it into (8.42b), we recover

$$G^{rec} = X_B X_Y \Delta G^*_{12}. \qquad (8.42)$$

It should be noted that ΔG^*_{12} has a unique value at a specified P and T; by contrast, G^{rec} has four possible values depending upon the choice of the independent components. If we settle for Set (a) (AX, AY, and BX), G^{rec} is as indicated in Eq.(8.42). For the other sets, Wood and Nicholls (1978) derived the following relations

$$G^{rec}(b) = - X_B X_X \Delta G^*_{12},$$
$$G^{rec}(c) = - X_A X_Y \Delta G^*_{12}, \text{ and}$$
$$G^{rec}(d) = X_A X_X \Delta G^*_{12}. \qquad (8.43)$$

Substituting the appropriate expression for G^{rec} into that for G, Wood and Nicholls (1978) derived the following equations for the chemical potentials

$$\mu_{AX} = \mu_{AX}^* + RT\ln[(X_A)^{q_1}(X_X)^{q_2}] - X_B X_Y \Delta G_{12}^*,$$
$$\mu_{AY} = \mu_{AY}^* + RT\ln[(X_A)^{q_1}(X_Y)^{q_2}] + X_B X_X \Delta G_{12}^*,$$
$$\mu_{BX} = \mu_{BX}^* + RT\ln[(X_B)^{q_1}(X_X)^{q_2}] + X_A X_Y \Delta G_{12}^*, \quad \text{and}$$
$$\mu_{BY} = \mu_{BY}^* + RT\ln[(X_B)^{q_1}(X_Y)^{q_2}] - X_A X_X \Delta G_{12}^*. \tag{8.44}$$

Comparing (8.44) with the general expression for μ_i,

$$\mu_i = \mu_i^* + RT\ln X_i + RT\ln\gamma_i, \tag{1.62}$$

it is evident that the last terms on the RHS of (8.44) measure the deviation from ideality. Because site mixing on both sublattices was assumed to be ideal in deriving (8.44), this deviation must be linked to the reciprocal interaction. Consequently, we write

$$RT\ln\gamma_{AX}^{rec} = -X_B X_Y \Delta G_{12}^*,$$
$$RT\ln\gamma_{AY}^{rec} = X_B X_X \Delta G_{12}^*,$$
$$RT\ln\gamma_{BX}^{rec} = X_A X_Y \Delta G_{12}^*, \text{ and}$$
$$RT\ln\gamma_{BY}^{rec} = -X_A X_X \Delta G_{12}^*. \tag{8.45}$$

These equations, of fundamental significance to the thermodynamics of reciprocal solutions, were first derived by Flood et al. (1954, p.309) in the context of molten salt solutions.

Yet other nonideality terms may arise in reciprocal solutions owing to nonideal site mixing. They can be treated by introducing appropriate γ^{site}_i terms, in addition to γ^{rec}_i. Denoting γ^{site}_i as γ_i for short, (8.44) may be extended to

$$\mu_{AX} = \mu_{AX}^* + RT\ln[(X_A\gamma_a)^{q_1}(X_X\gamma_x)^{q_2}] + RT\ln\gamma_{AX}^{rec},$$
$$\mu_{AY} = \mu_{AY}^* + RT\ln[(X_A\gamma_a)^{q_1}(X_Y\gamma_y)^{q_2}] + RT\ln\gamma_{AY}^{rec},$$
$$\mu_{BX} = \mu_{BX}^* + RT\ln[(X_B\gamma_b)^{q_1}(X_X\gamma_x)^{q_2}] + RT\ln\gamma_{BX}^{rec}, \quad \text{and}$$
$$\mu_{BY} = \mu_{BY}^* + RT\ln[(X_B\gamma_b)^{q_1}(X_Y\gamma_y)^{q_2}] + RT\ln\gamma_{BY}^{rec}. \tag{8.46}$$

The sum of the last two terms on the RHS of (8.46) is identical to $RT\ln a_i$. Consequently,

$$a_{AX} = (X_A\gamma_a)^{q_1}(X_X\gamma_x)^{q_2}\gamma_{AX}^{rec},$$
$$a_{AY} = (X_A\gamma_a)^{q_1}(X_Y\gamma_y)^{q_2}\gamma_{AY}^{rec},$$
$$a_{BX} = (X_B\gamma_b)^{q_1}(X_X\gamma_x)^{q_2}\gamma_{BX}^{rec}, \quad \text{and}$$
$$a_{BY} = (X_B\gamma_b)^{q_1}(X_Y\gamma_y)^{q_2}\gamma_{BY}^{rec}. \tag{8.47}$$

An important lesson to emerge from (8.47) is that a reciprocal solution does *not* reduce to an ideal one *unless* both γ^{site}_i and γ^{rec}_i are simultaneously unity.

Indeed, the magnitude of ΔG^*_{12} can already be substantial, leading to a considerable nonideality of mixing. For instance, ΔG^*_{12} of the $(Ca,Mg)_3(Al,Cr)_2[SiO_4]_3$ garnet has been estimated by Wood and Nicholls (1978) to be on the order of 75 kJ. Unfortunately, the ΔG^*_{12} of most minerals is unknown.

In the context of the thermodynamics of reciprocal solutions, attention may be drawn to the effect of long-range ordering on the thermodynamic properties of minerals. To give just one example, we may cite the fractionation of Fe into the M2-site of $(Mg,Fe)_2Si_2O_6$ orthopyroxene (Saxena and Ghose 1971, Fig. 1). Were it not for this effect, one could have handled the orthopyroxenes as a binary solution of $Mg_2Si_2O_6$ and $Fe_2Si_2O_6$. The Fe-Mg ordering proves that M1 and M2 are energetically distinct. Thus, the orthopyroxenes can be considered in terms of the reciprocal reaction (see Ganguly and Saxena 1987),

$$Mg(M1)Mg(M2)Si_2O_6 + Fe(M1)Fe(M2)Si_2O_6$$
$$= Fe(M2)Mg(M1)Si_2O_6 + Mg(M2)Fe(M1)Si_2O_6. \tag{8.48}$$

Before closing this section, the reader's attention is drawn to an alternative aproach to thermodynamic treatment of reciprocal crystalline solutions. It requires that the species mixing to give rise to a crystalline solution be identified first on the basis of crystallochemical considerations, followed by application of the technique of minimization of Gibbs energy (subject to appropriate constraints) to solve for the equilibrium distribution of species at any set of T, P, and bulk composition. Engi (1983) applied this technique to the reciprocal ternary spinels, $(Mg,Fe^{+2})(Al,Cr)_2O_4$, and concluded that these spinels show only a slight deviation from ideality. He observed that the very same spinels, modeled in terms of γ^{rec}_i and γ^{site}_i, reveal a significantly stronger nonideality. This difference is due to the alternative formulation of the S^{id}_m term in the two approaches.

8.3 Some Equations for Excess Properties of Crystalline Solutions

As discussed in Section 1.2.2.6, the composition dependence of Y^{ex}_m ($Y \equiv G$, H, S, or V) of solutions at a specified T and P is expressed empirically by solution models. The number of empirical parameters used to model the solutions depends upon the nature of the deviation from ideality and the quality of experimental data; the model with the smallest number of parameters accounting for a set of data is regarded as the best model. For mineral crystalline solutions, the asymmetric Margules equation,

$$Y^{ex}_m = X_A X_B [W_{Y,BA} + (W_{Y,AB}\text{-}W_{Y,BA})X_B], \tag{8.49}$$

generally suffices. Its functional form is most easily grasped by recasting (8.49) to

$$\frac{Y^{ex}}{X_A X_B} = W_{Y,BA} + (W_{Y,AB} - W_{Y,BA})X_B. \tag{8.50}$$

In other words, Y^{ex}_m/X_AX_B plotted against X_B gives a straight line with a slope equivalent to $(W_{Y,AB}-W_{Y,BA})$. If, within the limits of uncertainty of the data, a higher order fit appears indispensable, the Redlich-Kister equation (1.82) may be used. The uncertainties of most direct measurements of the excess mixing quantities, such as H^{ex}_m, S^{ex}_m, or V^{ex}_m, *in mineral systems* rarely justify use of the Redlich-Kister equation. On the contrary, in many cases the noise of the data is such that $(W_{Y,AB}-W_{Y,BA})$ is effectively zero. In such cases, one parameter, W_Y, emerges, and is defined as

$$W_Y \equiv W_{Y,AB} = W_{Y,BA}. \tag{8.51}$$

Note that the subscript of W_Y has been dropped altogether, because there is now only one parameter in a binary, and the subscript has become redundant. Substituting (8.51) into (8.49), the symmetric Margules equation may be recovered,

$$Y^{ex}_m = X_AX_BW_Y. \tag{8.52}$$

Numerous mineral solutions comprise more than two end-members. Solution modeling then requires that the binary models be extended to them. The formalism for extending the asymmetric Margules model to a multicomponent solution has been discussed by Andersen and Lindsley (1981), and more recently, by Helffrich and Wood (1989). Using the generalized expression given by the latter authors, Y^{ex}_m of a symmetric ternary solution A-B-C is

$$Y^{ex}_m = X_AX_BW_{Y,\underline{AB}} + X_AX_CW_{Y,\underline{AC}} + X_BX_CW_{Y,\underline{BC}} + X_AX_BX_CW_{Y,\underline{ABC}}. \tag{8.53}$$

Note that the subscripts of W_Y ($W_{Y,\underline{AB}}$, $W_{Y,\underline{BC}}$, and $W_{Y,\underline{AC}}$) had to be reintroduced here to distinguish W_Y of the limiting binaries from each other. Note furthermore that $W_{Y,\underline{AB}}$ is not to be confused with $W_{Y,AB}$; the former being a symmetric Margules parameter for A-B, whereas the latter is one of its two asymmetric parameters. It is essential to realize that the new term $W_{Y,\underline{ABC}}$ is related to the ternary interaction between the triplet A-B-C. Helffrich and Wood (1989, p.1021) demonstrated that $W_{Y,\underline{ABC}}$ cannot be predicted from a knowledge of the pair interaction parameters $\overline{W_{Y,AB}}$, $W_{Y,\underline{BC}}$, and $W_{Y,\underline{AC}}$; it has to be experimentally determined. The expressions for the three $RT\ln\gamma_i$ terms for such a ternary are

$$\begin{aligned} RT\ln\gamma_A &= (1-X_A)X_BW_{G,\underline{AB}} + (1-X_A)X_CW_{G,\underline{AC}} - X_BX_CW_{G,\underline{BC}} \\ &\quad + (1-2X_A)X_BX_C\overline{W}_{G,\underline{ABC}}, \end{aligned} \tag{8.54a}$$

$$\begin{aligned} RT\ln\gamma_B &= (1-X_B)X_CW_{G,\underline{BC}} + (1-X_B)X_AW_{G,\underline{AB}} - X_AX_CW_{G,\underline{AC}} \\ &\quad + (1-2X_B)X_AX_C\overline{W}_{G,\underline{ABC}}, \text{ and} \end{aligned} \tag{8.54b}$$

$$\begin{aligned} RT\ln\gamma_C &= (1-X_C)X_AW_{G,\underline{AC}} + (1-X_C)X_BW_{G,\underline{BC}} - X_AX_BW_{G,\underline{AB}} \\ &\quad + (1-2X_C)X_AX_B\overline{W}_{G,\underline{ABC}}. \end{aligned} \tag{8.54c}$$

8.4 Phase Relations in a Binary Solution

8.4.1 Isostructural Solution

Let us now consider the phenomena of stability and exsolution in an isostructural binary, $(A,B)_qR$. To facilitate handling of the relevant formulae, let us normalize its components by q. Thus, the solution we consider is $(a,b)r$, or simply, a-b. Moreover, to make the case rather general, an asymmetric Margules equation (rather than a symmetric one) is used in order to express the excess molar properties. The molar Gibbs energy G of a-b is then obtained from Eqs.(8.25) and (8.12),

$$G = [X_a\mu_a^* + X_b\mu_b^*] + [RT(X_a\ln X_a + X_b\ln X_b)]$$
$$+ [X_aX_b\{W_{G,ba} + (W_{G,ab} - W_{G,ba})X_b\}]. \tag{8.55}$$

The sum of the terms in the first square brackets indicates G^{mech} of a-b, the succeeding square-bracketed terms denoting its G^{id}_m and G^{ex}_m, respectively. As emphasized in 1.2.2.6, a negative G^{ex}_m enhances the stability of the real solution relative to that of the ideal solution. A positive G^{ex}_m, by contrast, makes the real solution less stable than the ideal one. Consequently, to explore the problems of stability *and exsolution*, we need only to consider a solution with positive G^{ex}_m. We assume that our solution has the following properties at stated P, and one subcritical temperature, say 700 K:

$\mu_a^* = -1000$ J mol^{-1}, $\mu_b^* = -4000$ J mol^{-1}, and
$W_{G,ab} = 10000$ J mol^{-1}, $W_{G,ba} = 17000$ J mol^{-1}.

If the components a and b were totally immiscible, the molar Gibbs energy G of the binary system would be given by the dashed straight line in Fig. 8.6a (above). Addition of G^{id}_m to the G^{mech} yields G^{id} of a-b, depicted as a solid line in the same diagram. Because $0 \le X_b \le 1$, the G^{id} vs X_b curve is always concave upward; that is, an ideal solution has a lower energy and is always stable relative to a mechanical mixture of the end-members. The slope of the G^{id}-X_b curve is given by

$$\left(\frac{\partial G^{id}}{\partial X_b}\right)_{P,T} = (\mu_b^* - \mu_a^*) + RT\ln\frac{X_b}{X_a}. \tag{8.56}$$

For $X_b \to 0$, $(\partial G^{id}/\partial X_b)_{P,T}$ tends to $-\infty$, whereas for $X_b \to 1$, it tends to $+\infty$. This interesting property of G^{id} is readily seen in the G^{id}-X_b curve of Fig. 8.6a (above), which has been computed using the set of data indicated above. We shall return to it in the context of the G-X_b plot discussed below. The second derivative of G^{id} is

$$\left(\frac{\partial^2 G^{id}}{\partial X_b^2}\right)_{P,T} = RT\left(\frac{1}{X_a} + \frac{1}{X_b}\right). \tag{8.57}$$

Figure 8.6a (below) is a plot of $(\partial^2 G^{id}/\partial X^2_b)_{P,T}$ as a function of X_b. It is positive throughout, and is symmetrical with respect to 0.5 X_b. Evidently, whenever $\partial^2 G/\partial X^2_b > 0$, the function $G(X_b)$ must be concave upward.

Next, we turn to the G^{ex}_m term. The partial derivative of G^{ex}_m with respect to X_b is given by

$$\left(\frac{\partial G^{ex}_m}{\partial X_b}\right)_{P,T} = W_{G,ab}X_b(3X_a - 1) - W_{G,ba}X_a(3X_b - 1). \tag{8.58}$$

This function attains its maximum absolute values both as $X_b \to 0$ and $X_b \to 1$; however, in either case, it remains finite. This behavior may be recognized in the G^{ex}_m-X_b plot of Fig. 8.6b (above), computed from the same set of data given above. Its second partial derivative is

$$\left(\frac{\partial^2 G^{ex}_m}{\partial X^2_b}\right)_{P,T} = 2W_{G,ab}(1 - 3X_b) + 2W_{G,ba}(1 - 3X_a). \tag{8.59}$$

Figure 8.6b (below) shows a $(\partial^2 G^{ex}_m/\partial X^2_b)_{P,T}$ vs X_b plot. Contrary to $(\partial^2 G^{id}/\partial X^2_b)_{P,T}$, which is positive across the entire range of the composition, $(\partial^2 G^{ex}_m/\partial X^2_b)_{P,T}$ is invariably negative, and a linear function of X_b.

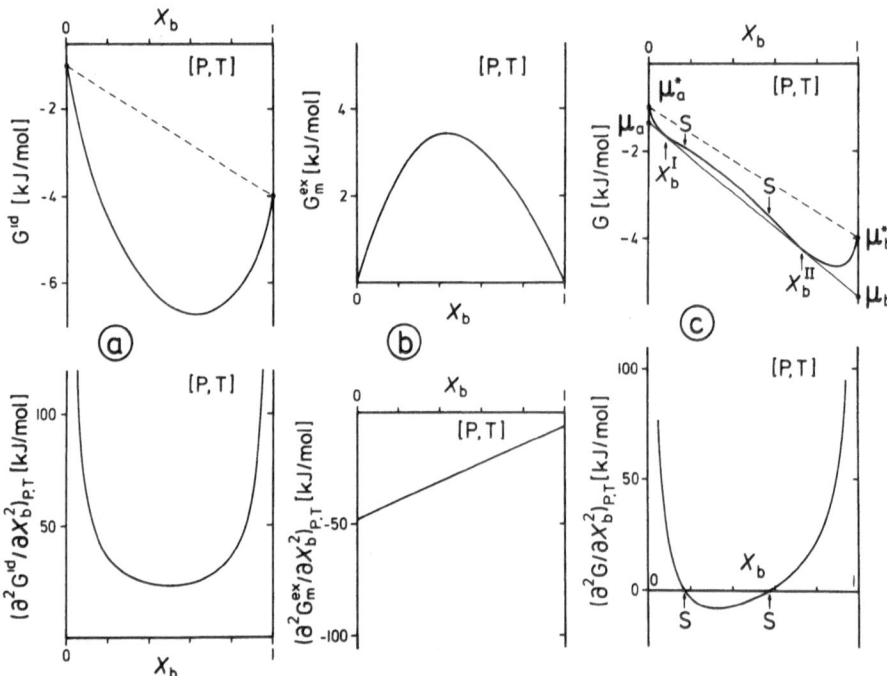

Fig. 8.6 The set of diagrams at the *top* shows the composition dependences of **a** G^{id}, **b** G^{ex}_m, and **c** G for a simple crystalline solution at 700 K and a given P. *Below* The second derivatives of those quantities with respect to X_b are plotted vs X_b. For further comments and input data used to compute these, see text

Having analyzed the properties of G^{id} and G^{ex}_m, we now focus on the molar Gibbs energy, G, of the solution a-b, which is the sum of G^{id} and G^{ex}_m. As demonstrated earlier, the slope of the G^{id}-X_b curve tends to -∞ and +∞ as X_b→0 and X_b→1, respectively. The slope of the G^{ex}_m-X_b curve remains finite at extreme dilution, however. Consequently, any positive deviation due to the G^{ex}_m term will be invariably overridden by the extremely strong negative deviation due to the G^{id} term in the limits of extreme dilution. An overall negative deviation thus results at either ends of composition, which is clearly mirrored in Fig. 8.6c (above). This explains why coexisting phases must show *some* mutual miscibility, even if this escapes detection by our analytical techniques.

There are some other features in the G-X_b curve of Fig. 8.6c (above) that warrant detailed comments. Although the G-X_b curve is concave upward at both ends, it has a very conspicuous downward concavity in the intervening range, whenever G^{ex}_m is large enough. Associating upward concavity of the $G(X_b)$ function with stability, and downward concavity with regions of metastability, it is clear that this solution is metastable in the intermediate range with respect to the phases I and II, whose compositions lie outside of the concave downward region. Thus, homogeneous solutions having a downward concavity in their G-X_b curve exsolve into two phases. This phenomenon is denoted as exsolution, and the exsolved phases are said to be located on the binodals (or the solvus). Rewriting Eq.(1.100) in terms of μ_i (i = a,b), the criteria for equilibrium between I and II are recovered as

$$\mu^I_a = \mu^{II}_a \text{ and} \qquad\qquad (8.60a)$$

$$\mu^I_b = \mu^{II}_b . \qquad\qquad (8.60b)$$

Simultaneous fulfillment of these two criteria, demanding equality of the chemical potential of each component in both phases, may be demonstrated by drawing a commmon tangent to the G-X_b curve. Note that the intercepts of the tangents to the G-X_b curve yield μ_a and μ_b, the *common* tangent being the coincidence of two such tangents, one for each phase I and II. How, in practice, the compositions of the two coexisting phases, denoted as X^I_b and X^{II}_b in Fig. 8.6c, are solved for, will be demonstrated in Chap. 9.

Next, we turn to Fig. 8.6c (below), where $(\partial^2 G/\partial X^2_b)_{P,T}$ is depicted as a function of X_b. This curve was obtained by summing $(\partial^2 G^{id}/\partial X^2_b)_{P,T}$ and $(\partial^2 G^{ex}_m/\partial X^2_b)_{P,T}$. Of these, the first term has been shown to be invariably positive (Fig. 8.6a, below), while the latter is a negative quantity (Fig. 8.6b, below), to the extent that G^{ex}_m is positive. The interplay of these two dictates the magnitude of $(\partial^2 G/\partial X^2_b)_{P,T}$. In the present case, $(\partial^2 G/\partial X^2_b)_{P,T}$ is positive at both ends of the composition range, but negative in between. Recalling that we can associate positive $(\partial^2 G/\partial X^2_b)_{P,T}$ with upward concavity of $G(X_b)$, and negative $(\partial^2 G/\partial X^2_b)_{P,T}$ with a downward concavity of $G(X_b)$, it is immediately apparent that $G(X_b)$ is concave upward in the dilute ranges, but concave downward in between. The segments of upward and downward concavity of

$G(X_b)$ meet each other at the inflection points of the G-X_b curve. These are better recognized on the $(\partial^2G/\partial X^2_b)_{P,T}$ vs X_b diagram, because $(\partial^2G/\partial X^2_b)_{P,T} = 0$ at these compositions. The inflection points on a $G(X_b)$ curve are the spinodes; they are labeled S in Fig. 8.6c.

Having established the stability relations of an isostructural solution at one temperature, let us concentrate on its isobaric-polythermal phase relations. To achieve our goal, we shall proceed in a slightly different manner. Rather than using G-X_b curves to monitor the extent of stability and exsolution on the individual isotherms, we shall use G_m-X_b curves. To comprehend this strategy, recall that only one equation of state is required to represent an isostructural solution. Therefore, when the criterion of equality of chemical potentials of each component [see Eqs.(8.60a,b)] is applied to both the phases, the standard chemical potentials cancel. Thus, we need to know only the equation of state to define the details of phase relations. Indeed, the representation of the polythermal G_m-X_b curve is also a lot simpler; regardless of the temperature, the intercepts of the G_m-X_b curve remain the same, since $G_m \rightarrow 0$ both as $X_b \rightarrow 0$ and $X_b \rightarrow 1$. Full advantage of this property of G_m will be taken by superposing all the G_m-X_b isotherms on one diagram.

Interpretation of the G_m-X_b isotherms requires knowledge of the functional nature of G_m, especially their partial derivatives with respect to X_b. These are summarized below:

$$\left(\frac{\partial G_m}{\partial X_b}\right)_{P,T} = RT\ln\frac{X_b}{X_a} + W_{G,ab}X_b(3X_a - 1) - W_{G,ba}X_a(3X_b - 1), \quad (8.61)$$

$$\left(\frac{\partial^2 G_m}{\partial X_b^2}\right)_{P,T} = RT\left(\frac{1}{X_a} + \frac{1}{X_b}\right) + 2W_{G,ab}(1 - 3X_b)$$

$$+ 2W_{G,ba}(1 - 3X_a), \quad \text{and} \qquad (8.62)$$

$$\left(\frac{\partial^3 G_m}{\partial X_b^3}\right)_{P,T} = RT\left(\frac{1}{X_a^2} + \frac{1}{X_b^2}\right) + 6(W_{G,ba} - W_{G,ab}). \qquad (8.63)$$

As before, the asymmetric Margules equation will be employed to represent the G^{ex}_m of the solution. Setting P at 1 bar, we use the following data to plot the G_m-X_b curves at several temperatures:

$W_{G,ab}$ (J mol^{-1}) = 24000 - 20 T and $W_{G,ba}$ (J mol^{-1}) = 38000 - 30 T.

Figure 8.7a depicts the G_m-X_b plots for a set of temperatures including 330, 380, 430, 480, and 507.6°C. The binodals, which are compositions satisfying the criterion of double-tangency, are shown as small circles on these curves. Indicated as dots are the spinodes, for which $(\partial^2 G_m/\partial X^2_b)_{P,T} = 0$. Note that both binodals and spinodes are further apart at low temperatures. As temperature increases, both the spinodes and the binodals approach each other, and finally coincide at the critical point, shown by the solid dot along with a small circle around it in Fig. 8.7a. The critical point is recognized by applying the dual criteria,

$$\left(\frac{\partial^2 G_m}{\partial X_b^2}\right)_{P,T} = 0 = \left(\frac{\partial^3 G_m}{\partial X_b^3}\right)_{P,T}. \tag{8.64}$$

In other words, $(\partial^2 G_m/\partial X_b^2)_{P,T}$ has two zeros (corresponding to two spinodes) at each subcritical temperature; as the composition is swept from one end to the other, $(\partial^2 G_m/\partial X_b^2)_{P,T}$ changes its sign twice (cf. Fig. 8.6c). With rising temperature, $(\partial^2 G_m/\partial X_b^2)_{P,T}$ becomes more and more positive, and the zeros (spinodes) approach each other. The spinodes coincide on the critical isotherm, and a unique zero ensues, satisfying the requirement $(\partial^3 G_m/\partial X_b^3)_{P,T}$ = 0. For supercritical isotherms, $(\partial^2 G_m/\partial X_b^2)_{P,T}$ is positive throughout the range of composition. The procedure used to solve for the critical conditions (for a given P, critical composition, X_c, and critical temperature, T_c) will be given in Chap. 9.

Figure 8.7b displays the isobaric temperature-composition (T-X_b) diagram generated simply by replotting the binodal, spinodal, and critical compositions at appropriate temperatures. The solvus, obtained by connecting the loci of the binodals, is depicted by a solid line. The spinodal curve, obtained from the

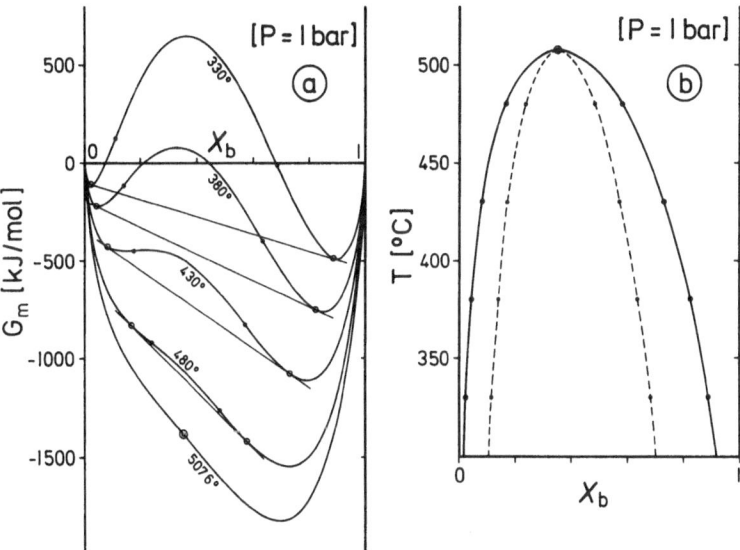

Fig. 8.7 a A G_m vs X_b plot for an isostructural solution at a set of temperatures (˚C) for a constant pressure of 1 bar, as obtained from the equation of state given in the text. The *dots* and *circles* show the locations of the spinodes and the binodals, respectively. Combination of these symbols depicts the critical point, which is the point of coincidence of the binodal and the spinode. **b** A T vs X_b plot for the same solution (P = 1 bar). The solvus is indicated by the *solid curve*, the spinodal by the *dashed curve*.

loci of all the spinodes, is shown as a dashed curve. As expected, they coincide at the critical point. Note that these curves are asymmetric with respect to 0.5 X_b, reflecting the use of the asymmetric Margules equation for generating them. Proceeding in this manner, and given data on the pressure dependence of G^{ex}_m, i.e. on V^{ex}_m, solvi may be calculated for any desired pressure.

The solvus and spinodal computed above apply to those binaries for which strain energy is of no consequence. This holds when the exsolved phases are non-coherent, or when they coexist as discrete grains. To emphasize this, they are called strain-free solvus and chemical spinodal, respectively. Coherent exsolution is, however, a common feature in many minerals. These are characterized by full continuity of the lattice across the exsolution lamellae. This can be achieved only by proper adjustment of the lattice spacings near the lamellar interface, which in turn, generates strain energy in the exsolving phase. This strain energy adds to the concave upward "part" of the molar Gibbs energy of a solution (Robin 1974). The net effect of the two is such that at any given temperature, the width of the solvus and the spinodal is reduced, and therefore, T_c is lowered. Robin (1974) gave an example of the computation of the coherent solvus and coherent spinodal. Readers willing to pursue this topic further may find Yund (1975) a good starting point.

8.4.2 Non-Isostructural Solution

Given the foregoing discussion, the non-isostructural case needs to be considered only very briefly. This time, the T-X_B diagram is derived from the condition of double-tangency applied to *two* G-X_B curves valid for the two structures at stated P and T. The system $Mg_2Si_2O_6$ (En)-$CaMgSi_2O_6$ (Di), explored by Lindsley et al. (1981), is a typical example of a non-isostructural binary solution among minerals. Figure 8.8a reproduces two G-X_{Di} curves for that system at 1200°C and 10 kbar pressure, one for the clinopyroxene (Cpx) and the other for the orthopyroxene (Opx) structure. A remarkable feature of this diagram is the clear downward concavity in the G-X_{Di} of the Cpx structure, implying that at the stated T and P, the Cpx solution is metastable in its intermediate composition range with respect to two discrete Cpx phases (indicated by the inverted arrows and the dashed double-tangent). However, such an exsolution is of no consequence for a phase diagram, which is a diagrammatic representation of the most stable state of a system. Indeed, as is clearly seen from this figure, the *most* stable state of the system at the intermediate composition range is the coexistence of an Opx with a Cpx, labeled En_{ss} and Di_{ss}. There are thus three stability levels implied in this G-X_{Di} diagram. A homogeneous Cpx (highest stability level) is metastable with respect to a pair of Cpx; this pair (intermediate stability level) is in turn metastable with respect to En_{ss} + Di_{ss} (the lowest stablility level). Figure 8.8b assembles a T-X_{Di} phase diagram based on a number of G-X_{Di} curves drawn at various temperatures between 800-1400°C, for a constant pressure of 10 kbar.

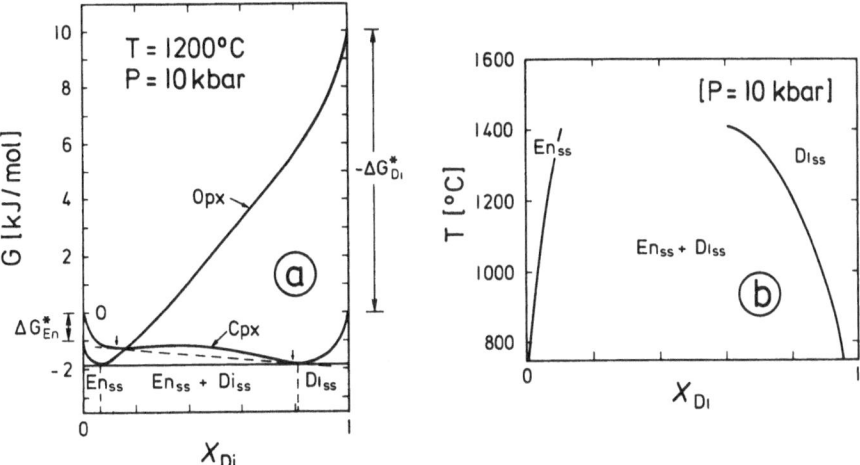

Fig. 8.8 a G-X_{Di} curves for the system $Mg_2Si_2O_6$(En)-$CaMgSi_2O_6$(Di) at 1200°C and 10 kbar. The condition of double tangency determines the compositions of the two coexisting pyroxenes with Opx and Cpx structures. **b** Temperature-dependent compositions of the coexisting phases define the two-phase field, En_{ss} + Di_{ss} (after Lindsley et al. 1981, Figs. 4a and 3b, respectively)

8.5 Formulation of Equations of State for Crystalline Solutions

8.5.1 Introduction

The objective of this section is to give a few examples of the derivation of equations of state for crystalline solutions. Stated otherwise, the goal is to express G^{ex}_m of crystalline solutions as a function of pressure, temperature, and composition. Availability of such data, along with a knowledge of the T-P-dependences of the molar Gibbs energies of the pure end-members, helps calculate the chemical potentials of the components of the crystalline solution. Data on chemical potentials are fundamental to all computations of mineral equilibria [cf. Eq.(1.100)].

A variety of experimental techniques is applied to obtain the requisite data, from which the equations of state of crystalline solutions are evaluated. In general, experimental data obtained by two or more methods are combined to formulate an equation of state. *Our examples* utilize data on binary crystalline solutions obtained by

1. calorimetric,
2. volumetric,
3. electrochemical, and
4. phase equilibria reversal experiments.

Some of these experiments measure the composition dependences of molar quantities, $Y(X_B)$, from which the excess molar quantities of mixing, Y^{ex}_m, are obtained. Calorimetric measurements of enthalpy of solution, $H°_{soln}(X_B)$, or molar entropy, $S°_{298}(X_B)$, belong to this category, as do X-ray data on molar volume, $V°_{298}(X_B)$. Other techniques, e.g. electrochemical cell measurements, obtain data on one partial molar quantity of mixing as a function of composition, $y_{m,B}(X_B)$, from which data for the other component, $y_{m,A}(X_B)$, are generally derived via Gibbs-Duhem integration. These two are then combined to obtain Y_m, or Y^{ex}_m, whichever is desired. The phase equilibria reversal experiments are widely employed to obtain excess mixing properties of crystalline solutions by analyzing the compositional data obtained at known P and T. This process calls for the use of models for S^{id}_m of the solution, and an analytical expression to render the composition dependence of the excess mixing quantities of interest. This type of data analysis is often called solution modeling.

To the extent that excess enthalpy and entropy are independent of T, and excess volume is independent of T and P, the combination of calorimetric data on H^{ex}_m and S^{ex}_m along with V^{ex}_m gives us the equation of state, $G^{ex}_m(P,T,X)$, of a crystalline solution. Because of the large uncertainty of this type of data, such equations of state may not be suitable for an accurate calculation of phase diagrams, however. The same statement might also apply to G^{ex}_m derived on the basis of measurement of partial molar properties. Despite that, both experimental techniques remain fundamental to the formulation of equations of state for crystalline solutions. The optimal use of such data would lie in their application as constraints for equations of state derived from solution modeling of phase equilibria reversal data. The equations of state of sanidine-analbite (Haselton et al. 1983) and diopside-jadeite (Cohen 1986b) crystalline solutions are good examples for such work (see also Sect.8.5.4).

8.5.2 Manipulation of Molar Quantity vs Composition Data

8.5.2.1 General Formalisms

Direct calorimetric measurements of $H°_{soln}$ vs X_B and $S°_{298}$ vs X_B belong to the category of molar quantity vs composition data, as do X-ray determination of $V°_{298}$ as a function of composition. The procedure for fitting such data will be given in this section in terms of the Margules equation, rather than the Redlich-Kister equation. Although the Redlich-Kister equation is more versatile, we prefer not to use it in the present context for two reasons. First, the quality of Y (Y ≡ H, S, and V) vs X_B data available at present for *mineral systems* seldom justifies fitting the three or more parameters of the Redlich-Kister equation (1.82). Secondly, using a one- or two-parameter Redlich-Kister equation is barely advantageous, because it may be easily recast to an asymmetric or a symmetric Margules equation [Eqs.(1.86),(1.92)], respectively.

An excess molar quantity of mixing, Y^{ex}_m, is expressed in the Margules equation in terms of the Margules parameter, $W_{Y,i}$, which is defined [cf.

Eq.(1.87)] as

$$W_{Y,i} \equiv y_{m,i}^{ex,\infty} ,\tag{8.65}$$

$y_{m,i}^{ex,\infty}$ being the excess partial molar quantity of mixing of i at infinite dilution. Given Y vs X_B data, the job at hand is thus to derive $W_{Y,i}$. This can be achieved in two ways. The straightforward of the two methods utilizes the equation

$$\frac{Y_m^{ex}}{X_A X_B} = W_{Y,BA} + (W_{Y,AB} - W_{Y,BA})X_B .\tag{8.50}$$

To evaluate $W_{Y,AB}$ and $W_{Y,BA}$, Y is first recast to Y^{ex}_m, using the relation

$$Y_m^{ex} = Y - (X_A Y_A + X_B Y_B) - Y_m^{id} .\tag{8.66}$$

After this, $Y^{ex}_m/X_A X_B$ is plotted against X_B. A linear relationship emerges (cf. Fig. 1.8). The intercepts of this plot for $X_B = 0$ and $X_B = 1$ yield $W_{Y,BA}$ and $W_{Y,AB}$, respectively.

However, in a routine data smoothing procedure, use of Eq. (8.66) is justified only if Y_A and Y_B have negligible uncertainties. In a real life situation, Y has the same uncertainty regardless of the composition, that is, Y_A and Y_B are known no better than Y at some intermediate X_B. For this reason, an alternative procedure of data smoothing, devised and advocated by Waldbaum and Thompson (1968), must be employed. In it, the experimentally observed data on Y are converted to Y', defined as

$$Y' \equiv Y - Y_m^{id} = (X_A Y_A + X_B Y_B) + Y_m^{ex} .\tag{8.67}$$

To the extent that Y is enthalpy or volume, for both of which $Y^{id}_m = 0$, Y' is identical to Y. If Y happens to be entropy, deducting S^{id}_m from S to derive S' is also simple, because S^{id}_m is dependent only on composition and site multiplicity. The next step is to fit Y' by the method of least squares to a polynomial in X_B,

$$Y' = (X_A Y_A + X_B Y_B) + Y_m^{ex} = a + bX_B + cX_B^2 + dX_B^3,\tag{8.68}$$

where a, b, c, and d denote empirical fit parameters. From (8.68), $W_{Y,i}$ is recovered by the following algebraic manipulations.

Consider the properties of the function $Y'(X_B)$ in Eq.(8.68). For $X_B = 0$,

$$Y' = a = Y_A,\tag{8.69}$$

and the slope of the tangent to the Y' vs X_B curve at $X_B = 0$ is

$$\left(\frac{\partial Y'}{\partial X_B}\right)_{P,T,X_B} = b .\tag{8.70}$$

Similarly, for $X_B = 1$,

$$Y' = a + b + c + d = Y_B, \tag{8.71}$$

and the slope of the tangent at $X_B = 1$ is found to be

$$\left(\frac{\partial Y'}{\partial X_B}\right)_{P,T,X_B=1} = b + 2c + 3d. \tag{8.72}$$

Note that (8.69) and (8.71) yield one intercept each, Y_A and Y_B, for the two tangents drawn at $X_B = 0$ and $X_B = 1$. To calculate the intercepts at the opposite end of the composition range, giving us $y^{ex,\infty}{}_B$ and $y^{ex,\infty}{}_A$, the excess partial molar quantities for B and A at infinite dilution, respectively (cf. Fig. 8.9), we may apply

$$y_A^{ex,\infty} = Y_B - 1 \cdot \left(\frac{\partial Y'}{\partial X_B}\right)_{P,T,X_B=1}, \quad \text{and} \tag{8.73a}$$

$$y_B^{ex,\infty} = Y_A + 1 \cdot \left(\frac{\partial Y'}{\partial X_B}\right)_{P,T,X_B=0} \tag{8.73b}$$

Substituting appropriate expressions for the quantities on the RHS of the last two equations, we have

$$y_A^{ex,\infty} = a - c - 2d, \text{ and} \tag{8.74a}$$

$$y_B^{ex,\infty} = a + b. \tag{8.74b}$$

The Margules parameters now follow from the above equations:

$$W_{Y,AB} \equiv y_{m,A}^{ex,\infty} = y_A^{ex,\infty} - Y_A = -(c + 2d) \text{ and} \tag{8.75a}$$

$$W_{Y,BA} \equiv y_{m,B}^{ex,\infty} = y_B^{ex,\infty} - Y_B = -(c + d). \tag{8.75b}$$

If the parameter d in (8.68) equals zero, both (8.75a) and (8.75b) simplify to

$$W_{Y,AB} = W_{Y,BA} = -c \equiv W_Y, \tag{8.76}$$

implying that we are now dealing with a symmetric solution. Thus, the empirical fit parameters of (8.68) are all we need to know to compute $W_{Y,AB}$ and $W_{Y,BA}$. Both methods for calculating the Margules parameters are illustrated below by suitable numerical examples.

8.5.2.2 An Example of Processing Calorimetric Data

Direct data on H^{ex}_m and S^{ex}_m can be obtained from measurements of enthalpy of solution, $H°_{soln}$, and $C°_p(T)$ in appropriate ranges of temperature. Though these are classical techniques of obtaining such data, their use in geochemistry has surfaced only during the last three or four decades. Measurements of $H°_{soln}$ of the (Na,K)Cl crystalline solution by Bunk and Tichelaar (1953) and by Barrett and Wallace (1954) may be cited as an early example, where water was used as the solvent. With the advent of HF-calorimetry, other crystalline solutions became accessible to this type of work (Waldbaum and Robie 1971; Hovis and Waldbaum 1977; Hovis 1988). Highly refractory minerals are now routinely studied by the technique of high-temperature, oxide-melt solution calorimetry (see Wood et al. 1980; Newton et al. 1981, Geiger et al. 1987). Although measurement of $H°_{soln}$ of mineral crystalline solutions has proliferated, calorimetric measurements of the composition dependence of $C°_p(T)$ remain deplorably scarce. The primary reason for this is that $C°_p(T)$ measurement in the low-temperature range still requires too large an amount of well characterized material, which is difficult to synthesize in sufficient quantity. Examples of such studies are provided by Haselton and Westrum (1980) and Haselton et al. (1983).

The objective of this section is to describe the procedure for fitting $H°_{soln}$ data, and to extract $W_{H,i}$ from that fit. The $H°_{soln}$ data on (Na,K)Cl crystalline solution will serve as our example. The fundamental requirement for such experiments is that the final solutions at the end of the dissolution process approach infinite dilution, so that solute-solute interactions can be ignored. This condition was fulfilled by the experiments of Bunk and Tichelaar (1953), who used 600 g of water at 25 °C to dissolve 0.03 mol of (Na,K)Cl in all their experiments. Because they listed $H°_{soln}$, our data-fitting procedure can operate directly. There is a second set of enthalpy measurements on (Na,K)Cl (Barrett and Wallace 1954). Unfortunately, the latter results were listed as H^{ex}_m; therefore, they will not be amenable to a similar treatment. We can, however, utilize them for comparison with H^{ex}_m computed from $W_{H,i}$ extracted from the fit to Bunk and Tichelaar's (1953) data.

Table 8.1 reproduces the raw calorimetric data on $-H°_{soln}$ vs X_{KCl} (Bunk and Tichelaar 1953). To fit this type of data by least squares, it is good to observe the general rule that roughly three times as many data points are necessary as there are parameters to be fitted. Given 11 data points, we may then introduce up to four parameters. A least squares fit to a cubic polynomial in X_{KCl} gave

$$-H°_{soln} \text{ (J mol}^{-1}) = -4091 \ (\pm 84) + 5711 \ (\pm 767) \ X_{KCl} - 20856 \ (\pm 1838) \ X^2_{KCl}$$
$$+ 1703 \ (\pm 1206) \ X^3_{KCl}; \ (\sigma_F = \pm 94.8 \text{ J mol}^{-1}). \tag{8.77}$$

Alternatively, a quadratic polynomial fit in X_{KCl} resulted in

$$-H°_{soln} \text{ (J mol}^{-1}) = -4029 \ (\pm 77) + 4737 \ (\pm 356) \ X_{KCl} - 18301 \ (\pm 343) \ X^2_{KCl};$$
$$(\sigma_F = \pm 100.5 \text{ J mol}^{-1}). \tag{8.78}$$

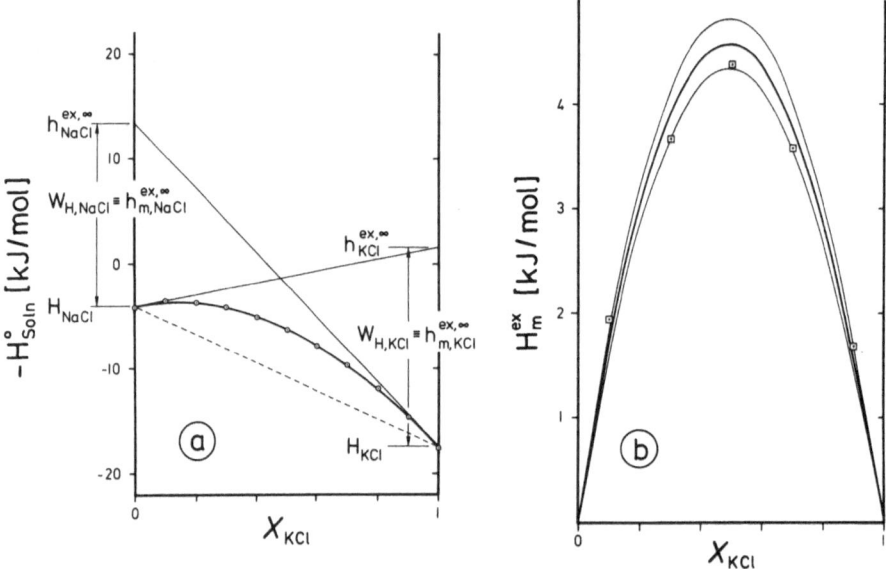

Fig. 8.9 a A plot of $-H^\circ_{soln}$ vs X_{KCl} for (Na,K)Cl crystalline solutions. The *encircled dots* represent the raw data of Bunk and Tichelaar (1953). The cubic fit to the data is indicated by a *bold line*. For further details, see text. **b** A plot of H^{ex}_m vs X_{KCl}, and its 2σ-uncertainty envelope, computed in the text. The *box symbols* show the independent data on H^{ex}_m by Barrett and Wallace (1954); these are compatible with Bunk and Tichelaar's (1953) results

The coefficients of the polynomial in (8.77) are the empirical fit parameters a, b, c, and d of (8.68), likewise, those of (8.78) are a, b, and c; the uncertainty of each of these parameters is given in parentheses. Also quoted is the overall uncertainty of the fit, σ_F, which is needed to derive $\sigma_{WH,i}$. Figure 8.9a displays the measured set of $-H^\circ_{soln}$ vs X_{KCl} data, and their fit to the cubic polynomial. The dashed straight line shows the calculated $-H^\circ_{soln}$ for the mechanical mixture of the two pure end-members. Evidently, the difference between the two reflects the extent of H^{ex}_m, which is positive in this case. Also depicted in Fig. 8.9a are the two tangents drawn at infinite dilution, which permit locating $h^{ex,\infty}_i$, whence $W_{H,i}$ for each component is obtained.

Evaluation of $W_{H,i}$ from the cubic fit follows from Eq.(8.75a,b):

$$W_{H,NaCl} = -(c + 2d) = 17450 \text{ J mol}^{-1} \text{ and} \tag{8.79a'}$$

$$W_{H,KCl} = -(c + d) = 19153 \text{ J mol}^{-1}. \tag{8.79b'}$$

Alternatively, the quadratic fit may be utilized for obtaining W_H from Eq.(8.76),

$$W_H = -c = 18301 \text{ J mol}^{-1}. \tag{8.80a}$$

In order to compute the uncertainties of $W_{Y,i}$, we may follow a procedure outlined by Meyer (1975), which utilizes the variance-covariance matrix obtained during the least squares fit, together with the overall uncertainty of the fit, σ_F. For a four parameter fit, the general form of the 4 x 4 variance-covariance matrix is

$$
\begin{array}{c|cccc}
 & A & B & C & D \\
\hline
a & y_{aA} & y_{aB} & y_{aC} & y_{aD} \\
b & y_{bA} & y_{bB} & y_{bC} & y_{bD} \\
c & y_{cA} & y_{cB} & y_{cC} & y_{cD} \\
d & y_{dA} & y_{dB} & y_{dC} & y_{dD}
\end{array}
\quad . \tag{8.81}
$$

Considering the coefficients of c and d in Eq.(8.75), $\sigma_{WY,i}$ may be obtained as follows

$$
\sigma_{WY,AB} = (y_{cC} + 4y_{cD} + 4y_{dD})^{\frac{1}{2}}\sigma_F \text{ and} \tag{8.82a}
$$

$$
\sigma_{WY,BA} = (y_{cC} + 2y_{cD} + y_{dD})^{\frac{1}{2}}\sigma_F. \tag{8.82b}
$$

For a quadratic fit involving three parameters, we likewise obtain a 3 x 3 matrix, and Eqs.(8.82a,b) reduce to

$$
\sigma_{WY} = (y_{cC})^{\frac{1}{2}}\sigma_F. \tag{8.83}
$$

Let us calculate $\sigma_{WH,i}$ from our fits to the Bunk and Tichelaar (1953) data. For the specific case of the cubic fit in (8.77), the variance-covariance matrix is

$$
\begin{array}{c|cccc}
 & A & B & C & D \\
\hline
a & 0.790 & -5.536 & 10.490 & -5.828 \\
b & -5.536 & 65.530 & -150.552 & 92.598 \\
c & 10.490 & -150.552 & 375.895 & -242.826 \\
d & -5.828 & 92.598 & -242.826 & 161.884
\end{array}
\quad . \tag{8.84}
$$

Employing these numbers and the appropriate σ_F given in (8.77), we have

$$
\sigma_{WH,NaCl} = 684 \text{ J mol}^{-1} \text{ and} \tag{8.85a}
$$

$$
\sigma_{WH,KCl} = 684 \text{ J mol}^{-1}. \tag{8.85b}
$$

Combining these with our earlier data on $W_{H,i}$, we obtain

$$
W_{H,NaCl} = 17450 \pm 684 \text{ J mol}^{-1} \text{ and} \tag{8.79a}
$$

$$
W_{H,KCl} = 19153 \pm 684 \text{ J mol}^{-1}. \tag{8.79b}
$$

Proceeding in an analogous manner, we may calculate σ_{WH}. The 3×3 variance-covariance matrix for the pertinent quadratic fit is

$$
\begin{array}{c|ccc}
 & A & B & C \\
\hline
a & 0.580 & -2.203 & 1.748 \\
b & -2.203 & 12.564 & -11.655 \\
c & 1.748 & -11.655 & 11.655
\end{array}
\tag{8.86}
$$

Inserting y_{cC} from this matrix and the appropriate σ_F from (8.78) into (8.83),

$$
\sigma_{W_H} = 343 \text{ J mol}^{-1}.
\tag{8.87}
$$

It is of interest, if trivial, to note that σ_{WH} is indeed the same as the uncertainty of the c parameter given in (8.78). This was to be expected, since $W_H = -c$ (8.76). Combining σ_{WH} with Eq.(8.80a), we have

$$
W_H = 18301 \pm 343 \text{ J mol}^{-1}.
\tag{8.80}
$$

Given $W_{H,i}$, the composition dependence of H^{ex}_m may be computed immediately from (8.49). The uncertainty of H^{ex}_m follows from the appropriate error propagation equation

$$
\sigma_{H^{ex}_m} = X_{NaCl} X_{KCl} [\sigma^2_{W_{H,NaCl}} X^2_{KCl} + \sigma^2_{W_{H,KCl}} X^2_{NaCl}]^{\frac{1}{2}} .
\tag{8.88}
$$

The computed H^{ex}_m and σ_{Hexm} are reproduced in Table 8.1. Figure 8.9b displays H^{ex}_m and the $2\sigma_{Hexm}$ error band as a function of X_{KCl}. Also shown are the experimental data on H^{ex}_m, given by Barrett and Wallace (1954). The

Table 8.1. The experimental data on $-H^\circ_{soln}$ for the halite-sylvite crystalline solution (Bunk and Tichelaar 1953), as a function of X_{KCl}. Also indicated are H^{ex}_m and $\sigma_{H^{ex}_m}$, derived by processing the $-H^\circ_{soln}$ data (see text)

X_{KCl}	$-H^\circ_{soln}$(J mol^{-1})	H^{ex}_m (J mol^{-1})	$\sigma_{H^{ex}_m}$(J mol^{-1})
0.0	-4197	0	0
0.1	-3573	1708	56
0.2	-3732	3010	90
0.3	-4201	3915	110
0.4	-5109	4433	119
0.5	-6339	4575	121
0.6	-7816	4351	119
0.7	-9661	3772	110
0.8	-11933	2846	90
0.9	-14598	1586	56
1.0	-17573	0	0

independent measurements of the latter authors are compatible with Bunk and Tichelaar's (1953) $H°_{soln}$ data.

8.5.2.3 Examples of Fitting V(X) Data for Crystalline Solutions

The objective of this section is to demonstrate how the molar quantity vs composition data can be alternatively processed from a plot of $Y^{ex}_m/X_A X_B$ vs X_B (see Sect.8.5.2.1). For this purpose, let us first consider the $V(X)$ data for (Na,K)Cl crystalline solution as reported by Bunk and Tichelaar (1953) and Barrett and Wallace (1954). Next, we shall briefly explore a complex set of $V(X)$ data provided by Newton et al. (1977) for the pyrope-grossular garnets, which calls for a more complicated data-fitting procedure.

The lattice parameter, $\underline{a}$, of the (Na,K)Cl crystalline solution has been measured by Bunk and Tichelaar (1953) and Barrett and Wallace (1954). Examination of the data shows that the cell edges reported by Barrett and Wallace (1954) are more precise; moreover, their cell edge data for the end-members are compatible with the molar volumes quoted by Robie et al. (1979). Bunk and Tichelaar's (1953) data look systematically low across the entire composition range. For this reason, we shall exclusively utilize the data from Barrett and Wallace (1954). Table 8.2 reproduces both $\underline{a}$, and the computed molar volume, V. The uncertainty of $\underline{a}$ was recast to that of V, using the error propagation equation[5],

$$\sigma_V = 3\underline{a}^2\sigma_{\underline{a}}. \tag{8.89}$$

Table 8.2 also summarizes V^{ex}_m, obtained from Eq.(8.66), bearing in mind that $V^{id}_m = 0$. Its uncertainty, σ_{Vexm}, was calculated from

$$\sigma_{V^{ex}_m} = [\sigma_V^2 + \sigma_{V_{NaCl}}^2 X_{NaCl}^2 + \sigma_{V_{KCl}}^2 X_{KCl}^2]^{\frac{1}{2}}. \tag{8.90}$$

And finally, Table 8.2 documents $V^{ex}_m/X_{NaCl}X_{KCl}$, along with its uncertainty, which follows from (8.90). Figure 8.10a is a plot of $V^{ex}_m/X_{NaCl}X_{KCl}$ vs X_{KCl}, the error bars shown being two times the uncertainties. It is clear that the whole set of data may be viewed as linear in X_{KCl}, with a zero slope. Referring to (8.50), it is then evident that (Na,K)Cl volumes may be handled in terms of a symmetrical Margules equation. The Margules volume parameter, W_V, is obtained simply by averaging the individual $V^{ex}_m/X_{NaCl}X_{KCl}$ data. The uncertainty quoted below is twice the standard deviation of the mean of $V^{ex}_m/X_{NaCl}X_{KCl}$,

$$W_V = 0.052 \pm 0.005 \text{ J bar}^{-1} \text{ mol}^{-1}. \tag{8.91}$$

---

[5] Bearing in mind that $V = \underline{a}^3$, application of (4.16) yields
$\sigma^2_V = (\partial V/\partial\underline{a})^2\sigma^2_{\underline{a}} = (3\underline{a}^2)^2\sigma^2_{\underline{a}}$, whence $\sigma_V = 3\underline{a}^2\sigma_{\underline{a}}$.

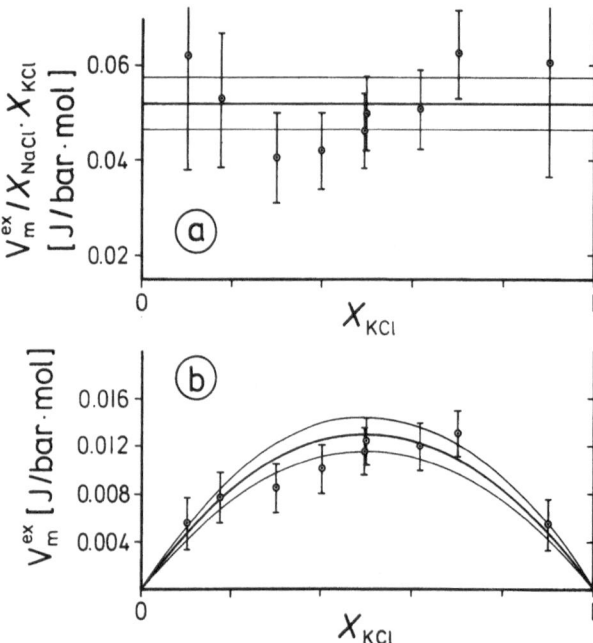

Fig. 8.10 a $V_m^{ex}/X_{NaCl}X_{KCl}$ of (Na,K)Cl crystalline solution plotted against X_{KCl}. The individual data points, and the 2σ error bars, are from Barrett and Wallace (1954). The *heavy line* with a zero slope corresponds to the value of W_V given in Eq.(8.91), the *thinner lines* depict the 2σ uncertainty envelope. **b** The same data replotted in terms of V_m^{ex}

Table 8.2. Calculations on the basis of the $V(X)$ data of (Na,K)Cl crystalline solution from Barrett and Wallace (1954)

X_{KCl}	$\underline{a}$ (Å)	V (J bar^{-1} mol^{-1})	V_m^{ex} (J bar^{-1} mol^{-1})	$V_m^{ex}/X_{NaCl}X_{KCl}$ (J bar^{-1} mol^{-1})
0.0000	5.6400(5)[a]	2.7010(8)	0.00000	0.00000
0.0997	5.7156(5)	2.8111(8)	0.00556(108)	0.06194(1203)
0.1757	5.7705(5)	2.8929(8)	0.00768(105)	0.05303(722)
0.3010	5.8571(5)	3.0251(8)	0.00850(101)	0.04040(478)
0.4019	5.9256(5)	3.1325(8)	0.01011(99)	0.04206(410)
0.4963	5.9883(5)	3.2329(8)	0.01153(98)	0.04612(392)
0.5001	5.9913(5)	3.2378(8)	0.01245(98)	0.04980(392)
0.6165	6.0654(5)	3.3594(8)	0.01200(99)	0.05076(418)
0.7003	6.1185(5)	3.4484(8)	0.01313(101)	0.06256(479)
0.8998	6.2354(5)	3.6499(8)	0.00546(108)	0.06056(1179)
1.0000	6.2916(5)	3.7495(8)	0.00000	0.00000

[a] Uncertainties indicated in parentheses always apply to the last digit(s) of the relevant quantities.

The W_V agrees with that obtained, when the $V(X)$ data are fitted to a quadratic polynomial in X_{KCl}, and W_V derived following (8.76). Figure 8.10b displays a V^{ex}_m vs X_{KCl} plot of the original data and compares it with V^{ex}_m computed from W_V of (8.91). A good agreement between the two underscores the correct derivation of W_V.

Given a simple $V(X)$ behavior for a crystalline solution, as in (Na,K)Cl, data manipulation may be accomplished with either method described above. As emphasized by Newton and Wood (1980), however, some oxides and silicates show a more complicated volume behavior. For example, a positive V^{ex}_m observed for the better part of the composition range may become negative as the end-member with the smaller molar volume is approached. This type of volume behavior requires an entirely different approach. Consider, for example, the $V(X)$ data for the pyrope-grossular garnets (Newton et al. 1977). The nature of the problem involved is best appreciated from a $V^{ex}_m/X_{Py}X_{Gr}$ vs X_{Gr} plot of those data (Fig. 8.11a). Note that all $V^{ex}_m/X_{Py}X_{Gr}$ in the range $0.18 \leq X_{Gr} \leq 1.00$ are easily fitted to a symmetrical Margules equation; the appropriate W_V is indicated by the dashed horizontal line, and its 2σ range by the pair of dotted lines. However, it does not explain the $V^{ex}_m/X_{Py}X_{Gr}$ datum for $0.1\ X_{Gr}$, whose 2σ range is far removed. Expanding its uncertainty to 3σ does not solve the problem either. However, by adding an exponential term to the symmetric Margules equation, the $0.1\ X_{Gr}$ datum may be better accounted for. This is the strategy adopted by

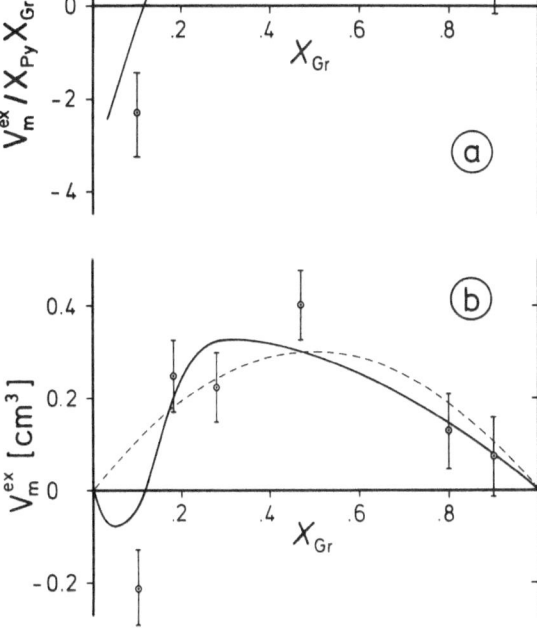

Fig. 8.11 Analysis of the molar volume vs composition data for pyrope-grossular garnets (Newton et al. 1977). a A $V^{ex}_m/X_{Py}X_{Gr}$ vs X_{Gr} plot of the raw data and their 2σ error band. The significance of the various curves is explained in the text. b A V^{ex}_m vs X_{Gr} plot of the same set of data

Haselton and Newton (1980) to fit the V-X_{Gr} data. The function employed by them has the form

$$V = [a + bX_{Py} + cX_{Py}^2] + D \cdot \exp\left[-\left(\frac{X_{Py} - E}{F}\right)^2/2\right].$$
(8.92)

The first square-bracketed term is a quadratic polynomial in X_{Py}, corresponding to W_V in Fig. 8.11a, whereas the exponential part accounts for the molar volume measured at 0.1 X_{Gr}. The $V^{ex}_m/X_{Py}X_{Gr}$ calculated from (8.92) is shown in this diagram by the solid line. Figure 8.11b depicts the experimental V^{ex}_m data and their 2σ uncertainties. Illustrated as a solid line in this diagram is V^{ex}_m computed from (8.92). Now, a somewhat better agreement between the experimental data and their computed counterpart emerges. For the sake of comparison, V^{ex}_m obtained from W_V only, is also delineated (dashed curve).

8.5.3 Handling Experimental Data on the Composition Dependence of Partial Molar Quantities of Mixing of a Component

Data on the partial molar Gibbs energy of mixing of component B in a binary crystalline solution A-B, $g_{m,B}$, may be obtained at T (and P) as a function of X_B by several experimental techniques:
1. *emf* measurements using solid oxide electrolyte sensors (e.g. Cameron and Unger 1970; Jacob 1978),
2. Measurement of the equilibrium partial pressure of B in the end-member phase and the crystalline solution by Knudsen cell mass spectrometry (Rammensee and Fraser 1981),
3. Phase equilibria reversal experiments under well defined f_{O2} and T (Hahn and Muan 1961), or at known P and T (Cressey et al. 1978).
   On the basis of data available at present, $g_{m,B}(X_B)$ generally proves to be linear in T, such that $h^{ex}_{m,B}(X_B)$ and $s_{m,B}(X_B)$ may be regarded as independent of T and calculated from Eqs.(1.11) and (1.8), respectively. The next step is to rewrite $g_{m,B}(X_B,T)$ and $s_{m,B}(X_B)$ as $g^{ex}_{m,B}(X_B,T)$ and $s^{ex}_{m,B}(X_B)$, followed by the derivation of $g^{ex}_{m,A}(X_B,T)$, $h^{ex}_{m,A}(X_B)$, and $s^{ex}_{m,A}(X_B)$. The latter part of the task is traditionally accomplished by integrating the Gibbs-Duhem equation (1.42). Different formats have been advocated to execute the Gibbs-Duhem integration. Worked examples will be provided to illustrate two of them (Sect. 8.5.3.2). An alternative to performing the Gibbs-Duhem integration will be demonstrated in a subsequent section (see 8.5.3.3).

8.5.3.1 Formalisms of Gibbs-Duhem Integration

For any excess partial molar quantity of mixing, $y^{ex}_{m,i}$, (Y ≡ G, H, S, V), the Gibbs-Duhem equation is written as [cf. Eq.(1.42)]

$$X_A \frac{dy^{ex}_{m,A}}{dX_B} + X_B \frac{dy^{ex}_{m,B}}{dX_B} = 0.$$
(8.93)

Because $y^{ex}_{m,A} = 0$ when $X_B = 0$, integration of (8.93) between the limits $y^{ex}_{m,B}$ at $X_B = 0$ to $y^{ex}_{m,B}$ at $X_B = X_B$ yields (as elaborated in Sect. 1.2.2.1)

$$y^{ex}_{m,A}(\text{at } X_B = X_B) = - \int_{y^{ex}_{m,B}(\text{at } X_B=0)}^{y^{ex}_{m,B}(\text{at } X_B=X_B)} \frac{X_B}{X_A} dy^{ex}_{m,B} . \tag{8.94}$$

The RHS of (8.94) is usually evaluated graphically to come up with $y^{ex}_{m,A}(X_B)$. Having obtained $y^{ex}_{m,A}$, the integral molar quantity of mixing, Y^{ex}_m, may be put together as a function of X_B from

$$Y^{ex}_m = (1-X_B)y^{ex}_{m,A} + X_B y^{ex}_{m,B}. \tag{8.95}$$

Since $y^{ex}_{m,B}$ is experimentally measured over an intermediate range of X_B, this integration requires that $y^{ex}_{m,B}$ be extrapolated to $y^{ex}_{m,B}$ at $X_B = 0$, the lower limit of integration. This may give rise to a perceptible error in the integration (see example given below). Extrapolating $y^{ex}_{m,B}$ for $X_B \rightarrow 1$ is problematic also, since X_B/X_A becomes indeterminate at $X_B = 1$.

To achieve a better control on the extrapolation of $y^{ex}_{m,B}$ to the dilute ranges, alternative formats of integration have been proposed (Wagner 1940; Darken and Gurry 1953; Chiotti 1972). The procedure advocated by Chiotti (1972) is recommended if a very accurate result is desired, though data processing for this method (Sect. 8.5.3.2) is more arduous. The accuracy of extrapolation of $y^{ex}_{m,B}$ to the dilute range is achieved here by introducing the two boundary constraints, $X_B y^{ex}_{m,B} = 0$ for $X_B = 0$ and $X_B = 1$.[6] This requires that the term $X_B y^{ex}_{m,B}$ be employed in this integration. To satisfy this need, Chiotti (1972) started with the derivative of $X^2_B y^{ex}_{m,B}$,

$$d(X^2_B y^{ex}_{m,B}) = 2X_B y^{ex}_{m,B} dX_B + X^2_B dy^{ex}_{m,B}. \tag{8.96}$$

Now, from Eq.(1.41), we have $X_B dy^{ex}_{m,B} = -X_A dy^{ex}_{m,A}$. Substituting it into the RHS of (8.96), it may be rearranged to

$$X_B X_A dy^{ex}_{m,A} = -d(X^2_B y^{ex}_{m,B}) + 2X_B y^{ex}_{m,B} dX_B, \tag{8.97}$$

which, when integrated, yields

$$\int_{y^{ex}_{m,A}(\text{at } X_B=0)}^{y^{ex}_{m,A}(\text{at } X_B=X_B)} X_B X_A dy^{ex}_{m,A} = -X^2_B y^{ex}_{m,B} + 2 \int_{X_B=0}^{X_B=X_B} X_B y^{ex}_{m,B} dX_B . \tag{8.98}$$

---

[6] If $X_B = 0$, $y^{ex}_{m,B}$ is a finite quantity equaling $y^{ex,\infty}_{m,B}$; thus, $X_B y^{ex}_{m,B} = 0$. On the other hand, $y^{ex}_{m,B} = 0$ if $X_B = 1$, so that we again end up with $X_B y^{ex}_{m,B} = 0$.

Given experimental data for $y^{ex}_{m,B}(X_B)$, the RHS of (8.98) may be evaluated. Note that the RHS of (8.98) provides *the integral of the LHS*, from which $y^{ex}_{m,A}(X_B)$ must be calculated employing the trapezoidal rule (see Chiotti 1972, p.2912). Fitting $X_B y^{ex}_{m,B}(X_B)$ data prior to numerical integration of the RHS of (8.98) must take into account two additional constraints (Chiotti 1972):

$$\frac{d(X_B y^{ex}_{m,B})}{dX_B} = y^{ex,\infty}_{m,B}, \quad \text{as} \quad X_B \to 0, \quad \text{and} \tag{8.99a}$$

$$\frac{d(X_B y^{ex}_{m,B})}{dX_B} = 0, \quad \text{when} \quad X_B \to 1. \tag{8.99b}$$

As indicated by Chiotti (1972, p.2913), $y^{ex,\infty}_{m,B}$ is estimated for this purpose by plotting $y^{ex}_{m,B}$ vs X_B and $y^{ex}_{m,B}/(1-X_B)^2$ vs X_B. At $X_B = 0$, both curves meet at a common intercept, yielding $y^{ex,\infty}_{m,B}$.

8.5.3.2 Worked Examples of Gibbs-Duhem Integration

Let us now turn to the execution of the Gibbs-Duhem integration, utilizing each of the above methods. The results will be compared with each other by plotting $Y^{ex}_m/(1-X_B)X_B$ vs X_B. The experimental data to be used for this purpose were reported by Jacob (1978), who determined $g_{m,Cr2O3}$ by measuring the reversible *emf* for the following electrochemical cell,

$$Pt,Cr+Cr_2O_3 \,|\, Y_2O_3\text{-}ThO_2 \,|\, Cr+(Al,Cr)_2O_3,Pt.$$

Without going into the details of the theory (cf. Seetharaman and Abraham 1980), it will suffice to say that the observed *emf*, E, at any T and X_i, is related to $g_{m,Cr2O3}$ by

$$g_{m,Cr_2O_3} = - zFE = RT\ln a_{Cr_2O_3}, \tag{8.100}$$

where z denotes the number of electrons transferred through the cell (6 per mol of Cr_2O_3), and F is the Faraday constant (= 96487 J Volt^{-1} mol^{-1}). The *emf* data were obtained between 800-1320°C for seven compositions spanning $0.1 \le X_{Cr2O3} \le 0.9$ (see Jacob 1978, Fig.1). For each composition, E was found to be linear in T within the limits of uncertainty,

$$E \text{ (in mV)} = a + bT, \tag{8.101}$$

with T given in K (Jacob 1978, Table 1). The overall uncertainty of E, σ_E was reported, not however σ_a and σ_b. Because E is linear in T, we may apply Eqs.

(1.8), (1.11), and (8.101) to solve for $s_{m,B}$ and $h^{ex}_{m,B}$:

$$s_{m,Cr_2O_3} = -\left(\frac{\partial g_{m,Cr_2O_3}}{\partial T}\right)_P = zF\left(\frac{\partial E}{\partial T}\right)_P = zFb, \quad \text{and} \tag{8.102}$$

$$h^{ex}_{m,Cr_2O_3} = g_{m,Cr_2O_3} + Ts_{m,Cr_2O_3} = -zFa. ^{7)} \tag{8.103}$$

Let us now evaluate $G^{ex}_m/(1-X_{Cr2O3})X_{Cr2O3}$ for the $(Al,Cr)_2O_3$ crystalline solution, using Jacob's (1978) *emf* data. The initial step in data processing is to convert $g_{m,Cr2O3}$ to $ln\gamma_{Cr2O3}$ (which equals $g^{ex}_{m,Cr2O3}/RT$). Bearing in mind that the site multiplicity, q, in this crystalline solution is two, we have

$$ln\gamma_{Cr_2O_3} = lna_{Cr_2O_3} - 2lnX_{Cr_2O_3}. \tag{8.104}$$

The resulting data on $ln\gamma_{Cr2O3}$ are listed in Table 8.3. To employ the format of integration in Eq. (8.94), we have to plot $ln\gamma_{Cr2O3}$ against $-X_{Cr2O3}/X_{Al2O3}$ at a specified T, and to determine the area under this curve graphically. Figure 8.12 depicts the plot for T = 1273.15 K. That $ln\gamma_{Cr2O3}$ has been plotted against X_{Cr2O3}/X_{Al2O3}, rather than $-X_{Cr2O3}/X_{Al2O3}$, is of no consequence; it merely requires us to switch the limits of integration. We integrate from $ln\gamma_{Cr2O3}$ at $X_{Cr2O3} = 0.5$ (A in Fig. 8.12) to $ln\gamma_{Cr2O3}$ at $X_{Cr2O3} = 0$ (O in Fig. 8.12). Graphical determination of this area (hatched in Fig. 8.12) yields $ln\gamma_{Al2O3} = 1.006$. Proceeding in this fashion, $ln\gamma_{Al2O3}$ has been computed for every composition explored by Jacob (1978), and listed in Table 8.3.

---

[7] $h^{ex}_{m,Cr2O3}$ $= g_{m,Cr2O3} + Ts_{m,Cr2O3} = -zFE + T[zF(\partial E/\partial T)_P]$
$= -zF[E-T(\partial E/\partial T)_P]$ $= -zF[a + bT - Tb] = -zFa.$

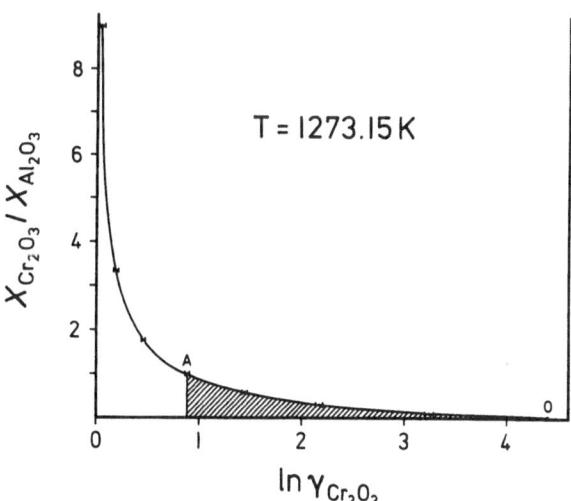

Fig. 8.12. A plot of $ln\gamma_{Cr_2O_3}$ vs $X_{Cr_2O_3}/X_{Al_2O_3}$ at 1273.15 K. The *shaded area* yields $ln\gamma_{Al_2O_3}$ for $X_{Cr_2O_3} = 0.5$. The significance of the points A and O is explained in the text

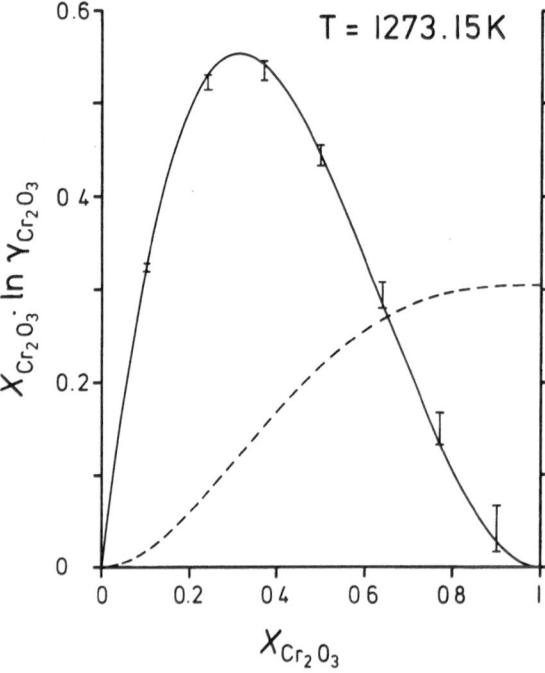

Fig. 8.13 A plot of $X_{Cr_2O_3}\ln\gamma_{Cr_2O_3}$ vs $X_{Cr_2O_3}$ at 1273.15 K, using data listed in Table 8.3. Note that the *solid curve* fitted through the uncertainty brackets of $X_{Cr_2O_3}\ln\gamma_{Cr_2O_3}$ shows a zero slope as $X_{Cr_2O_3}\to1$; by contrast, the slope equals $\ln\gamma_{Cr_2O_3}^{\infty}$ as $X_{Cr_2O_3}\to0$. The *dashed curve* represents the integral of the LHS of Eq.(8.98), from which $\ln\gamma_{Al_2O_3}$ follows.

Table 8.3. Results of Gibbs-Duhem integration following Eq. (8.94) for 1273.15 K

$X_{Cr_2O_3}$	E (mV)	g_{m,Cr_2O_3} (J mol^{-1})	$\ln\gamma_{Cr_2O_3}$	$\ln\gamma_{Al_2O_3}$	$\dfrac{G_m^{ex}}{(1-X_{Cr_2O_3})X_{Cr_2O_3}}$ (J mol^{-1})
0.10	24.952(700)	-14445(405)	3.2405(383)[a]	0.063	44782
0.24	12.379(600)	-7166(347)	2.1772(328)	0.269	42189
0.37	9.920(500)	-5743(289)	1.4460(273)	0.575	40746
0.50	9.089(400)	-5262(232)	0.8892(219)	1.006	40123
0.64	7.911(400)	-4580(232)	0.4599(219)	1.547	39110
0.77	5.996(400)	-3471(232)	0.1948(219)	2.152	38549
0.90	2.993(500)	-1733(289)	0.0470(273)	2.908	39178

[a] Uncertainties indicated in parentheses always apply to the last digit(s) of the relevant quantities.

For comparison, we now apply the method of Chiotti (1972) to integrate the Gibbs-Duhem equation. The first step is to integrate the second term on the RHS of Eq. (8.98). Figure 8.13 shows a plot of $X_{Cr2O3}\ln\gamma_{Cr2O3}$ (and its uncertainties) against X_{Cr2O3}, using $\ln\gamma_{Cr2O3}$ data of Table 8.3. Prior to integration, $X_{Cr2O3}\ln\gamma_{Cr2O3}$ vs X_{Cr2O3} data were fitted (Chatterjee et al. 1982 give details of the data fitting procedure) to the following polynomial, such that both constraints (8.99a) and (8.99b) are simultaneously satisfied

$$X_{Cr_2O_3}\ln\gamma_{Cr_2O_3} = 4.05964X_{Cr_2O_3} - 9.10853X^2_{Cr_2O_3}$$
$$+ 6.03814X^3_{Cr_2O_3} - 0.98925X^4_{Cr_2O_3}. \tag{8.105}$$

The integration was then accomplished numerically, and the result added to the first term on the RHS of (8.98). The sum of these two terms gives the *integral* of $X_{Cr2O3}X_{Al2O3}d\ln\gamma_{Al2O3}$, from which $\ln\gamma_{Al2O3}$ was derived by the trapezoidal rule. These results are listed in Table 8.4. To compute G^{ex}_m from (8.95), $\ln\gamma_{Al2O3}$, thus derived, was combined with $\ln\gamma_{Cr2O3}$ recalculated from (8.105). The G^{ex}_m, recast to $G^{ex}_m/(1-X_{Cr2O3})X_{Cr2O3}$, are summarized in Table 8.4 and juxtaposed with $G^{ex}_m/(1-X_{Cr2O3})X_{Cr2O3}$ data given earlier in Table 8.3.

Comparison of $G^{ex}_m/(1-X_{Cr2O3})X_{Cr2O3}$ calculated by the two contrasting methods of Gibbs-Duhem integration is facilitated by a knowledge of the uncertainties involved. A rough estimation may be attempted by postulating $\sigma_{RT\ln aCr2O3} \equiv \sigma_{RT\ln\gamma Cr2O3} = \sigma_{RT\ln\gamma Al2O3}$. Error propagation formalism would then yield,

$$\frac{\sigma_{G^{ex}_m}}{(1 - X_{Cr_2O_3})X_{Cr_2O_3}} = \frac{\sigma_{RT\ln aCr_2O_3}[X^2_{Cr_2O_3} + X^2_{Al_2O_3}]^{\frac{1}{2}}}{(1 - X_{Cr_2O_3})X_{Cr_2O_3}}. \tag{8.106}$$

Employing $\sigma_{RT\ln aCr2O3}$ as given in Table 8.3, the uncertainties of $G^{ex}_m/(1-X_{Cr2O3})X_{Cr2O3}$ were computed and listed in Table 8.4.

Table 8.4. The results of Gibbs-Duhem integration based on Eq. (8.98), compared to those obtained from Eq. (8.94); T = 1273.15 K

		Eq. (8.98)		Eq. (8.94)
$X_{Cr_2O_3}$	$\ln\gamma_{Cr_2O_3}$[a]	$\ln\gamma_{Al_2O_3}$	$\dfrac{G^{ex}_m \text{ (J mol}^{-1})}{(1-X_{Cr_2O_3})\cdot X_{Cr_2O_3}}$	$\dfrac{G^{ex}_m \text{ (J mol}^{-1})}{(1-X_{Cr_2O_3})\cdot X_{Cr_2O_3}}$
0.10	3.2082	0.0449	42486(4075)	44782(4075)
0.24	2.2077	0.2490	41731(1516)	42189(1516)
0.37	1.4660	0.5738	41048(906)	40746(906)
0.50	0.8913	1.0154	40365(656)	40123(656)
0.64	0.4441	1.6067	39632(739)	39110(739)
0.77	0.1745	2.2491	38948(1053)	38549(1053)
0.90	0.0317	2.9684	38268(2908)	39178(2908)

[a] Recalculated from Eq. (8.105).

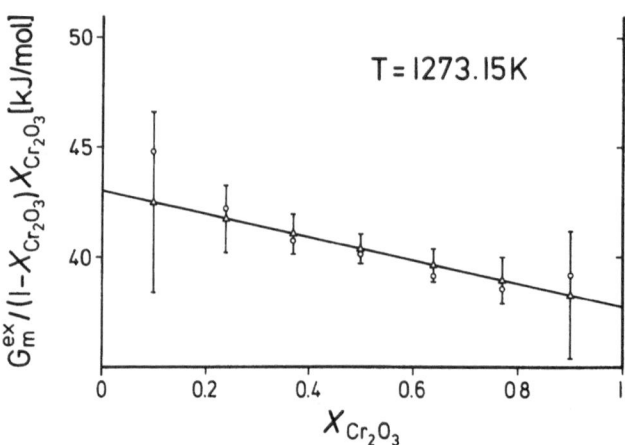

Fig. 8.14 A $G_m^{ex}/(1-X_{Cr_2O_3})X_{Cr_2O_3}$ vs $X_{Cr_2O_3}$ plot contrasting the results obtained by different methods of Gibbs-Duhem integration. The results obtained from Eq.(8.94) are shown as *circles*, those from Eq.(8.98), and their uncertainties, as *triangles*. The linear fit to the latter set of data is also indicated

Figure 8.14 depicts a $G^{ex}_m/(1-X_{Cr2O3})X_{Cr2O3}$ vs X_{Cr2O3} plot based on the identical set of *emf* data processed by two alternate formats of the Gibbs-Duhem integration. The results obtained by Chiotti's (1972) method of integration (8.98) are shown by small triangles; the error bars are the appropriate uncertainties. Note the linear relationship between the plotted quantities, implying that the $(Al,Cr)_2O_3$ crystalline solution indeed obeys the Margules equation (8.50). A linear fit to these data gives

$$\frac{G_m^{ex}}{(1-X_{Cr_2O_3})X_{Cr_2O_3}}[J\ mol^{-1}] = 43002 - 5265X_{Cr_2O_3}\ . \tag{8.107}$$

Using Eq.(8.50), from (8.107) we recover

$$W_{G,Al_2O_3} = 37737\ J\ mol^{-1}\ \text{and}\ W_{G,Cr_2O_3} = 43002\ J\ mol^{-1}. \tag{8.108}$$

The $G^{ex}_m/(1-X_{Cr2O3})X_{Cr2O3}$ vs X_{Cr2O3} data computed from Eq.(8.94) are shown in Fig. 8.14 by small circles. Although they deviate from linearity, the deviation is not significant in the light of the uncertainties involved. We may then conclude that the Margules equation is indeed adequate to formulate an equation of state for $(Al,Cr)_2O_3$ crystalline solution. In addition to evaluating Jacob's (1978) $g^{ex}_{m,Cr2O3}$ data, Chatterjee et al. (1982) used (8.98) also to integrate $h^{ex}_{m,Cr2O3}$. For this purpose, $X_{Cr2O3}h^{ex}_{m,Cr2O3}$ were fitted to the following polynomial in X_{Cr2O3},

$$X_{Cr_2O_3}h^{ex}_{m,Cr_2O_3} = 37458X_{Cr_2O_3} - 86375X^2_{Cr_2O_3} + 60375X^3_{Cr_2O_3} - 11458X^4_{Cr_2O_3}. \tag{8.109}$$

The $h^{ex}_{m,Al2O3}$ values obtained by the Gibbs-Duhem integration were then combined with $h^{ex}_{m,Cr2O3}$ to derive

$$W_{H,Al_2O_3} = 31729 \text{ J mol}^{-1} \text{ and } W_{H,Cr_2O_3} = 37484 \text{ J mol}^{-1}. \tag{8.110}$$

Noting that E is linear in T, (8.108) and (8.110) were combined to deduce

$$W_{S,Al_2O_3} = -4.719 \text{ J K}^{-1} \text{ mol}^{-1} \text{ and } W_{S,Cr_2O_3} = -4.334 \text{ J K}^{-1} \text{ mol}^{-1}. \tag{8.111}$$

To be applicable to geochemical calculations, the above data must be supplemented by $W_{V,Al2O3}$ and $W_{V,Cr2O3}$. These were obtained from experimental V-X_{Cr2O3} data (Chatterjee et al. 1982). Following Eqs. (1.86) and (1.97), the equation of state for $(Al,Cr)_2O_3$ was then put together as

$$G^{ex}_m \text{ (J mol}^{-1}) = (1-X_{Cr_2O_3})X_{Cr_2O_3} \cdot$$
$$[W_{G,Cr_2O_3} + (W_{G,Al_2O_3} - W_{G,Cr_2O_3})X_{Cr_2O_3}], \tag{8.112a}$$

$$W_{G,Al_2O_3} \text{ (J mol}^{-1}) = 31729 + 4.719 \text{ T} + 0.0006 \text{ P, and} \tag{8.112b}$$

$$W_{G,Cr_2O_3} \text{ (J mol}^{-1}) = 37484 + 4.334 \text{ T} + 0.0386 \text{ P}, \tag{8.112c}$$

with T expressed in K and P in bar. We shall utilize this equation of state on a later occasion to model the thermodynamic properties of $Zn(Al,Cr)_2O_4$ spinels (see Sect. 8.5.4.3).

8.5.3.3 An Alternative to Gibbs-Duhem Integration

As an alternative to integrating the Gibbs-Duhem equation, one may start by ˙ rearranging Eq.(8.14b) to

$$\frac{RT\ln\gamma_B}{(1 - X_B)^2} = W_{G,BA} + 2X_B(W_{G,AB} - W_{G,BA}). \tag{8.113}$$

All we require is to recast the experimental data on $g_{m,B}$ (that is $RT\ln a_B$) at any T and X_B to $RT\ln\gamma_B/(1-X_B)^2$, and plot them against X_B. To the extent that the solution obeys the Margules equation, a linear relationship emerges, from which the $W_{G,i}$ are obtained. If a quadratic polynomial is required to fit these data within their limits of uncertainties, a three-parameter Redlich-Kister equation can be used. To do this, we simply rewrite Eq.(1.84) as

$$\frac{RT\ln\gamma_B}{(1 - X_B)^2} = (A + B + C) + (-4B - 8C)X_B + (12C)X_B^2. \tag{8.114}$$

Defining the bracketed quantities on the RHS of (8.114) as a_0, a_1, and a_2, respectively, we solve for the Redlich-Kister parameters:

$A = a_0 + a_1/4 + a_2/12$, $B = -a_1/4 - a_2/6$, and $C = a_2/12$.

Evaluation of $\ln\gamma_{Cr2O3}(X_B)$ data at 1273.15 K (listed in Table 8.3) by this technique leads to the following results,

$$W_{G,Al_2O_3} = 37602 \pm 4468 \text{ J mol}^{-1} \text{ and } W_{G,Cr_2O_3} = 43046 \pm 4569 \text{ J mol}^{-1},$$

(8.115)

which are in excellent agreement with those obtained earlier [see (8.108)] by Gibbs-Duhem integration.[8] This method has the added advantage that $\sigma_{WG,i}$ follows from the linear fit of $RT\ln\gamma_{Cr2O3}/(1-X_{Cr2O3})^2$ vs X_{Cr2O3} data in a straightforward manner.

Experimental data on many crystalline solutions do not permit resolution of two $W_{G,i}$, let alone the A, B, and C parameters of the Redlich-Kister equation. The manganosite-bunsenite solution, (Mn,Ni)O, studied by Hahn and Muan (1961), is a case in point. In order to establish the a_i-X_i relations in (Mn,Ni)O, Hahn and Muan (1961) performed two sets of reversal experiments. In the first set, they equilibrated the NiO end-member with Ni metal at 1 atm P_{tot} at known T and f_{O2} (specified by mixing CO_2 and H_2). For the relevant reaction

$$Ni + \frac{1}{2}O_2 = NiO \,,$$

(8.116)

the condition of equilibrium may be written as

$$0 = \Delta G_T^\circ + \int_{P^\circ=1\,\text{bar}}^{P=1\,\text{atm}} \Delta V_{T,s}(P)dP - \frac{1}{2}RT\ln\frac{f_{O_2}(1)}{f_{O_2}^\circ} \,.$$

(8.117)

In a second series of experiments, Ni metal was equilibrated with (Mn,Ni)O crystalline solutions at desired T and f_{O2} at a P_{tot} of 1 atm, followed by the determination of the compositions of (Mn,Ni)O solutions. To the extent that Ni remains a pure phase, which holds true for this set of experiments, the condition of equilibrium is

$$0 = \Delta G_T^\circ + \int_{P^\circ=1\,\text{bar}}^{P=1\,\text{atm}} \Delta V_{T,s}(P)dP - \frac{1}{2}RT\ln\frac{f_{O_2}(2)}{f_{O_2}^\circ} + RT\ln a_{NiO} \,.$$

(8.118)

Equating (8.118) with (8.117), we obtain

$$a_{NiO} = \left[\frac{f_{O_2}(2)}{f_{O_2}(1)}\right]^{\frac{1}{2}} \,.$$

(8.119)

---

[8] Readers seeking to verify this procedure may note that the uncertainties of $RT\ln\gamma_{Cr2O3}/$ $(1-X_{Cr2O3})^2$ increase dramatically as X_{Cr2O3} goes from 0.10 to 0.90. Consequently, fitting these data by regression requires that they weighted by the reciprocals of the squares of their uncertainties.

Table 8.5. Evaluation of a_i-X_i relations for manganosite-bunsenite solutions at 1373.15 K, based on reversal experiments by Hahn and Muan (1961)

$\log f_{O_2}(1)$ (atm)	$\log f_{O_2}(2)$ (atm)	X_{NiO}	a_{NiO}	γ_{NiO}	$RT\ln\gamma_{NiO}/(1-X_{NiO})^2$ (J)
-8.9736	-10.996	0.045	0.0975	2.1657	9673
-8.9736	-10.996	0.047	0.0975	2.0736	9167
-8.9736	-10.297	0.112	0.2179	1.9458	9638
-8.9736	-10.297	0.114	0.2179	1.9117	9424
-8.9736	-10.088	0.145	0.2772	1.9118	10121
-8.9736	-10.088	0.159	0.2772	1.7435	8973
-8.9736	-9.781	0.250	0.3947	1.5790	9271
-8.9736	-9.781	0.250	0.3947	1.5790	9271
-8.9736	-9.625	0.294	0.4724	1.6068	10863
-8.9736	-9.625	0.321	0.4724	1.4717	9568
-8.9736	-9.533	0.370	0.5252	1.4194	10075
-8.9736	-9.533	0.387	0.5252	1.3571	9277
-8.9736	-9.462	0.440	0.5699	1.2953	9419
-8.9736	-9.462	0.447	0.5699	1.2750	9070
-8.9736	-9.358	0.512	0.6424	1.2547	10878
-8.9736	-9.358	0.550	0.6424	1.1680	8757
-8.9736	-9.317	0.601	0.6735	1.1206	8164
-8.9736	-9.317	0.608	0.6735	1.1077	7598
-8.9736	-9.233	0.668	0.7418	1.1106	10861
-8.9736	-9.233	0.680	0.7418	1.0910	9706
-8.9736	-9.118	0.808	0.8469	1.0481	14550
-8.9736	-9.118	0.853	0.8469	0.9928	-3812

Table 8.5 lists the results of evaluation of the a_i-X_i data of the manganosite-bunsenite solution, based on Hahn and Muan (1961) experiments at 1373.15 K. The last column of this table reproduces $RT\ln\gamma_{NiO}/(1-X_{NiO})^2$ data, which, appropriately weighted and fitted against X_{Nio}, yields a linear function having a zero slope within the limits of uncertainties of the data. The zero slope implies a symmetric solution, whose W_G is found to be

$$W_G = 9645 \pm 995 \text{ J mol}^{-1}. \tag{8.120}$$

8.5.4 More on Equations of State for Crystalline Solutions: Some Worked Examples

A variety of phase equilibria experiments may be utilized to formulate equations of state for crystalline solutions. Only three of these will be considered below:

1. Solvus in a binary system;
2. Cation-exchange in a reciprocal ternary; and
3. Isothermal displacement of the equilibrium P for a solid-solid reaction as a function of composition of the crystalline solution. To the extent that calorimetrically measured data on H^{ex}_m are also available for the relevant crystalline solution, they can be used to place further constraints on the derived equation of state.

8.5.4.1 An Equation of State for the Halite-Sylvite Crystalline Solution, (Na,K)Cl

A number of workers (Nacken 1918; Bunk and Tichelaar 1953; Barrett and Wallace 1954) experimentally obtained isobaric solvus data of the halite-sylvite crystalline solution (Table 8.6). These data will be combined with its $V(X)$ data given above (Sect. 8.5.2.3), to formulate a polybaric-polythermal equation of state for the halite-sylvite crystalline solutions.

Derivation of the excess mixing properties from solvus data may follow alternative formats of evaluation. One method of data smoothing, devised originally by Thompson and Waldbaum (1969), has been quite popular among geochemists. Two other possibilities will be demonstrated below. One of these permits a direct solution for the Margules parameter, $W_{G,i}$, given a pair of solvus compositions at any T. This method will be used to process the halite-sylvite solvus data at 1 bar. The other technique shall be elucidated in Section 8.5.4.2, using polybaric solvi (Warner and Luth 1973) on the forsterite-monticellite join.

The condition of two-phase equilibrium, a halite-rich (H) and a sylvite-rich (S) solution, at any given temperature (at constant pressure), is the simultaneous fulfillment of equality of chemical potentials of the components NaCl (h) and KCl (s) in H and S,

$$\mu_h^H = \mu_h^S \text{ and} \tag{8.121a}$$

$$\mu_s^H = \mu_s^S. \tag{8.121b}$$

The standard state chosen is the pure component with appropriate structure having unit activity at system T and P. Since both NaCl and KCl are isostructural, $\mu^{*H}_h = \mu^{*S}_h$ and $\mu^{*H}_s = \mu^{*S}_s$; therefore, (8.121a,b) reduce to

$$RT\ln(1-X_s^H) + RT\ln\gamma_h^H = RT\ln(1-X_s^S) + RT\ln\gamma_h^S \text{ and} \tag{8.122a}$$

$$RT\ln X_s^H + RT\ln\gamma_s^H = RT\ln X_s^S + RT\ln\gamma_s^S. \tag{8.122b}$$

Table 8.6. Experimental data on the compositions of the halite and sylvite crystalline solutions coexisting on the 1 atm solvus, and the derived values of $W_{G,i}$

T (K)	X_s^H	X_s^S	Source of data[a]	$W_{G,NaCl}$ (J mol^{-1})	$W_{G,KCl}$ (J mol^{-1})
582.15	0.021	0.889	2	11646	19310
608.15	0.020	0.880	1	11636	20347
640.15(a)	0.044	0.830	3	11090	17914
640.15(b)	0.044	0.830	3	11090	17914
664.15	0.061	0.771	2	9987	17122
673.15	0.060	0.740	1	8969	17292
695.15(a)	0.096	0.709	3	9698	16098
695.15(b)	0.096	0.712	3	9788	16106
720.15	0.137	0.634	2	9170	15365
735.15(a)	0.139	0.612	3	8793	15587
735.15(b)	0.138	0.612	3	8759	15606
738.15	0.150	0.560	1	7638	15284
739.15	0.195	0.542	2	8637	14684
745.15	0.191	0.521	2	8003	14802
769.15	0.292	0.436	3	8881	14464

[a] The sources of experimental data are: 1, Nacken (1918); 2, Bunk and Tichelaar (1953); 3, Barrett and Wallace (1954).

Employing the asymmetric Margules equation to express the $RT\ln\gamma_i$ terms in these equations, they may be rewritten as

$$0 = RT\ln\left[\frac{(1-X_s^H)}{(1-X_s^S)}\right] + W_{G,s}[2\{(X_s^H)^2(1-X_s^H) - (X_s^S)^2(1-X_s^S)\}]$$

$$+ W_{G,h}[(X_s^H)^2(2X_s^H - 1) - (X_s^S)^2(2X_s^S - 1)] \quad \text{and} \quad (8.123a)$$

$$0 = RT\ln\left[\frac{X_s^H}{X_s^S}\right] + W_{G,s}[\{(2X_s^S - 5)X_s^S + 4\}X_s^S - \{(2X_s^H - 5)X_s^H + 4\}X_s^H]$$

$$+ W_{G,h}[2\{(X_s^H - 1)^2 X_s^H - (X_s^S - 1)^2 X_s^S\}]. \quad (8.123b)$$

Defining the square-bracketed terms on the RHS of the last two equations as b_1, b_2, b_3, and a_1, a_2, a_3, respectively, they can be rewritten as

$$0 = b_1 + b_2 W_{G,s} + b_3 W_{G,h} \quad \text{and} \quad (8.123a')$$

$$0 = a_1 + a_2 W_{G,s} + a_3 W_{G,h}. \quad (8.123b')$$

Given experimental data on X^H_s and X^S_s at some T (and P), b_i and a_i are known, and we are left with two equations in two unknowns, $W_{G,h}$ and $W_{G,s}$. Solving for them, we have

$$W_{G,h} = \frac{\left(\dfrac{b_1}{b_2} - \dfrac{a_1}{a_2}\right)}{\left(\dfrac{a_3}{a_2} - \dfrac{b_3}{b_2}\right)} \quad \text{and} \tag{8.124a}$$

$$W_{G,s} = -\left(\frac{b_1}{b_2}\right) - W_{G,h}\left(\frac{b_3}{b_2}\right). \tag{8.124b}$$

Let us now address the numerical solution for $W_{G,i}$, starting from the pairs of raw X^H_s and X^S_s data listed in Table 8.6. These data were summarized earlier by Thompson and Waldbaum (1969), and were taken from Nacken (1918), Bunk and Tichelaar (1953), and Barrett and Wallace (1954). Although we shall evaluate the $W_{G,i}$ parameters, we shall not go into a rigorous estimation of their uncertainties, since only 3 out of 15 data points (those by Nacken 1918) were reversed, and in none of these cases was the width of the composition brackets unambiguously reported. To give just one example of evaluation of $W_{G,i}$, consider the pair of X^H_s and X^S_s data at 673.15 K (see Table 8.6): $X^H_s = 0.060$ and $X^S_s = 0.740$. Employing these,

$a_1 = -14060.8$ J, $a_2 = 0.810016$,   and $a_3 = 0.005984$, and
$b_1 = 7193.0$ J, $b_2 = -0.277984$,   and $b_3 = -0.266016$.

Inserting them into Eqs. (8.124a) and (8.124b), it follows that

$$W_{G,NaCl} = 8969.3 \text{ J mol}^{-1} \text{ and} \tag{8.125a}$$

$$W_{G,KCl} = 17292.4 \text{ J mol}^{-1}. \tag{8.125b}$$

The complete set of $W_{G,i}$ data is displayed in Table 8.6. The $W_{G,i}$, thus obtained, may be combined with W_V to formulate a polybaric-polythermal equation of state for (Na,K)Cl. To accomplish this, $W_{G,i}$ must be expressed as a function of T. Prior to least squares fitting of the data, duplicate $W_{G,i}$ values for three temperatures [flagged (a) and (b) in Table 8.6] were averaged. A least squares fit to the remaining 12 data points gave,

$$W_{G,NaCl} = 23730 - 20.527 \text{ T } (\pm 585) \text{ J mol}^{-1} \text{ and} \tag{8.126a}$$

$$W_{G,KCl} = 37959 - 30.954 \text{ T } (\pm 518) \text{ J mol}^{-1}, \tag{8.126b}$$

where the quantities on the RHS are $W_{H,i}$ and $W_{S,i}$, respectively. These results are in agreement with Thompson and Waldbaum's (1969) $W_{G,i}$ data. They imply that the (Na,K)Cl solution is asymmetric and both H^{ex}_m and S^{ex}_m are positive. It is striking, however, that the $W_{H,i}$ suggested by the solvus is far greater than those obtained by calorimetric measurements of $H°_{soln}$ ($W_{H,NaCl}$ = 17450 ± 684 J mol^{-1} and $W_{H,KCl}$ = 19153 ± 684 J mol^{-1}; see Eqs. (8.79a) and (8.79b). Although no calorimetric data are available on $W_{S,i}$ of (Na,K)Cl, comparison with calorimetric $W_{S,i}$ data on Na-K mixing in alkali feldspars [W_S = 10.3 ± 0.3 J K^{-1} mol^{-1} (Haselton et al. 1983)] would suggest that they are also too large. These discrepancies notwithstanding, Eqs. (8.126a) and (8.126b) may be used for phase equilibria calculations *if* we restrict ourselves to the range of T explored experimentally, that is, for interpolation. They are not recommended for the range of T beyond that of the original experiments, however. Bearing this in mind, we combine Eq. (8.126) with (8.91) to derive a polybaric-polythermal equation of state,

$$W_{G,NaCl} \text{ (J mol}^{-1}) = 23730 - 20.527\ T + 0.052\ P \text{ and} \tag{8.127a}$$

$$W_{G,KCl} \text{ (J mol}^{-1}) = 37959 - 30.954\ T + 0.052\ P, \tag{8.127b}$$

with T expressed in K and P in bar. Note that the numbers on the RHS of [(8.127a) and (8.127b)] refer to $W_{U,i}$, $W_{S,i}$, and $W_{V,i}$, $W_{U,i}$ being practically

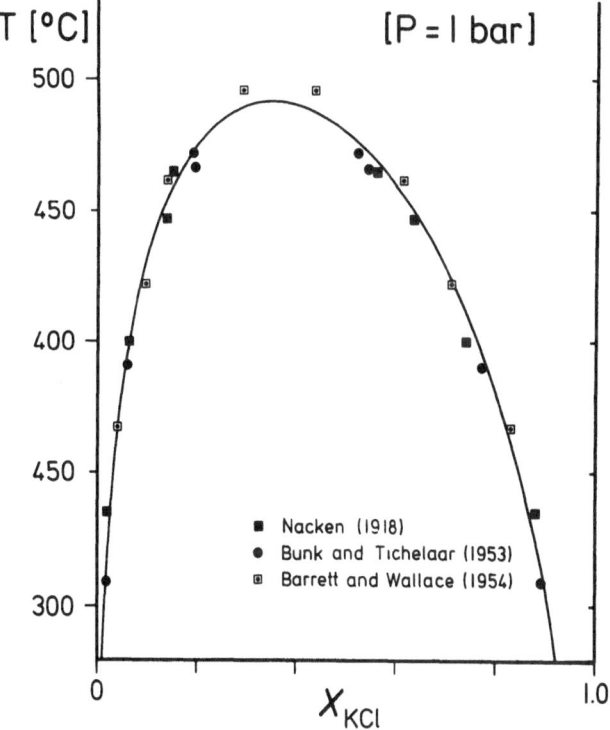

Fig. 8.15 The (Na,K)Cl solvus at 1 bar pressure, recalculated from its equation of state, Eqs.(8.127a) and (8.127b). The experimental solvus data are reproduced for comparison

identical to $W_{H,i}$ at 1 bar pressure $(W_{H,i} \equiv W_{U,i} + PW_{V,i})$. The reliability of this equation of state for smoothing the solvus data may be demonstrated by recalculating the solvus at 1 bar (for the method of computing the solvus, see Chapt. 9), and comparing it with the original experimental data (Fig. 8.15).

It has been emphasized above that this equation of state will no longer be valid at temperatures outside the range of T of the experiments on which they are based, nor will it indicate the true magnitudes of H^{ex}_m and S^{ex}_m for (Na,K)Cl solution. To deduce the correct magnitudes for H^{ex}_m and S^{ex}_m we need to apply calorimetric constraints to these data. Given independent (calorimetric) $W_{H,i}$ for (Na,K)Cl solution, we can use these to put constraints on the above equation of state. To do that, we must also postulate that $W_{H,i}$ and $W_{S,i}$ are temperature-dependent above 298.15 K. Recalling the relations,

$$\left(\frac{\partial W_{H,i}}{\partial T}\right)_P = W_{C_{p,i}} \quad \text{and} \tag{8.128a}$$

$$\left(\frac{\partial W_{S,i}}{\partial T}\right)_P = \frac{W_{C_{p,i}}}{T}, \tag{8.128b}$$

and assuming for simplicity that $W_{Cp,i}$ is constant, $W_{G,i}$ is recast to

$$W_{G,i} = [W_{H,298,i} + (T - 298)W_{C_{p,i}}] - T\left[W_{S,298,i} + W_{C_{p,i}}\ln\frac{T}{298}\right]. \tag{8.129}$$

Supposing $\sigma_{WG,i} = \sigma_{WH,i}$ and $\sigma_T = 5$ K, a least squares fit to $W_{G,i}$ data of Table 8.6, using $W_{H,i}$ [from (8.79)] as constraints, yields

$$W_{S,298,NaCl} = 7.29 \pm 1.60 \text{ J K}^{-1} \text{ mol}^{-1}, W_{C_{p,NaCl}} = 15.13 \pm 5.55 \text{ J K}^{-1} \text{ mol}^{-1}. \tag{8.130a}$$

$$W_{S,298,KCl} = 10.92 \pm 1.60 \text{ J K}^{-1} \text{ mol}^{-1}, W_{C_{p,KCl}} = 53.10 \pm 5.55 \text{ J K}^{-1} \text{ mol}^{-1}. \tag{8.130b}$$

Although $W_{S,298,i}$, thus derived, is comparable in magnitude to $W_{S,298,i}$ observed for Na-K mixing in sanidine-analbite (Haselton et al. 1983), $W_{Cp,i}$ appears to be exorbitant. Evidently, a direct calorimetric measurement of $C_p{}^{ex}_m$ at T > 298.15 is needed to solve the problem.

8.5.4.2 Excess Mixing Properties of the Monticellite-Forsterite Crystalline Solution, $(Ca,Mg)MgSiO_4$

Warner and Luth (1973) experimentally determined the solvus of the monticellite $(CaMgSiO_4, \text{mo})$-forsterite (Mg_2SiO_4, fo) binary at 1 atm, and at 1, 2, 5, and 10 kbar pressure. A lack of independent H^{ex}_m data in this system will require us to process only the phase equilibria data to arrive at the polybaric-polythermal equation of state for this crystalline solution. Before proceeding further, we note that the Ca-Mg mixing in the monticellite-forsterite solution involves the M2 site only (Warner and Luth 1973, p.1003), so that we may treat it as a single-site solution ($q = 1$).

The coexistence of a monticellite-rich (M) and a forsterite-rich (Ol) phase on the solvus, at any T and P, demands that

$$\mu_{mo}^{M} = \mu_{mo}^{Ol} \text{ and} \tag{8.131a}$$

$$\mu_{fo}^{M} = \mu_{fo}^{Ol}. \tag{8.131b}$$

Recalling that monticellite and olivine are isostructural phases, whose compositions will be expressed by the master variable X_{fo}, and choosing the pure components with appropriate structure having unit activity at system T and P as the standard state, (8.131a,b) reduce to

$$RT\ln(1-X_{fo}^{M}) + RT\ln\gamma_{mo}^{M} = RT\ln(1-X_{fo}^{Ol}) + RT\ln\gamma_{mo}^{Ol} \text{ and} \tag{8.132a}$$

$$RT\ln X_{fo}^{M} + RT\ln\gamma_{fo}^{M} = RT\ln X_{fo}^{Ol} + RT\ln\gamma_{fo}^{Ol}. \tag{8.132b}$$

To keep our treatment perfectly general at this stage, let us introduce the asymmetric Margules equation to express the $RT\ln\gamma_i$ terms of these equations. This yields,

$$RT\ln(1-X_{fo}^{M}) - RT\ln(1-X_{fo}^{Ol}) + [W_{G,mo} + 2(1-X_{fo}^{Ol})(W_{G,fo}-W_{G,mo})](X_{fo}^{M})^2$$
$$- [W_{G,mo} + 2(1-X_{fo}^{Ol})(W_{G,fo}-W_{G,mo})](X_{fo}^{Ol})^2 = 0 \text{ and} \tag{8.133a}$$

$$RT\ln X_{fo}^{M} - RT\ln X_{fo}^{Ol} + [W_{G,fo} + 2X_{fo}^{Ol}(W_{G,mo}-W_{G,fo})](1-X_{fo}^{M})^2$$
$$- [W_{G,fo} + 2X_{fo}^{Ol}(W_{G,mo}-W_{G,fo})](1-X_{fo}^{Ol})^2 = 0. \tag{8.133b}$$

Given experimental data on $X^M{}_{fo}$ and $X^{Ol}{}_{fo}$ at specified T and P, we once again have two equations in two unknowns, $W_{G,fo}$ and $W_{G,mo}$. Rather than solving for these unknowns at each pair of T and P, as was done in Sect. 8.5.4.1, we may split $W_{G,i}$ into $W_{U,i}$, $W_{S,i}$, and $W_{V,i}$, and solve them simultaneously for the six unknowns: $W_{U,mo}$, $W_{S,mo}$, $W_{V,mo}$, $W_{U,fo}$, $W_{S,fo}$, and $W_{V,fo}$ by least squares. To achieve this, Eqs. (8.133a) and (8.133b) are equated to each other and rewritten as

$$RT\ln\left[\frac{(1-X_{fo}^{M})X_{fo}^{Ol}}{(1-X_{fo}^{Ol})X_{fo}^{M}}\right]$$
$$+(W_{U,fo} - TW_{S,fo} + PW_{V,fo})\{(1-X_{fo}^{M})(3X_{fo}^{M}-1) - (1-X_{fo}^{Ol})(3X_{fo}^{Ol}-1)\}$$
$$+(W_{U,mo} - TW_{S,mo} + PW_{V,mo})\{X_{fo}^{Ol}(2-3X_{fo}^{Ol}) - X_{fo}^{M}(2-3X_{fo}^{M})\} = 0 . \tag{8.134}$$

Warner and Luth (1973, Table 4) reported 64 pairs of solvus compositions, $X^M{}_{fo}$ and $X^{Ol}{}_{fo}$, for temperatures between 800-1450°C at 1 bar and at 1, 2, 5, and 10 kbar pressure. Using them, (8.134) may be solved for the six

Table 8.7. Composition dependence of molar volume, V, on the join monticellite-forsterite, computed from cell volume measurements by Warner and Luth (1973)

X_{fo}	$V_{cell}(\text{Å}^3)$	Number of observations	$V(\text{J bar}^{-1} \text{ mol}^{-1})$
0.010	340.740(40)	1	5.1299(6)
0.055	338.590(50)	1	5.0976(8)
0.065	338.340(70)	1	5.0938(11)
0.110	335.840(40)	1	5.0562(6)
0.164	332.960(60)	1	5.0128(9)
0.218	330.560(70)	1	4.9767(11)
0.820	297.000(40)	1	4.4714(6)
0.863	295.480(40)	1	4.4485(6)
0.909	292.815(67)	2	4.4084(10)
0.955	291.400(58)	3	4.3871(9)
1.000	290.073(69)	3	4.3671(10)

unknowns. Before starting with that, we note that the experimental solvus is highly symmetrical, although the V-X_{fo} data (Warner and Luth 1973, Table 8.7) do show a slight asymmetry. The observed symmetrical solvus may be accounted for by entering the following two constraints into Eq. (8.134) prior to data reduction:

$$W_{U,mo} = W_{U,fo} \text{ and } W_{S,mo} = W_{S,fo}. \tag{8.135}$$

In other words, the least squares reduction was carried through to yield four parameters: $W_{U,mo}(= W_{U,fo})$, $W_{S,mo}(= W_{S,fo})$, $W_{V,mo}$ and $W_{V,fo}$. The results, termed here Model 1, are listed in Table 8.8. Though the uncertainties of $W_{U,i}$ and $W_{S,i}$ are reasonable, those of $W_{V,i}$ are huge. Indeed, $W_{V,i}$ itself is too large to be reconcilable with the slight deviation from ideality observed (Warner and Luth 1973, p.1000).

To analyze the cause of the discrepancy between the calculated and the observed V-X_{fo} data, we focus on the cell volumes given by Warner and Luth (1973, Table 3). Table 8.7 reproduces their molar volume (V)-composition (X_{fo}) data, averaging each set of multiple observations. Using the data fitting procedure outlined in Sections 8.5.2.1 and 8.5.2.3, the V-X_{fo} data from Table 8.7 give

$$W_{V,mo} = -0.2740 \pm 0.0308 \text{ J bar}^{-1} \text{ mol}^{-1} \text{ and} \tag{8.136a}$$

$$W_{V,fo} = 0.1029 \pm 0.0291 \text{ J bar}^{-1} \text{ mol}^{-1}. \tag{8.136b}$$

These results corroborate our earlier suspicion; the true $W_{V,i}$ are much smaller than those obtained by least squares analysis of the compositional data (see Table 8.8, Model 1). The $W_{V,i}$ from Eqs. (8.136a,b) were used in the next

Table 8.8. The Margules parameters for the monticellite-forsterite solution, computed by least squares technique

Parameters	Model 1	Model 2
$W_{U,mo}$ (J mol^{-1})	4.7185E4 (0.1634E4)	5.4200E4 (0.1893E4)
$W_{S,mo}$ (J K^{-1} mol^{-1})	8.606 (1.005)	12.513 (1.225)
$W_{V,mo}$ (J bar^{-1} mol^{-1})	2.6580 (1.0890)	-0.2740 (0.0308)
$W_{U,fo}$ (J mol^{-1})	4.7185E4 (0.1634E4)	5.4200E4 (0.1893E4)
$W_{S,fo}$ (J K^{-1} mol^{-1})	8.606 (1.005)	12.513 (1.225)
$W_{V,fo}$ (J bar^{-1} mol^{-1})	-2.0505 (1.1670)	0.1029 (0.0291)

cycle of least squares data processing. That is, a total of 64 data points were used to solve for two unknowns, $W_{U,mo}(=W_{U,fo})$ and $W_{S,mo}(=W_{S,fo})$. These results are listed in Table 8.8 as Model 2. Using them, the polybaric-polythermal equation of state for the monticellite-forsterite solution may be written as

$$W_{G,mo}(P,T) = 54200 - 12.513\ T - 0.2740\ P\ \text{and} \qquad (8.137a)$$

$$W_{G,fo}(P,T) = 54200 - 12.513\ T + 0.1029\ P, \qquad (8.137b)$$

with T given in K and P in bar. It should be quite clear that the solution is very slightly asymmetric, because the molar volume is slightly asymmetric. As in the case of (Na,K)Cl, the veracity of this equation of state was also affirmed by computing the solvus.

8.5.4.3 Thermodynamic Mixing Properties of $Zn(Al,Cr)_2O_4$ Spinel

The second variety of phase equilibria experiments, often used to obtain the thermodynamic mixing properties of binary solutions, studies cation (or anion) fractionation between a suitably chosen pair of binary solutions at a specified T and P. The thermodynamic theory underlying such experiments, though developed below in the context of Al-Cr exchange experiments between $Zn(Al,Cr)_2O_4$ spinel (Sp) and $(Al,Cr)_2O_3$ corundum (Co), is of general validity, and may be applied to any other reciprocal ternary system comprising two solutions.

Thermodynamic treatment of the reciprocal system Al_2O_3-Cr_2O_3-$ZnAl_2O_4$-$ZnCr_2O_4$ requires a knowledge of the cation-mixing process in both the solutions, Sp and Co. The corundum structure has two equivalent [6]-sites (q = 2) on which Al and Cr mix in a random manner. The structure of spinel, by contrast, comprises both [6]- and [4]-sites. However, due to a high tetrahedral site-preference energy of Zn and a high octahedral site-preference energy of Cr (see O'Neil and Navrotsky 1984), little Zn enters the [6]-site and practically no Cr goes into the [4]-site (Bruckmann-Benke et al. 1988, Fig.1). Therefore, the $Zn(Al,Cr)_2O_4$ spinel may also be treated as a simple two-site (q = 2) solution involving mixing of Al and Cr on [6]-sites. Writing an Al-Cr exchange reaction between Sp and Co in terms of the q-normalized mixing units, $Zn_{0.5}AlO_2$ (gahnite, ga), $Zn_{0.5}CrO_2$ (zincochromite[9], zc), $AlO_{1.5}$ (corundum, co), and $CrO_{1.5}$ (eskolaite, es),

$$Zn_{0.5}AlO_2 + CrO_{1.5} = Zn_{0.5}CrO_2 + AlO_{1.5}, \qquad (8.138)$$
(ga in Sp)   (es in Co)   (zc in Sp)   (co in Co)

the condition of equilibrium for reaction (8.138) is written as

$$0 = \mu_{zc}^{Sp} + \mu_{co}^{Co} - \mu_{ga}^{Sp} - \mu_{es}^{Co}. \qquad (8.139)$$

Note that the subscripts in this equation unambiguously identify the phases, making the superscripts redundant; therefore, for the sake of brevity, their use will be discontinued here. Choosing the pure components with the appropriate structures (normal spinel for ga and zc, and corundum for co and es) having unit activity at any T and P as the standard state and splitting the chemical potential terms into their constituent parts, (8.139) may be expanded to

$$0 = [\mu_{zc}^* + \mu_{co}^* - \mu_{ga}^* - \mu_{es}^*] + RT\ln\left[\frac{X_{zc}(1 - X_{es})}{(1 - X_{zc})X_{es}}\right]$$
$$+ [RT\ln\gamma_{co} - RT\ln\gamma_{es}] + [RT\ln\gamma_{zc} - RT\ln\gamma_{ga}]. \qquad (8.140)$$

The last two square-bracketed quantities on the RHS of (8.140) sum the nonideality terms due to Co and Sp, respectively. Thus, if we wish to derive the mixing properties of spinel, as Bruckmann-Benke et al. (1988) did, the equation of state for the corundum solution should be known. Bruckmann-Benke et al. (1988) used the equation of state for corundum from Chatterjee et al. (1982), to solve for the properties of the spinel, $Zn(Al,Cr)_2O_4$. That is, Co served as a "known", and spinel as an "unknown" quantity. A general strategy to pick a "known" solution follows from the above requirement.[10]

    Bruckmann-Benke et al. (1988) experimentally obtained nine tie lines between Sp and Co at 900, 1100, and 1300°C for a constant pressure of 25 kbar. Figure 8.16 shows their data at 1100°C. Thus, at each isobar-isotherm, X_{zc} and X_{es} are known for each tie line. Also known is $[RT\ln\gamma_{co} - RT\ln\gamma_{es}]$ for

---

[9] Zincochromite is a naturally occurring spinel of near end-member $ZnCr_2O_4$ composition (Nesterov and Rumyantseva 1987).

[10] If an ideally behaving "known" solution can be found, we are rid of at least one of the terms contributing to the uncertainty in the derived mixing properties of the "unknown" phase. Because the dilute aqueous solutions of NaCl and KCl are virtually ideal (see Thompson and Waldbaum 1968), they have been employed as the "known" phase for Na-K ion-exchange studies with alkali feldspar (Orville 1963) or with nepheline-kalsilite solution (Zyrianov et al. 1978).

each tie line, since it can be evaluated from the equation of state for Co given by Chatterjee et al. (1982). Noting that the first square-bracketed term in (8.140) is ΔG^* of the reaction (8.138), and that the second one is K_D, the distribution coefficient, we may rearrange Eq. (8.140) as

$$-RT\ln K_D - [RT\ln\gamma_{co} - RT\ln\gamma_{es}] = \Delta G^* + [RT\ln\gamma_{zc} - RT\ln\gamma_{ga}]. \qquad (8.141)$$

This equation sums the known quantities on the LHS, retaining the unknowns on the RHS. Note that although, in principle, it is also possible that ΔG^* is independently known, this is not true in this case. Prior to solving for the unknowns, we insert the expressions for $RT\ln\gamma_i$ from Eqs.(8.15a,b) into the RHS of (8.141),

$$\begin{aligned} -RT\ln K_D - [RT\ln\gamma_{co} - RT\ln\gamma_{es}] &= \Delta G^* + [W_{G,zc} + 2X_{zc}(W_{G,ga} - W_{G,zc})](1-X_{zc})^2 \\ &\quad - [W_{G,ga} + 2(1-X_{zc})(W_{G,zc} - W_{G,ga})]X_{zc}^2, \qquad (8.142) \end{aligned}$$

and recast it to

$$\begin{aligned} -[RT\ln K_D + (RT\ln\gamma_{co} - RT\ln\gamma_{es})] \\ = (\Delta G^* + W_{G,zc}) + 2(W_{G,ga} - 2W_{G,zc})X_{zc} + 3(W_{G,zc} - W_{G,ga})X_{zc}^2. \qquad (8.143) \end{aligned}$$

Defining

$$(\Delta G^* + W_{G,zc}) \equiv a_0, \qquad (8.144a)$$

$$2(W_{G,ga} - 2W_{G,zc}) \equiv a_1, \text{ and} \qquad (8.144b)$$

$$3(W_{G,zc} - W_{G,ga}) \equiv a_2, \qquad (8.144c)$$

Eq.(8.143) may be rewritten as

$$-[RT\ln K_D + (RT\ln\gamma_{co} - RT\ln\gamma_{es})] = a_0 + a_1X_{zc} + a_2X_{zc}^2. \qquad (8.145)$$

It is evident from (8.145) that, given data on the LHS for any set of T and P, we may fit them to a polynomial in X_{zc}. Should a_2 be nonzero, we are dealing with an asymmetric solution; if a_2 is zero but a_1 is nonzero, the solution is symmetric. If both a_1 and a_2 are zero, the solution is ideal. Solving Eqs.(8.144b) and (8.144c) simultaneously, we obtain

$$W_{G,ga} = -\frac{a_1}{2} - \frac{2a_2}{3} \quad \text{and} \qquad (8.146a)$$

$$W_{G,zc} = -\frac{a_1}{2} - \frac{a_2}{3}. \qquad (8.146b)$$

The above format for *isothermwise* evaluating compositional data of a reciprocal ternary system was introduced and popularized by Thompson and

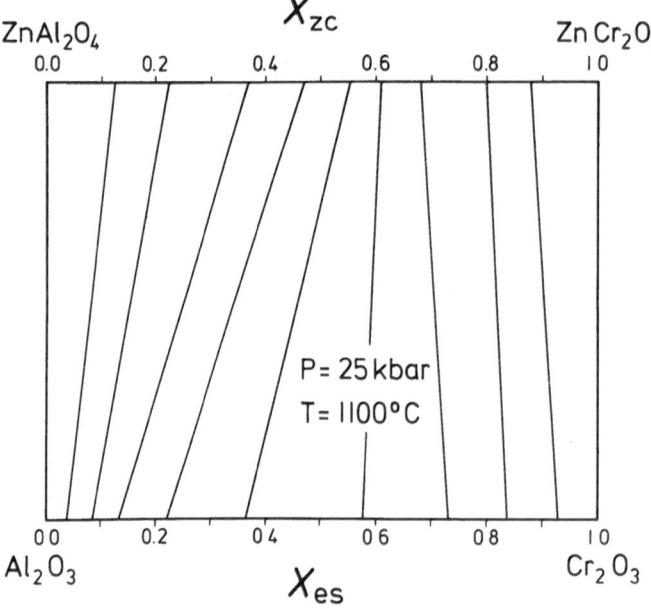

Fig. 8.16 Experimental Sp-Co tie lines in the reciprocal ternary $ZnAl_2O_4$-$ZnCr_2O_4$-Al_2O_3-Cr_2O_3 (after Bruckmann-Benke et al. 1988) at 25 kbar and 1100°C

Waldbaum (1968). We shall apply it to calculate the mixing properties for $Zn(Al,Cr)_2O_4$ spinels. The compositional data by Bruckmann-Benke et al. (1988) are reproduced in Table 8.9. Also given are $-RTlnK_D$ ($\equiv$ A), σ_A, $-[RTln\gamma_{co}-RTln\gamma_{es}]$ ($\equiv$ B), σ_B, A+B, and $\sigma_{(A+B)}$. The σ_A was obtained by propagating σ_T, σ_{Xzc}, and σ_{Xes}, whereas only σ_T and σ_{Xes} was propagated to calculate σ_B, ignoring any $\sigma_{WG,co}$, and $\sigma_{WG,es}$.

The results of the least-squares processing of the 1100°C, 25 kbar data, are reproduced in Table 8.10. In the quadratic fit, σ_{a2} turns out to be larger than a_2 (left column), suggesting that the $Zn(Al,Cr)_2O_4$ spinel must be a symmetric solution. The linear fit to the data (right column) supports this notion. The evaluation of the 900°C Al-Cr exchange-isotherm also vindicates this conclusion, although the 1300°C data do suggest a marginal asymmetry. Chances are that these spinels are slightly asymmetric, but that asymmetry is drowned by the noise of the compositional data. In the absence of more compelling evidence to the contrary, we shall handle the $Zn(Al,Cr)_2O_4$ spinels as a symmetric solution. Having established the symmetric nature of the $Zn(Al,Cr)_2O_4$ spinel, and substituting $W_{G,zc} = W_{G,ga} \equiv W_G$ into (8.143), this equation simplifies to

$$0 = [RTlnK_D + (RTln\gamma_{co}-RTln\gamma_{es})] + \Delta G^* + W_G(1-2X_{zc}). \tag{8.147}$$

In order to explore any temperature dependence of ΔG^* and W_G, the former is split into ΔH^* and $T\Delta S^*$, and the latter into W_H and TW_S,

$$0 = [RTlnK_D + (RTln\gamma_{co}-RTln\gamma_{es})] + \Delta H^* - T\Delta S^* + (W_H - TW_S)(1-2X_{zc}). \tag{8.148}$$

Table 8.9. Experimental data on the compositions of the coexisting spinel and corundum crystalline solutions at 25 kbar and at 900, 1100, and 1300°C (Bruckmann-Benke et al. 1988). Abbreviations: $-RT\ln K_D$ ($\equiv A$) and $-[RT\ln\gamma_{co}-RT\ln\gamma_{es}]$ ($\equiv B$)

T (K)	X_{zc}	X_{es}	A J mol^{-1}	σ_A J mol^{-1}	B J mol^{-1}	σ_B J mol^{-1}	A + B J mol^{-1}	$\sigma_{(A+B)}$ J mol^{-1}
1173.15	0.1178(18)	0.0312(16)	-13872	556	20222	75	6350	561
1173.15	0.2228(13)	0.0554(27)	-15477	525	19037	122	3560	539
1173.15	0.3767(17)	0.1198(19)	-14541	226	15937	84	1396	242
1173.15	0.4727(17)	0.2095(22)	-11886	177	11748	93	-138	200
1173.15	0.5462(17)	0.3576(55)	-7521	251	5162	215	-2359	331
1173.15	0.5710(17)	0.5797(33)	347	148	-3943	115	-3595	188
1173.15	0.6387(28)	0.7348(54)	4383	297	-9752	173	-5369	344
1173.15	0.7273(38)	0.8485(12)	7237	217	-13724	39	-6488	220
1173.15	0.8577(51)	0.9348(32)	8452	658	-16577	93	-8125	665
1373.15	0.1224(10)	0.0383(05)	-14309	229	20277	34	5967	231
1373.15	0.2228(13)	0.0844(14)	-12953	253	17997	67	5044	262
1373.15	0.3677(14)	0.1331(11)	-15204	189	15633	53	428	196
1373.15	0.4686(11)	0.2207(17)	-12967	171	11489	74	-1478	186
1373.15	0.5537(12)	0.3657(18)	-8749	131	4943	72	-3806	150
1373.15	0.6071(06)	0.5776(15)	-1395	77	-3922	54	-5317	94
1373.15	0.6794(17)	0.7318(27)	2886	182	-9850	90	-6965	203
1373.15	0.7992(15)	0.8387(09)	3052	134	-13702	34	-10650	138
1373.15	0.8797(27)	0.9298(15)	6782	397	-16817	51	-10035	400
1573.15	0.1110(03)	0.0375(11)	-15233	426	20720	60	5487	430
1573.15	0.2314(05)	0.0833(11)	-15668	243	18418	57	2750	250
1573.15	0.3588(08)	0.1519(14)	-14900	206	15042	67	142	217
1573.15	0.4621(05)	0.2414(06)	-12990	134	10767	31	-2223	137
1573.15	0.5305(06)	0.4054(29)	-6607	172	3313	116	-3294	208
1573.15	0.5981(12)	0.5732(57)	-1342	312	-3804	210	-5146	376
1573.15	0.6754(09)	0.7264(12)	3188	100	-9851	44	-6664	109
1573.15	0.7740(06)	0.8384(09)	5432	111	-14001	37	-8569	117
1573.15	0.8863(09)	0.9345(09)	7906	237	-17379	41	-9472	241

Table 8.10. Results of least squares reduction of the 1100°C, 25 kbar data presented in Table 8.9, using the format of Eq. (8.145)

Fit parameters	Quadratic fit to X_{zc}	Linear fit to X_{zc}
a_0 [$\equiv \Delta G^* + W_{G,zc}$]	9919 ± 1219 J mol^{-1}	9308 ± 653 J mol^{-1}
a_1 [$\equiv 2(W_{G,ga}-2W_{G,zc})$]	-26666 ± 5373 J mol^{-1}	-23525 ± 1138 J mol^{-1}
a_2 [$\equiv 3(W_{G,zc}-W_{G,ga})$]	3143 ± 5240 J mol^{-1}	-

Table 8.11. The results of simultaneous least-squares processing of the polythermal compositional data summarized in Table 8.9. For the nature of the solution models A and B, see text

Parameters	Model A	Model B
ΔH^* (J)	1.629E3 ± 1.280E3	1.618E3 ± 1.253E3
ΔS^* (J K^{-1})	2.830 ± 0.926	2.821 ± 0.901
W_H (J mol^{-1})	1.138E4 ± 0.280E4	1.0682E4 ± 0.0317E4
W_S (J K^{-1} mol^{-1})	0.506 ± 2.013	-

Given data on $[RT\ln K_D + (RT\ln \gamma_{co} - RT\ln \gamma_{es})]$ [$\equiv$ -(A+B) of Table 8.9] as a function of X_{zc} at 900, 1100, and 1300°C, all at 25 kbar, we may solve (8.148) for the four unknowns ΔH^*, ΔS^*, W_H, and W_S by least squares. The results of processing the polythermal data are summarized in Table 8.11 (Model A). It is readily seen that this model is inadequate, the computed uncertainty of W_S being larger than W_S itself. Therefore, the subsequent cycle of data reduction (Model B) constrained W_S to zero. These results are also indicated in Table 8.11. Clearly, the latter model is more satisfactory.

Utilizing Model B (Table 8.11) and the volume data provided by Bruckmann-Benke et al. (1988), and bearing in mind that ΔG^* and W_G refer to the q-normalized mixing units, the thermodynamic properties derived above may be summarized as

$$\Delta G^*(P,T) = 1618 - 2.821\ T + 0.0081\ P \text{ and} \tag{8.149}$$

$$W_G(P,T) = 10682 + 0.0193\ P, \tag{8.150}$$

with T given in K and P in bar.

8.5.4.4 Solution Modeling of Grossular-Almandine Garnets

A third category of phase equilibria experiments, frequently used to obtain the equation of state for a binary crystalline solution, is exemplified by the work of Cressey et al. (1978) on almandine-grossular garnet, $(Fe,Ca)_3Al_2[SiO_4]_3$. At a given T, they measured the displacement of the equilibrium P of the reaction

$$\underset{\text{(Gr in Gt)}}{1\ Ca_3Al_2[SiO_4]_3} + \underset{\text{(Ky/Sill)}}{2\ Al_2SiO_5} + \underset{\text{(Qtz)}}{1\ SiO_2} = \underset{\text{(An)}}{3\ Ca[Al_2Si_2O_8]}, \tag{8.151}$$

as a function of the mole fraction of the $Ca_3Al_2[SiO_4]_3$ component, X_{Gr}, in garnet. At a specified T, the equilibrium P of the solid-solid reaction was first measured for the end-member phases. Next, an Al_2SiO_5 (Ky/Sill), quartz, and anorthite were equilibrated with the grossular-almandine crystalline solution at some other (lower) pressure. Following termination of the run, the final composition of the garnet was determined by X-rays.

The condition of equilibrium for the *end-members* reaction is

$$0 = \Delta G_T^\circ + \int_1^P \Delta V_T(P)dP.$$ (8.152)

For the same equilibrium involving a *garnet crystalline solution*, and provided that the other phases remain pure end-members, the condition of equilibrium becomes

$$0 = \Delta G_T^\circ + \int_1^{P_x} \Delta V_T(P)dP - RTln a_{Gr}(P_x, T).$$ (8.153)

Note that P and P_x in these equations are equilibrium P for the reactions involving the end-member phases and the garnet solution, respectively. Moreover, little or no Fe^{2+} enters the structures of Al_2SiO_5, quartz, and anorthite in equilibrium with the garnet; and thus, we may regard them as essentially pure end-members. Equating (8.152) with (8.153), an expression for a_{Gr} is obtained,

$$RTln a_{Gr}(P_x, T) = - \int_{P_x}^P \Delta V_T(P)dP.$$ (8.154)

Recalling that Ca-Fe mixing involves three sites in a garnet (q = 3), we may rewrite the activity as a_{gr}, where gr is the normalized mixing unit $CaAl_{2/3}[SiO_4]$. Splitting a_i into X_i and γ_i, (8.154) is recast to

$$3RTln\gamma_{gr}(P_x, T) = - \int_{P_x}^P \Delta V_T(P)dP - 3RTlnX_{gr}.$$ (8.155)

It is important to realize at this stage that γ_{gr}, thus obtained, refers to the equilibrium pressure P_x (and T), at which the garnet crystalline solution was equilibrated with the other phases. To handle all γ_{gr} data simultaneously, they need to be recalculated to some standard pressure, which will be 1 bar in this case. To do this, we proceed as follows. Recalling Eqs. (1.67) and (1.9), and considering $v^{ex}_{m,i}$ as independent of T and P, we may write

$$3 RTln\gamma_{gr}(P_x,T) = 3 RTln\gamma_{gr}(1 \text{ bar},T) + (P_x-1)v^{ex}_{m,Gr},$$ (8.156)

which, upon rearranging, yields

$$3 RTln\gamma_{gr}(1 \text{ bar},T) = 3 RTln\gamma_{gr}(P_x,T) - (P_x-1)v^{ex}_{m,Gr}.$$ (8.157)

The $v^{ex}_{m,Gr}$ data, needed to recast γ_{gr} to 1 bar pressure, follows from V-X_{Gr} data of Cressey et al. (1978), which were expressed in the Margules form by Newton et al. (1989),

$$V^{ex}_m = (1-X_{Gr})X_{Gr}[W_{V,Alm}X_{Gr} + W_{V,Gr}(1-X_{Gr})], \tag{8.158}$$

with $W_{V,Alm} = 0.30$ J bar^{-1} mol^{-1} and $W_{V,Gr} = 0.00$ J bar^{-1} mol^{-1}. Applying the tangent intercept formalism to (8.158), $v^{ex}_{m,Gr}$ turns out to be:

$$v^{ex}_{m,Gr} = 2W_{V,Alm}X_{Gr}(1-X_{Gr})^2 + W_{V,Gr}(1-2X_{Gr})(1-X_{Gr})^2. \tag{8.159}$$

The experimental data points, documented in Table 2 of Cressey et al. (1978), were used to compute 1 bar $RT\ln\gamma_{gr}$. These were then converted to $RT\ln\gamma_{gr}/(1-X_{gr})^2$, and listed in Table 8.12. From an inspection of these data it is clear that $RT\ln\gamma_{gr}/(1-X_{gr})^2$ becomes negative as X_{gr} goes to zero, implying that $W_{G,gr}$ is negative and $W_{G,alm}$ is positive. This is compatible with the calorimetric data of Geiger et al. (1987), according to which, $W_{H,Gr} = -9080$ J mol^{-1} and $W_{H,Alm}$

Table 8.12. The calculation of $RT\ln\gamma_{gr}/(1-X_{gr})^2$ at 1 bar pressure, based on the experimental data by Cressey et al. (1978). Note that only those data, used later in obtaining the equation of state for the garnet, have been listed here

T (K)	P_x (bar)	Stable Al$_2$SiO$_5$ is	X_{gr}	P (bar)	$-\int_{P_x}^{P}\Delta VdP$ (J)	$RT\ln\gamma_{gr}/(1-X_{gr})^2$ (J)
1373.15	11600	Sill	0.12	27407	-76245	-1839
1373.15	15000	Sill	0.18	27407	-59337	-813
1373.15	20400	Ky	0.35	25474	-29671	3531
1373.15	20800	Ky	0.37	25474	-27309	4125
1373.15	21300	Ky	0.40	25474	-24361	4798
1373.15	22000	Ky	0.48	25474	-20246	3912
1373.15	22300	Ky	0.52	25474	-18485	3340
1373.15	22700	Ky	0.57	25474	-16142	3020
1373.15	22700	Ky	0.59	25474	-16142	1148
1273.15	12000	Sill	0.14	25074	-63366	-755
1273.15	12500	Sill	0.15	25074	-60867	-662
1273.15	14000	Sill	0.20	25074	-53406	-1756
1273.15	14600	Sill	0.20	25074	-50437	-234
1273.15	17900	Ky	0.30	23178	-31179	3725
1273.15	17900	Ky	0.31	23178	-31179	3100
1273.15	18500	Ky	0.34	23178	-27600	3837
1273.15	19500	Ky	0.40	23178	-21654	5332
1273.15	20000	Ky	0.44	23178	-18691	6085
1273.15	20300	Ky	0.45	23178	-16915	7476
1273.15	22000	Ky	0.62	23178	-6899	16390

= 13680 J mol^{-1}. Thus, instead of handling Cressey et al. (1978) data isothermwise, they can be reduced simultaneously, using the data of Geiger et al. (1987) on $W_{H,i}$ as constraints. For this purpose, $RT\ln\gamma_{gr}/(1-X_{gr})^2$ must be expressed as a function of X_{gr} and T. This is achieved in two steps. First, $RT\ln\gamma_{gr}/(1-X_{gr})^2$ is recast as a function of X_{gr},

$$\frac{RT\ln\gamma_{gr}}{(1-X_{gr})^2} = W_{G,gr} + 2X_{gr}(W_{G,alm} - W_{G,gr}),$$ (8.160)

and then, $W_{G,i} = W_{Hi} - TW_{S,i}$ is substituted into this equation,

$$\frac{RT\ln\gamma_{gr}}{(1-X_{gr})^2} = (W_{H,gr} - TW_{S,gr}) + 2X_{gr}[(W_{H,alm} - TW_{S,alm})$$
$$- (W_{H,gr} - TW_{S,gr})].$$ (8.161a)

The final equation, to be used for a least squares data reduction, is obtained by rearranging (8.161a) to

$$\frac{RT\ln\gamma_{gr}}{(1-X_{gr})^2} = W_{H,gr}(1 - 2X_{gr}) - TW_{S,gr}(1 - 2X_{gr})$$
$$+ 2W_{H,alm}X_{gr} - 2TW_{S,alm}X_{gr}.$$ (8.161)

Constraining $W_{H,gr}$ to -3027 J mol^{-1}, and $W_{H,alm}$ to 4560 J mol^{-1}, and using the 31 data points from Cressey et al. (1978, Table 2), of which only 20 are reproduced in Table 8.12 (for an explanation, see later), a least squares treatment gave

$W_{S,alm} = 1.985 \pm 0.716$ J K^{-1} mol^{-1} and $W_{S,gr} = 1.096 \pm 1.222$ J K^{-1} mol^{-1}. (8.162a)

Newton et al. (1989), however, noted that 11 data points from the 1173.15 and 1123.15 K isotherms contradict their H^{ex}_m data. Thus, in a second cycle of data reduction, these 11 were eliminated, and the remaining 20 (reproduced in Table 8.12) processed. That gave,

$W_{S,alm} = 0.924 \pm 0.691$ J K^{-1} mol^{-1} and $W_{S,gr} = -0.301 \pm 1.297$ J K^{-1} mol^{-1}. (8.162)

In other words, the excess entropy on the join almandine-grossular is practically zero, as already concluded by Newton et al. (1989). Noting that the V^{ex}_m given earlier applies to the formula unit $(Ca,Fe)_3Al_2[SiO_4]_3$, the equation of state for $(Ca,Fe)Al_{2/3}SiO_4$ can be written as

$W_{G,alm}(P,T)$ [J mol^{-1}] = 4560 + 0.1 P, and (8.163a)

$W_{G,gr}(P,T)$ [J mol^{-1}] = -3026 + 0.0 P, (8.163b)

with P expressed in bar.

8.6 An Epilogue

In concluding this chapter, we emphasize that although we have analyzed only a few types of reversal experiments to come up with equations of state for crystalline solutions, many others can, and have been, devised to achieve this goal. Indeed, the only sensible approach to design experiments is on the basis of thermodynamic reasoning. Experiments are commonly expensive and time-consuming. Theoretical considerations prior to experimenting often help prune redundant efforts.

Chapter 9

Phase Equilibria Involving Nonideal Solutions and Outlines of Geothermometry and Geobarometry

9.1 Introduction and Scope

Given the thermodynamic data for pure substances (mineral end-members or pure fluid species) and the equations of state for the solutions, the chemical potentials of the components in any phase, solid or fluid, are readily computed at any pressure, temperature, and composition. The knowledge of the chemical potentials sets the stage for the computation of phase diagrams comprising nonideal solutions. A phase diagram predicted from thermodynamic data may be compared with mineral assemblages observed in nature to gain insight into the conditions of the formation of rocks and mineral deposits. The computed phase diagram may also be utilized to plan new phase equilibria experiments. Alternatively, knowledge of the chemical potentials permits recovery of intensive parameters (T, P, a_i, etc.) from compositional data on the coexisting minerals of rocks, providing the observed compositions reflect the condition of equilibrium. The extraction of the temperature of equilibration is geothermometry; geobarometry is the calculation of the pressure of equilibrium for any assemblage. Because the T and P of solid-solid reactions are independent of the activity or fugacity of fluids prevalent during equilibration, solid-solid equilibria form the backbone of geothermometry and geobarometry. Once temperature and pressure of equilibration of an assemblage are known, a solid-fluid equilibrium may be used to evaluate a_i (or f_i) of the fluid i present during equilibration.

Both the above aspects of the problem will be addressed in the following, albeit in broad outline. A detailed treatment of either of them is beyond the scope of this book. A few examples will be given for the calculation of phase equilibria, emphasizing the principles inherent to them. This will be followed by a treatment of geothermometry and geobarometry. The emphasis will again be on the basic principles. Worked examples will be given to wrap up the chapter.

9.2 Calculation of Heterogeneous Phase Equilibria Involving Nonideal Solutions

9.2.1 General Considerations

Depending upon the thermodynamic database to be employed for computing heterogeneous phase relations among solution phases in a multicomponent system, we may adopt two alternative strategies to calculate geological phase diagrams. To the extent that we have a complete set of internally consistent thermodynamic data for the end-members *and* the solutions, the elegant and versatile strategy is to apply the technique of Gibbs energy minimization. This type of database (for an example, see Sack and Ghiorso 1989), is rather rare at present. For this reason, the worked examples given below (Sect. 9.2.3) will have to do with whatever data are available at the present state of development of mineralogical thermodynamics. Therefore, we need not venture into the details of the technique of Gibbs energy minimization, but acquaint ourselves with it in a qualitative manner. Because of its versatility, however, it is the method of choice for the future; accordingly, current research is abuzz with attempts to refine the algorithms, and make them available for routine use by geologists (e.g. Brown and Skinner 1974; Wood and Holloway 1984; de Capitani and Brown 1987).

9.2.2 A Qualitative Look at the Gibbs Energy Minimization Method

9.2.2.1 Theoretical Basis

From basic thermodynamics, it is well known that the criterion of spontaneity and equilibrium at any T and P may be written as (e.g. Klotz and Rosenberg 1972, p.130)

$$dG_{sys} \leq 0, \tag{9.1}$$

where G_{sys} denotes the Gibbs energy of the system. Here, the sign of dG_{sys} indicates the direction of a spontaneous change, whereas the equality applies to the state of equilibrium. By implication, at constant T and P, an equilibrium is characterized by a minimum in G_{sys}.

Consider a c-component (i = 1,2,...c) N-phase (j = 1,2,...N) system. At any T and P, the molar Gibbs energy of the phase j, G^j, may be written by analogy to Eq.(1.35) as

$$nG^j = \sum_{i=1}^{c} n_i^j \mu_i^j, \tag{9.2}$$

n_i^j being the number of moles of the component i in the j-th phase and μ_i^j the chemical potential of i in phase j. The Gibbs energy of the system, G_{sys}, is then obtained by summing over N phases of the system,

$$G_{sys} = \sum_{j=1}^{N} (\sum_{i=1}^{c} n_i^j \mu_i^j). \qquad (9.3)$$

This G_{sys} is minimized at constant T and P to identify the stable assemblage. In actual practice, G_{sys} is expressed in terms of X_i^j, not n_i^j, so that the compositions of the phases in equilibrium may also be derived in terms of the mole fractions of i.

The computation of a phase diagram by the method of Gibbs energy minimization may be envisioned as a two-step process. The first step considers the G-X section of the system at a specified set of P and T, and solves for the stable phase assemblage and the phase compositions on that isobar-isotherm. In a second step, the phase composition and phase assemblage data from a series of isobar-isotherms are put together to generate the P-T-X phase diagram (P-T, T-X, and P-X sections).

9.2.2.2 Graphical Analysis of Isobaric-Isothermal G-X Sections

From the theoretical exposition given above, it is immediately apparent that a G-X section through the system of interest at any specified T and P is best suited to illustrate what the technique of Gibbs energy minimization strives to achieve. Figure 9.1 shows two schematic isobaric-isothermal G-X_B diagrams for a binary A-B, having two non-isostructural phases 1 and 2. A pair of G-X_B curves appears on the binary, labeled 1 and 2, corresponding to those two phases. The stable state of the pure component A is the structure 2, that of the pure component B is the structure 1.

The two questions we need to answer are: (1) How do we predict the stable assemblage for a specified bulk composition at a given P and T? (2) How do we obtain the compositions of the coexisting phases in the stable assemblage? To the extent that we are dealing with a binary system, we may, in principle, solve our problems by very simple graphical means. Recalling that at equilibrium G_{sys} attains a minimum, we *graphically* determine the *lowest level* of G_{sys} for the specified *bulk* composition. This may be accomplished by starting from any initial energy level, which is below both the G-X curves. This level has been depicted by the line (a) in Fig. 9.1A. The bulk composition of the system is indicated by an open circle on that line. The next step is to raise this energy level parallel to itself, until it hits either of the two G-X curves. In Fig. 9.1A, this state is shown by the line (b). Note that (b) is now tangential to the G-X curve of the phase 2, X^b_{B2} denoting the point of contact of the two curves. If X^b_{B2} were the specified bulk composition of the system, line (b) would already be the lowest energy level we are looking for. Since the bulk composition is not identical to X^b_{B2}, it will be possible to lift the energy level (b) still further, although no longer parallel to itself. In fact,

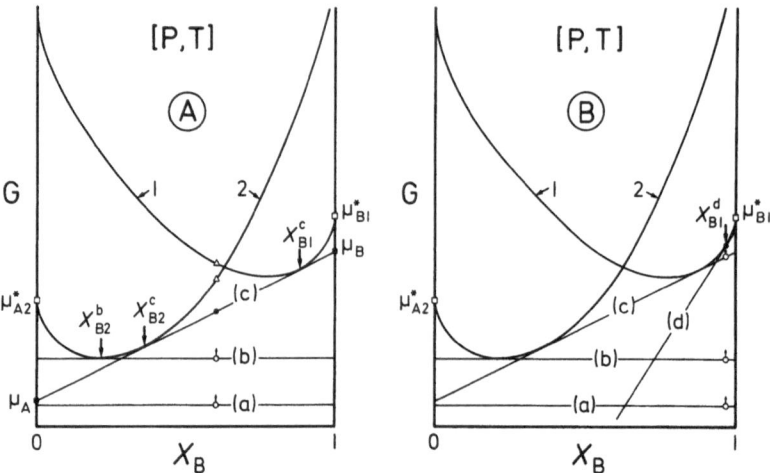

Fig. 9.1 Schematic G-X diagrams for the binary system A-B at constant P and T, illustrating the graphical method of minimizing the Gibbs energy for a specified bulk composition. The G-X curves apply to two non-isostructural phases 1 and 2. **A** depicts how the lowest energy level for a bulk composition (denoted by a *small circle*) is identified by gradually lifting the energy level from the initial state (a) to the final state (c). The energy of the final state has been indicated by a *solid dot*. The equilibrium assemblage comprises two phases, 1 and 2, having the compositions X_{B1}^c and X_{B2}^c. In **B**, the same process has been repeated for a different bulk composition. This time the equilibrium assemblage comprises only phase 1, whose composition, X_{B1}^d, is identical to the bulk composition of the system

any further rise of this tangential curve will successively shift the point of contact, X^b_{B2}, to higher values of X_B. In other words, the tangent line will now rock around the G-X curve for the phase 2. This rocking movement stops the very moment the tangent line hits the second G-X curve, that of the phase 1. This state is depicted in our figure by the line (c). Note that line (c) is now cotangential to both G-X curves (i.e. $\mu^1_A = \mu^2_A$ and $\mu^1_B = \mu^2_B$) and the G_{sys} (denoted by the solid dot on that line) can no longer be raised unless we quit the stable equilibrium and pass over to the metastable states, indicated by two open triangles in Fig. 9.1A. Clearly, three stability levels are recognized in this G-X diagram for the given bulk composition. The lowest one is the two-phase assemblage 1-2, with the compositions X^c_{B1} and X^c_{B2}. This, then, is the stable assemblage, with a *minimum* for G_{sys}. Thus, the Gibbs energy minimization technique permits identification of the stable assemblage *and* the compositions of the coexisting phases in that assemblage.

Next, consider another bulk composition, indicated by the open circle on the intial energy level (a) in Fig. 9.1B. As we continue raising G_{sys}, we pass through the states depicted by lines (b) and (c). This time, curve (c) does not, however, represent the stable equilibrium. The G_{sys} may be raised further, until it corresponds to the solid dot, which is identical to the G of phase 1 for the system bulk composition. In other words, we end up with a single-phase assemblage. Proceeding in this manner, it is always possible to derive the phase relations for every composition on the binary system A-B.

The technique of Gibbs energy minimization, delineated above in terms of a binary system, can be extended to any multicomponent system. Note, however, that in a ternary system, a tangential line becomes a plane, and in systems having more than three components, a tangential plane becomes a tangential hyperplane. Consequently, G-X relations in a multicomponent system can no longer be analyzed graphically; it has to be handled mathematically. For details of the algorithms devised for treating that problem, the readers may refer to Brown and Skinner (1974), Wood and Holloway (1984), and de Capitani and Brown (1987), among others.

9.2.2.3 From Isobaric-Isothermal G-X Sections to Phase Diagrams

As indicated in Section 9.2.2.1, calculation of phase diagrams is a two-step process. In the first step, isobaric-isothermal G-X data are processed by Gibbs energy minimization to solve for the stable assemblages and the compositions

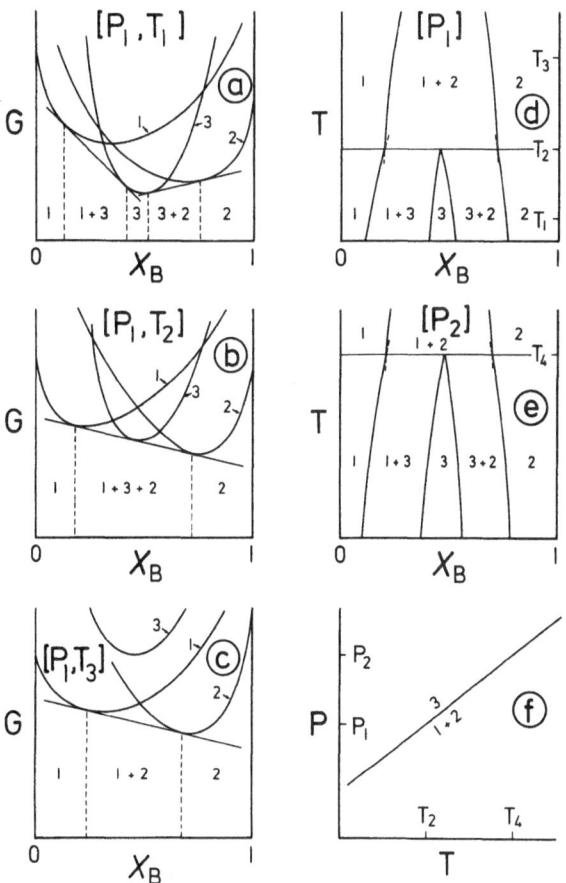

Fig. 9.2 Three schematic isobaric-isothermal G-X sections (**a-c**) for a binary system A-B allow construction of the T-X diagram (**d**). Two T-X diagrams at pressures P_1 and P_2, (**d**) and (**e**), leads to the P-T diagram shown in (**f**)

of the coexisting phases. The second step generates a P-T-X phase diagram by assembling this type of information from several isobar-isotherms. How the latter job is achieved will be explained commencing with a fresh batch of binary G-X curves.

Figure 9.2a-c depicts three schematic isobaric-isothermal G-X sections through the binary A-B. In the P-T range regarded here, three non-isostructural phases occur, designated 1, 2, and 3. The three G-X curves, labeled 1, 2, and 3, refer to them. Furthermore, note that all G-X sections relate to the *same* pressure P_1, though each one to a *different* temperature, with T rising from T_1 (Fig. 9.2a) to T_3 (Fig. 9.2c). At T_1, we have three one-phase fields (one for each of 1, 2, and 3) and two two-phase fields (for each of the assemblages 1+3 and 3+2) in between. With rising T, the G-X curves will shift relative to each other. Let the relative shifts of the G-X curves be such that at T_2, the G-X curves for all three phases are cotangential; at this temperature, 3 is at equilibrium with 1 and 2. At a higher temperature T_3, the relative positions of the G-X curves are such that 3 is now metastable with respect to the assemblage 1+2. Given such isobaric-polythermal G-X data at three temperatures, we may utilize them to draw the isobaric T-X section of Fig. 9.2d. Figure 9.2e depicts a similar T-X section through A-B at a pressure P_2 (for $P_2 > P_1$). The P-T diagram shown in Fig. 9.2f was generated from these T-X sections. From the above, it is clear that the technique of Gibbs energy minimization leads to complete phase diagrams in multicomponent-multiphase systems.

9.2.3 Alternative Techniques for Phase Diagram Calculations with Nonideal Solutions, and Some Worked Examples

Given adequate thermodynamic data for *each* participating phase of an equilibrium, phase relations are best computed by the Gibbs energy minimization technique. Although such data are accumulating rapidly, they are not always available at present. For example, we may conceive of a case where $\Delta G°_T$ of the reaction is known, not, however, $G°_{f,T}$ of the *individual* phases. Given additional data on $\Delta V°_{298}$ [or $\Delta V(P,T)$] and the mixing properties of the solutions, we may still calculate the phase diagram. The computational procedure utilized in such cases will be demonstrated below with some simple examples involving solid and fluid phases belonging to the system K_2O-Na_2O-CaO-Al_2O_3-SiO_2-H_2O-CO_2. The solids considered are quartz, jadeite, corundum, calcite, grossular, and wollastonite (each of which is treated as an end-member), in addition to the crystalline solutions such as alkali feldspars, plagioclases, and muscovite-paragonite micas. Both solid-solid and solid-fluid equilibria will be handled. To the extent that the equilibrium involves a fluid, it will be either H_2O or a binary solution of H_2O and CO_2. It may be stressed that these examples were chosen purely for the purpose of demonstrating the computational method, not because we really lack adequate thermodynamic data for those phases.

9.2.3.1 Calculation of P-T-X Phase Diagram for the $NaAlSi_3O_8$-$KAlSi_3O_8$ Binary

As a prelude to computing phase relations among solutions, let us calculate the phase diagram for the alkali feldspar binary. The thermodynamic properties of these feldspars depend on Al-Si-mixing on the four T-sites and on Na-K-mixing on the A-site. For the sake of simplicity, let us deal with two end-members with an identical state of Al-Si mixing, so that the properties of the solution will be a function of Na-K-mixing only. Analbite and sanidine are two such end-members (with random mixing of Al and Si on all T-sites), for which an equation of state is available (Haselton et al.1983). We shall use their data in our exercises, bearing in mind that the computed equilibria may not indicate the most stable state of the system at all T owing to the T-dependent Al/Si ordering. According to them, the analbite-sanidine solutions show a positive deviation from ideality, such that at some P-T conditions, they will exsolve into two phases. Therefore, an isobaric T-X_{kf} section will be used to display these phase relations. Note that the mole fraction of K-feldspar, X_{kf}, is used as the master variable, so that the mole fraction of albite is $(1$-$X_{kf})$.

9.2.3.1.1 Calculation of the Isobaric T-X_{kf} Section for the $NaAlSi_3O_8$-$KAlSi_3O_8$ Binary: the Analbite-Sanidine Solvus

Following Section 8.4, the conditions for two-phase equilibria on this solvus may be written as

$$\mu_{ab}^{Ab} = \mu_{ab}^{Kf}, \text{ and} \tag{9.4a}$$

$$\mu_{kf}^{Ab} = \mu_{kf}^{Kf}, \tag{9.4b}$$

where the superscripts Ab and Kf indicate the phases analbite and sanidine, respectively. The simultaneous fulfillment of (9.4a) and (9.4b) could, at least in principle, be demonstrated by drawing a common tangent to the two concave upward segments of the G-X_{kf} curve (cf. Fig. 8.6c). The two compositions corresponding to the common tangent would give us the compositions of the two phases on the solvus. Solving *exactly* for the solvus compositions by drawing a tangent turns out to be a difficult task in practice, especially at rather high temperatures, where the curvatures of G-X_{kf} become less pronounced (cf. the G-X_b curves for 330 and 480°C in Fig. 8.7a). Two alternative algorithms have been suggested to compute a solvus (Waldbaum and Thompson 1969, p.1280; Luth and Fenn 1973). Of these, the latter is simpler, and involves comparatively little computational work; it will be followed here. To appreciate how it works, recall that due to the equalities of the standard chemical potentials in an isostructural crystalline solution, the criteria of equilibrium [Eqs.(9.4a) and (9.4b)] on the solvus reduce to

$$a_{ab}^{Ab} = a_{ab}^{Kf}, \text{ and} \tag{9.5a}$$

$$a_{kf}^{Ab} = a_{kf}^{Kf}. \tag{9.5b}$$

Thus, at any P and T, obtaining the solvus compositions boils down to a search for two values of X_{kf}, for which (9.5a) and (9.5b) are simultaneously satisfied. This can be achieved graphically or by a stepwise search. The graphical method will be described below. The first step is to compute the activities from

$$a_{ab} = (1 - X_{kf}) \exp \left[\frac{\{W_{G,ab} + 2(1 - X_{kf})(W_{G,kf} - W_{G,ab})\} X_{kf}^2}{RT} \right], \text{ and} \tag{9.6a}$$

$$a_{kf} = X_{kf} \exp \left[\frac{\{W_{G,kf} + 2X_{kf}(W_{G,ab} - W_{G,kf})\}(1 - X_{kf})^2}{RT} \right], \tag{9.6b}$$

utilizing the $W_i(T,P)$ data given by Haselton et al. (1983),

$$W_{G,ab} \text{ (J mol}^{-1}) = 18810 - 10.3 \text{ T} + 0.364 \text{ P, and} \tag{9.7a}$$

$$W_{G,kf} \text{ (J mol}^{-1}) = 27320 - 10.3 \text{ T} + 0.364 \text{ P,} \tag{9.7b}$$

with T given in K and P in bar. Substituting Eqs. (9.7a,b) into (9.6a,b), a_{ab} and a_{kf} are obtained at any desired set of T and P. To demonstrate the graphical method of searching for the pair of compositions on the solvus, a_{ab} and a_{kf} were computed at 0.01 intervals of X_{kf} for a fixed P = 10 kbar and T = 700°C.

Table 9.1. Computed values of a_{ab} and a_{kf} of the analbite-sanidine crystalline solution at 700°C and 10 kbar pressure. A simultaneous fulfillment of the equal activity criteria, demanded by the solvus, may be noted for the two values of X_{kf} highlighted by asterisks

X_{kf}(Ab)	a_{ab}	a_{kf}	X_{kf}(Kf)	a_{ab}	a_{kf}
0.1310	0.92064	0.75082	0.6400	0.92091	0.75171
0.1311	0.92061	0.75097	0.6401	0.92085	0.75174
0.1312	0.92058	0.75112	0.6402	0.92078	0.75177
0.1313	0.92056	0.75128	0.6403	0.92072	0.75180
0.1314	0.92053	0.75143	0.6404	0.92065	0.75183
0.1315	0.92050	0.75158	0.6405	0.92059	0.75186
0.1316	0.92047	0.75173	0.6406	0.92052	0.75189
0.1317*	0.92044	0.75188	0.6407*	0.92046	0.75192
0.1318	0.92042	0.75203	0.6408	0.92039	0.75195
0.1319	0.92039	0.75218	0.6409	0.92033	0.75198

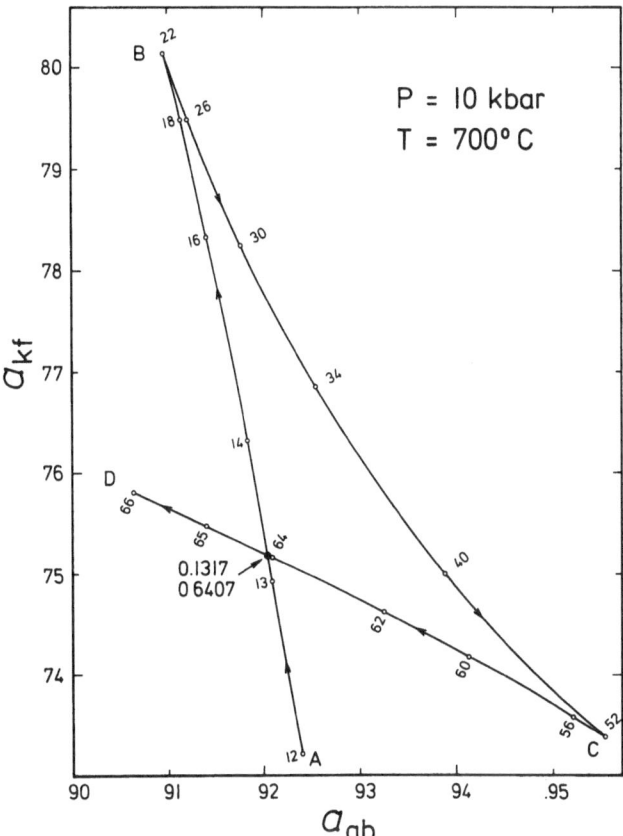

Fig. 9.3 Graphical solutions for the solvus compositions on the analbite-sanidine binary at 700°C and 10 kbar pressure from an a_{ab} vs a_{kf} plot. The *numbers* along the curve indicate X_{kf}. The equal activity criteria are fulfilled in the composition ranges 0.13-0.14 X_{kf} (Ab) and 0.64-0.65 X_{kf} (Kf), where the two limbs of the a_{ab} vs a_{kf} curves intersect; the point of intersection is shown by a *solid dot* (cf. Table 9.1)

Next, a_{ab} and a_{kf} were plotted against each other for each X_{kf}. Part of the plot is reproduced in Fig. 9.3. With increasing X_{kf}, the a_{ab} vs a_{kf} plot describes the loop A-B-C-D, the values of X_{kf} along the loop being indicated by the numbers. Two segments of the loop, AB and CD, intersect in the composition ranges 0.13-0.14 and 0.64-0.65, thereby satisfying the conditions of equality of activities demanded by Eqs. (9.5a,b). Alternatively, the solvus compositions can be deduced by matching a_{ab} and a_{kf} at close intervals of X_{kf} (see Table 9.1); at 10 kbar and 700°C, they are 0.1317 and 0.6407 X_{kf}. A repetition of these steps of calculation at a number of temperatures at constant pressures yields a series of points on the isobaric solvi of Fig. 9.4. Three solvi, for 1 bar and for 10 and 20 kbar, are displayed by solid lines in Fig. 9.4. Besides the solvi, it also indicates the spinodes (dotted lines). But how were the spinodes obtained?

To calculate the spinodes, recall their properties as outlined in Section
8.4.1. For the analbite-sanidine solution, the relevant equation (8.62) may be
recast to

$$0 = \left(\frac{\partial^2 G_m}{\partial X_{kf}^2}\right)_{P,T} = \frac{RT}{(1-X_{kf})X_{kf}} + 2W_{G,ab}(1-3X_{kf}) + 2W_{G,kf}(3X_{kf}-2). \quad (9.8)$$

At any P and T, this equation may be solved for X_{kf} by iteration. In so doing,
it should be borne in mind that at all subcritical T, two solutions (i.e. two
zeros) for spinodes must result. Table 9.2 is an example of the calculation of
the alkali feldspar spinodes at P = 10 kbar and T = 700°C. Repeating that
process over a number of T yields the entire spinodal curve.

Having obtained the spinodes, we are left with only one more quantity of
interest, the critical point. The critical point (see Sect.8.4.1) may be
recognized from the conditions

$$\left(\frac{\partial^2 G_m}{\partial X_{kf}^2}\right)_{P,T} = 0 = \left(\frac{\partial^3 G_m}{\partial X_{kf}^3}\right)_{P,T}. \quad (9.9)$$

Consequently, the critical point can be identified from a maximum on the
spinodal; it may be solved for by iteration loops of T. As T rises, the spinodes
approach each other, and become identical at the critical point. For the
analbite-sanidine crystalline solution the critical point for 10 kbar pressure is

Table 9.2. An example of the calculation of the spinodes for the analbite-sanidine
solution at 700°C and 10 kbar, from Eq. (9.8). The compositions on the spinodes are
shown by double asterisks

X_{kf}	F(J)[a]	X_{kf}	F(J)[a]
0.1	36113.6	0.4	-4756.4
0.2	1888.0	0.5	-998.9
0.3	-5046.3	0.6	5455.6
0.26	-3564.2	0.56	2537.6
0.22	-509.2	0.52	74.2
0.21	600.2	0.51	-475.3
0.212	364.8	0.512	-367.5
0.214	136.4	0.514	-258.6
0.216	-85.3	0.516	-148.7
0.215	24.7	0.518	-37.8
0.2152**	2.6	0.519	18.1
0.2153	-8.5	0.5188	6.9
		0.5186	-4.3
		0.5187**	1.3

[a] $F(X_{kf}) = RT/(1-X_{kf})X_{kf} + 2W_{G,ab}(1-3X_{kf}) + 2W_{G,kf}(3X_{kf}-2)$.

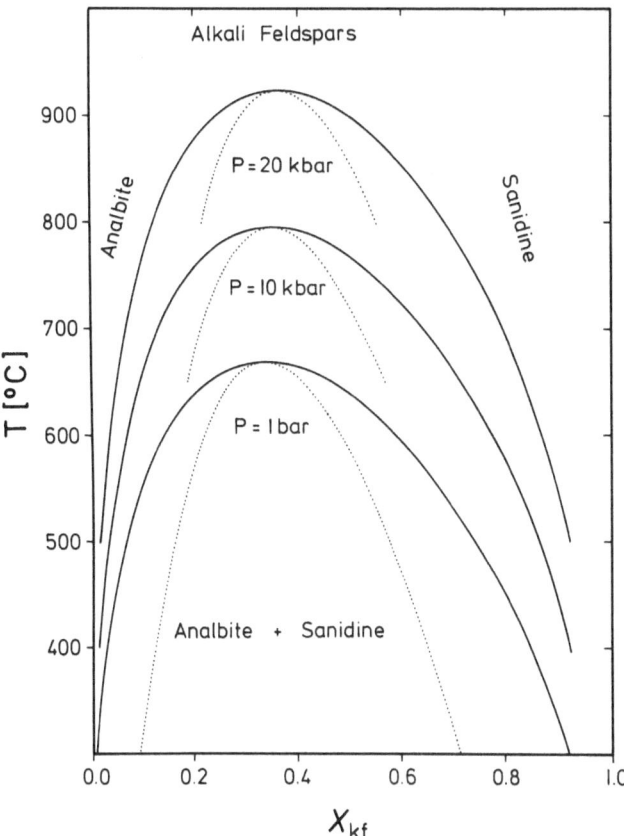

Fig. 9.4 A polybaric T-X_{kf} diagram showing the phase relations for the analbite-sanidine crystalline solutions. The solvi appear as *solid lines*, whereas the spinodes are *dotted*. The region within a solvus is that of a two-phase (Ab plus Kf) coexistence. The one-phase regions to the left and to the right of the solvus belong to Ab and Kf, respectively. At T > T_c, a homogeneous alkali feldspar is stable across the entire composition range

found to be

$$T_c = 795.5\,°C, X_c = 0.3509\,X_{kf}, \qquad (9.10)$$

with T_c and X_c denoting the critical temperature and the critical composition, respectively. Figure 9.4 summarizes the results of computation of the T-X_{kf} sections in a polybaric phase diagram. In the pressure range 1 bar to 20 kbar, T_c increases from 668.8 to 923.6°C; X_c, however, changes only from 0.3372 to 0.3629 X_{kf}.

With a further rise in temperature, the alkali feldspars will melt. The nature of the melting process, not treated in this book, is fairly complex, and is pressure- and composition-dependent (see Waldbaum and Thompson 1969).

Likewise, increasing pressure also destabilizes the alkali feldspars. These stability relations will be computed next, and displayed on P-X_{kf} sections. Finally, the T-X_{kf} and the P-X_{kf} sections will be combined to a P-T projection.

9.2.3.1.2 Calculation of the P-X_{kf} Section Through the $NaAlSi_3O_8$-$KAlSi_3O_8$ Pseudobinary, and the P-T Diagram

Although the analbite-sanidine solvus was computed above to 20 kbar pressure (Fig. 9.4), parts of the phase relations turn out to be metastable by virtue of the reaction

$$NaAlSi_2O_6 + SiO_2 = Na[AlSi_3O_8].$$ (9.11)
(jadeite)   (quartz)   (analbite)

For instance, at 600°C, the ab component of alkali feldspar would decompose to jadeite plus quartz as the pressure increases beyond 16 kbar. Thus, at those high pressures, the alkali feldspar phase relations must be treated in terms of the $NaAlSi_2O_6$-SiO_2-$KAlSi_3O_8$ ternary; that is, the $NaAlSi_3O_8$-$KAlSi_3O_8$ join is now pseudobinary. The objective of this section is to explore these phase relations by thermodynamic calculations. In so doing, we shall assume that alkali feldspar is the only crystalline solution involved, whereas quartz and jadeite retain their end-member compositions. This is a realistic picture, because the structure of jadeite, with its [8]-coordinated Na-site, would barely accomodate a considerably larger cation like K. Note, however, that in contrast to albite, sanidine remains stable to quite high pressures.

To compute the P-T curve for the reaction (9.11), we start with the general condition of heterogeneous equilibrium,

$$\Delta\mu = 0 = \Delta G_T^\circ + \int_1^P \Delta V_T(P)dP + RT\ln K .$$ (1.106)

In the specific case of reaction (9.11), involving two end-members and a solution, and recalling that the alkali feldspar is a one-site solution (q = 1) whose composition will have to be expressed in terms of X_{kf}, (1.106) may be explicitly rewritten as

$$\Delta\mu = 0 = \Delta G_T^\circ + \int_1^P \Delta V_T(P)dP + RT\ln(1 - X_{kf}) + RT\ln\gamma_{ab} .$$ (9.12)

Calculation of reaction (9.11) requires that data on ΔG°_T and $\Delta V(P,T)$ be available, in addition to the equation of state for the alkali feldspar. Using the experimental reversal brackets for reaction (9.11) from Holland (1980), and the $C^\circ_P(T)$, V°_{298}, α_o, and β_o data (Table 9.3), and employing the thermodynamic formalism put forward in Section 5.5, we may come up with

Table 9.3. A summary of input data employed to extract ΔH°_{298} and ΔS°_{298} of reaction (9.11) from the reversal brackets indicated by Holland (1980)

| Phases | $C^\circ_P(T)$(in J K^{-1} mol^{-1}) = a + b·T + c·T^{-2} + d·T$^{-0.5}$ + f·T^{-3} | | | | |
	a	b	c	d	f
Jadeite[a]	311.29	0.0	-5.3503E6	-2.0051E3	6.6257E8
Analbite[a]	391.87	0.0	-9.3975E6	-2.2696E3	1.32604E9
Quartz(T < 844)[b]	44.603	3.7754E-2	-1.0018E6	0.0	0.0
Quartz(T > 844)[b]	58.928	1.0031E-2	0.0	0.0	0.0

The standard state transition properties for quartz are[b]
T_{tr} = 844 K, ΔH°_{844} = 476 J, ΔS°_{844} = 0.55 J K^{-1}

Phases	V°_{298}(J bar^{-1} mol^{-1})	$\alpha(T)=\alpha_0$(K^{-1})	$\beta(P)=\beta_0$(bar^{-2})
Jadeite	6.0400[b]	2.84E-5[c]	0.75E-6[c]
Analbite	10.0430[b]	2.71E-5[c]	1.59E-6[c]
Quartz	2.2688[b]	5.36E-5[c]	2.28E-6[c]

[a] Berman and Brown (1985).
[b] Robie et al. (1979).
[c] Holland (1980).

$\Delta H°_{298}$ and $\Delta S°_{298}$ for reaction (9.11). Combining these with the properties of alkali feldspars (Haselton et al. 1983), the desired phase relations can be computed. To achieve the first step, extraction of $\Delta H°_{298}$ and $\Delta S°_{298}$, use is made of the relation

$$-\Delta H^\circ_{298} + T\Delta S^\circ_{298} = \int_{298}^{844} \Delta C^\circ_P(T)dT + \Delta H^\circ_{844} + \int_{844}^{T} \Delta C^\circ_P(T)dT$$

$$-T\left[\int_{298}^{844}\left\{\frac{\Delta C^\circ_P(T)}{T}\right\}dT + \Delta S^\circ_{844} + \int_{844}^{T}\left\{\frac{\Delta C^\circ_P(T)}{T}\right\}dT\right]$$

$$+ \int_{1}^{P} \Delta V_T(P)dP. \tag{9.13}$$

The RHS of (9.13) was calculated for the 12 P-T bracketing points for reaction (9.11) reported by Holland (1980). Treating that data as a linear function of T, the two quantities on the LHS of (9.13) are found to be,

$$\Delta H^\circ_{298} = 10469 \pm 2042 \text{ J and } \Delta S^\circ_{298} = 43.936 \pm 1.670 \text{ J K}^{-1}. \tag{9.14}$$

Table 9.4. Examples of computation of the P-X_{kf} sections for the alkali feldspar pseudobinary at 900 and 1000 °C

T (°C)	X_{kf} (J)	ΔG_T (J)	$\int_1^P \Delta V_T^\circ(P)dP$ (J)	RTln(1-X_{kf}) (J)	RTlnγ_{ab} (J)	$\Delta\mu$ (bar)	P
900	0.00	-39528.28	39527.61	0.00	0.00	-0.67	24098
900	0.02	-39528.28	39712.12	-197.06	12.89	-0.33	24216
900	0.05	-39528.28	39949.89	-500.31	79.41	0.71	24368
900	0.10	-39528.28	40246.78	-1027.68	309.84	0.66	24558
900	0.15	-39528.28	40434.18	-1585.20	678.96	-0.33	24678
900	0.20	-39528.28	40530.96	-2176.53	1173.92	0.07	24740
1000	0.00	-43595.89	43596.39	0.00	0.00	0.50	26748
1000	0.02	-43595.89	43797.34	-213.85	12.86	0.46	26878
1000	0.05	-43595.89	44059.96	-542.96	79.28	0.39	27048
1000	0.10	-43595.89	44401.09	-1115.28	309.40	-0.68	27269
1000	0.15	-43595.89	44637.08	-1720.32	678.27	-0.86	27426
1000	0.20	-43595.89	44785.08	-2362.05	1173.16	0.30	27518
1000	0.25	-43595.89	44860.59	-3045.22	1781.00	0.48	27567
1000	0.30	-43595.89	44882.17	-3775.53	2488.50	-0.75	27581

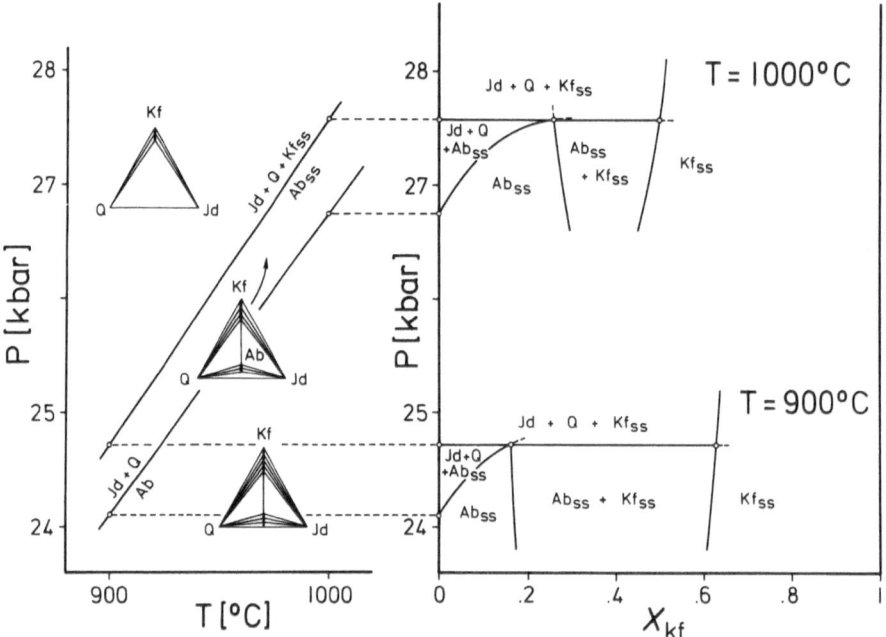

Fig. 9.5 Two P-X_{kf} isotherms (for 900 and 1000 °C) through the alkali feldpar pseudobinary (cf. Table 9.4) are shown on the *right side* of this figure. The P-T diagram, constructed from these P-X_{kf} sections, is shown on the *left side*

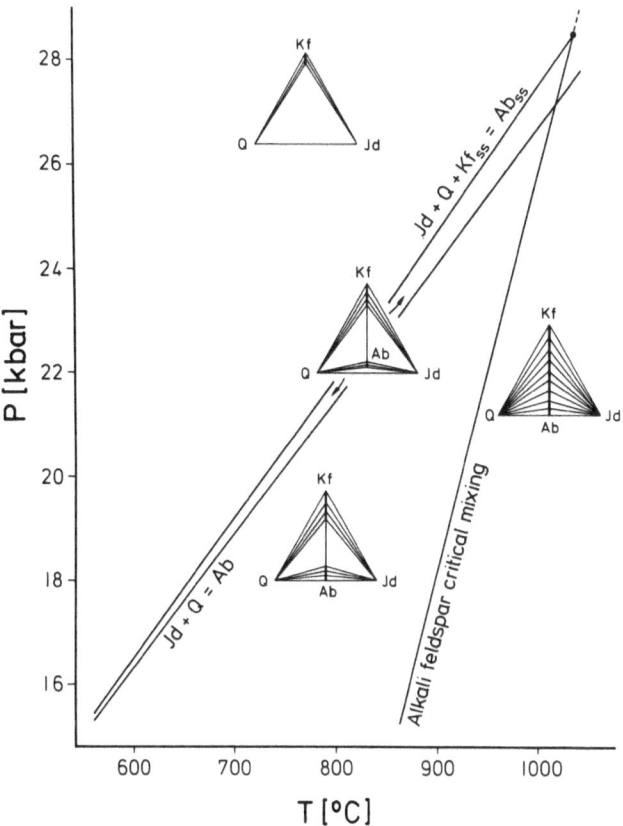

Fig. 9.6 A P-T projection depicting the phase relations for the $NaAlSi_2O_6$-SiO_2-$KAlSi_3O_8$ ternary

To compute the P-T curve of reaction (9.11), the $\Delta H°_{298}$ and $\Delta S°_{298}$ from (9.14) were input into (9.12), along with ancillary data from Table 9.3 and Haselton et al. (1983). In practice, at a specified T and X_{kf}, P was iterated until $\Delta\mu$ approached zero to within ±1 J. Table 9.4 summarizes the results for the 900 and 1000°C isotherms. They are also displayed on the isothermal P-X_{kf} plots of Fig. 9.5. Note that $X_{kf} = 0$ on these plots corresponds to the end-member reaction (9.11). With increasing X_{kf}, the equilibrium P for the coexistence of jadeite, quartz, and an albite-rich solution (Ab_{ss}) rises steadily, until the P-X_{kf} curve intersects the Ab_{ss} limb of the isothermal alkali feldspar solvus. The mutual intersection of these curves indicates the equilibrium pressure for the reaction

$$Ab_{ss} = \text{jadeite} + \text{quartz} + Kf_{ss}. \tag{9.15}$$

At this pressure, the most Kf-rich Ab_{ss} is in equilibrium with jadeite + quartz + Kf_{ss}; the composition of the latter is given by the Kf_{ss} limb of the alkali feldspar solvus. A further increase in pressure leads to the coexistence of jadeite + quartz + Kf_{ss}. A P-T projection for these phase relations has been constructed from the 900°C and 1000°C isotherms and depicted on the left half of Fig. 9.5. The results of computation of a number of these $P\text{-}X_{kf}$ isotherms were combined with data on the alkali feldspar critical curve (pressure dependence of the alkali feldspar critical point) to obtain a P-T projection delineating the phase relations for the $NaAlSi_3O_8\text{-}KAlSi_3O_8$ pseudobinary (Fig. 9.6).

9.2.3.2 $NaAl_2[AlSi_3O_{10}(OH)_2]\text{-}KAl_2[AlSi_3O_{10}(OH)_2]$ Pseudobinary: Computation of Phase Relations for Multiple Equilibria

The micas on the paragonite (pg)-muscovite (ms) join, like the alkali feldspars, show a partial miscibility; their intermediate range of composition being metastable with respect to alkali feldspar + corundum + H_2O. Thus, $NaAl_2[AlSi_3O_{10}(OH)_2]\text{-}KAl_2[AlSi_3O_{10}(OH)_2]$ is also a pseudobinary. With increasing T, both end-members decompose to the corresponding feldspar end-member, corundum, and H_2O,

$$NaAl_2[AlSi_3O_{10}(OH)_2] = Na[AlSi_3O_8] + Al_2O_3 + H_2O \text{ and} \qquad (9.16a)$$
$$\text{(pg)} \qquad\qquad\qquad \text{(ab)} \qquad\quad \text{(C)}$$

$$KAl_2[AlSi_3O_{10}(OH)_2] = K[AlSi_3O_8] + Al_2O_3 + H_2O. \qquad\qquad (9.16b)$$
$$\text{(ms)} \qquad\qquad\qquad \text{(kf)} \qquad\quad \text{(C)}$$

An equilibrium coexistence of a mica and a feldspar solution with corundum (C) and H_2O, thus, implies that both the above equilibria are simultaneously operative. This is one of the numerous examples of multiple equilibria of relevance to rock-forming processes. The objective of this section is to give an example of the computation of such equilibria.

In view of the above, the mica phase relations will be handled in terms of the $NaAlSi_3O_8\text{-}KAlSi_3O_8\text{-}Al_2O_3\text{-}H_2O$ quaternary, applying the dual criteria of equilibrium to the reactions (9.16a) and (9.16b), to be abbreviated below as reactions (a) and (b):

$$\Delta\mu_a = 0 = \Delta G^\circ_{T,a} + \int_1^P \Delta V_{T,s,a}(P)dP + RT\ln K_a, \quad \text{and} \qquad (9.17a)$$

$$\Delta\mu_b = 0 = \Delta G^\circ_{T,b} + \int_1^P \Delta V_{T,s,b}(P)dP + RT\ln K_b. \qquad\qquad (9.17b)$$

Note that the equilibrium constants, K_a and K_b, comprise f^*_{H2O} and a_i terms. Thus, ΔG°_T, $\Delta V(T,P)$, a_i, and f^*_{H2O} data will be required to compute the desired phase relations. The reversal brackets for the end-member reactions

(9.16a) and (9.16b), given by Chatterjee (1970) and Chatterjee and Johannes (1974), will be used to extract $\Delta G°_{T,a}$ and $\Delta G°_{T,b}$. Derivation of $\Delta G°_T$ from experimental reversal brackets for solid-fluid equilibria has been discussed in Section 5.5. It was demonstrated that to the extent $\Delta G°_T$ is needed merely to recalculate the equilibrium within the experimentally explored range of T, we may ignore $C°_p(T)$ of the participating phases and split the $\Delta G°_T$ into $\Delta H°_{<T>}$ and $T\Delta S°_{<T>}$ terms, $\Delta H°_{<T>}$ and $\Delta S°_{<T>}$ being the average standard state enthalpy and entropy differences for the experimentally covered range of T. We shall cut corners in the present exercise, and apply the simple format of computation. Furthermore, for the restricted P-T range of interest, the P-T dependences of ΔV_s will also be ignored. Due to these reasons, and recalling that mica and feldspar are single-site solutions, whose compositions may be expressed in terms of X_{ms} and X_{kf}, Eqs.(9.17a) and (9.17b) are recast to

$$\Delta\mu_a = 0 \;\; = \Delta G°_{T,a} + \Delta V°_{298,s,a}(P-1) + RT\ln f^*_{H_2O}$$
$$+ RT[\ln(1-X_{kf})-\ln(1-X_{ms})] + RT[\ln\gamma_{ab}-\ln\gamma_{pg})], \text{ and} \tag{9.18a}$$

$$\Delta\mu_b = 0 \;\; = \Delta G°_{T,b} + \Delta V°_{298,s,b}(P-1) + RT\ln f^*_{H_2O}$$
$$+ RT[\ln X_{kf}-\ln X_{ms}] + RT[\ln\gamma_{kf}-\ln\gamma_{ms})]. \tag{9.18b}$$

Given data on the excess mixing properties of alkali feldspars and micas, from which $RT\ln\gamma_i$ follows, these two equations are amenable to simultaneously solution for the two unknowns, X_{ms} and X_{kf}.

To achieve this goal, the $\Delta H°_{<T>}$, $\Delta S°_{<T>}$, and $\Delta V°_{298,s}$ listed in Table 9.5, were used. They were supplemented by the Haselton et al. (1983) data on the excess mixing properties of alkali feldspar, whereas those for the micas were taken from Chatterjee and Flux (1986). The Haselton et al. (1983) equation of state for the alkali feldspars leads to

$$RT\ln\gamma_{ab} = [W_{G,ab} + 2(1-X_{kf})(W_{G,kf}-W_{G,ab})]X^2_{kf} \text{ and} \tag{9.19a}$$

$$RT\ln\gamma_{kf} = [W_{G,kf} + 2X_{kf}(W_{G,ab}-W_{G,kf})](1-X_{kf})^2, \tag{9.19b}$$

with $W_{G,i}$ given by Eq.(9.7). Likewise, the Redlich-Kister equation of Chatterjee and Flux (1986) translates to [cf.(1.83) and (1.84)]

$$RT\ln\gamma_{pg} = [A + B(3-4X_{ms}) + C(1-2X_{ms})(5-6X_{ms})]X^2_{ms}, \tag{9.20a}$$

$$RT\ln\gamma_{ms} = [A + B(1-4X_{ms}) + C(1-2X_{ms})(1-6X_{ms})](1-X_{ms})^2. \tag{9.20b}$$

where

$A(\text{J mol}^{-1}) = 11222 + 1.389\ T + 0.2359\ P,$
$B(\text{J mol}^{-1}) = -1134 + 6.806\ T + 0.0840\ P, \text{ and}$
$C(\text{J mol}^{-1}) = -7305 + 9.043\ T,$

Table 9.5. $\Delta H^\circ_{<T>}$, $\Delta S^\circ_{<T>}$, and $\Delta V^\circ_{298,s}$ data used to calculate the phase relations in the system $NaAlSi_3O_8$-$KAlSi_3O_8$-Al_2O_3-H_2O

Reaction	$\Delta H^\circ_{<T>}(J)$	$\Delta S^\circ_{<T>}(J\ K^{-1})$	$\Delta V^\circ_{298,s}(J\ bar^{-1})$
$pg = ab + c + H_2O$	99550	174.407	-0.5875
$ms = kf + c + H_2O$	97692	163.545	-0.6195

Table 9.6. Results of simultaneous solution of Eqs. (9.18a) and (9.18b) for the two unknowns, X_{kf} and X_{ms}, at a series of temperatures for a constant pressure of 1000 bar

P(bar)	T(°C)	X_{kf}	X_{ms}
1000	536	0.0	0.0
1000	540	0.0149	0.0958
	Eq. (9.21) $Pg_{ss} = Ab_{ss} + Ms_{ss} + C + H_2O$		
1000	545	0.0200	0.6350
1000	550	0.0233	0.6866
1000	555	0.0271	0.7224
1000	560	0.0315	0.7504
1000	565	0.0366	0.7735
1000	570	0.0426	0.7933
1000	575	0.0499	0.8106
1000	580	0.0589	0.8261
1000	585	0.0705	0.8400
1000	590	0.0864	0.8528
1000	595	0.1105	0.8646
	Eq. (9.22) $Ab_{ss} + Ms_{ss} = Kf_{ss} + C + H_2O$		
1000	600	0.7218	0.8923
1000	605	0.8515	0.9362
1000	610	0.9664	0.9854
1000	612	1.0	1.0

with T given in K and P in bar. Inserting (9.19a,b) and (9.20a,b) into (9.18a,b), the two final equations were eventually obtained. To solve them simultaneously at the desired T and P for the two unknowns, X_{kf} and X_{ms}, they are equated to each other, and solved by iteration. The results have been reproduced for 1000 bar pressure and a series of temperatures in Table 9.6. The first and the last lines of this table indicate equilibria of the end-member reactions (9.16a) and (9.16b), respectively. Note that with rising T, the compositions of the coexisting feldspar and mica become progressively K-rich until, at a T between 540 and 545°C, the mica limb of this coexistence

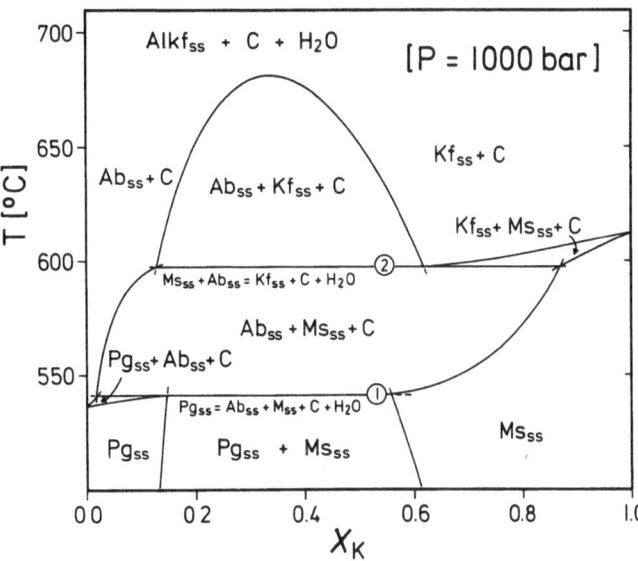

Fig. 9.7 An isobaric T-X_K section showing the phase relations on the $NaAl_2[AlSi_3O_{10}(OH)_2]$-$KAl_2[AlSi_3O_{10}(OH)_2]$ pseudobinary join. The *horizontal lines*, indicated by 1 and 2, refer to the reactions (9.21) and (9.22)

curve intersects with the paragonite limb of the mica solvus (see Fig. 9.7). This is followed by an abrupt break in X_{ms}, which jumps from 0.0958 (for 540°C) to 0.6350 (for 545°C). This is in response to the discontinuous reaction

$$Pg_{ss} = Ab_{ss} + Ms_{ss} + C + H_2O, \tag{9.21}$$

by which the paragonite solution with a maximum X_{ms} decomposes to Ab_{ss} + Ms_{ss} + C + H_2O. With increasing temperature, both feldspar and mica continue to evolve to more K-rich compositions, until the feldspar limb cuts the albite limb of the feldspar solvus between 595 and 600°C (Fig. 9.7). Now, the feldspar composition jumps from 0.1105 (at 595°C) to 0.7218 (at 600°C), this time in response to the reaction

$$Ab_{ss} + Ms_{ss} = Kf_{ss} + C + H_2O. \tag{9.22}$$

Further up the temperature scale, both solutions become more and more K-rich, until the end-member reaction (9.16b) takes place at 612°C.

Figure 9.7 depicts the isobaric T-X section calculated at 1000 bar pressure. Note that the mole fractions of both solutions have been denoted as X_K in this figure, rather than as X_{kf} and X_{ms}. It also indicates the stable segments of the mica and feldspar solvi, computed from the respective equations of state. The discontinuous reactions (9.21) and (9.22) are labeled by the numbers 1 and 2.

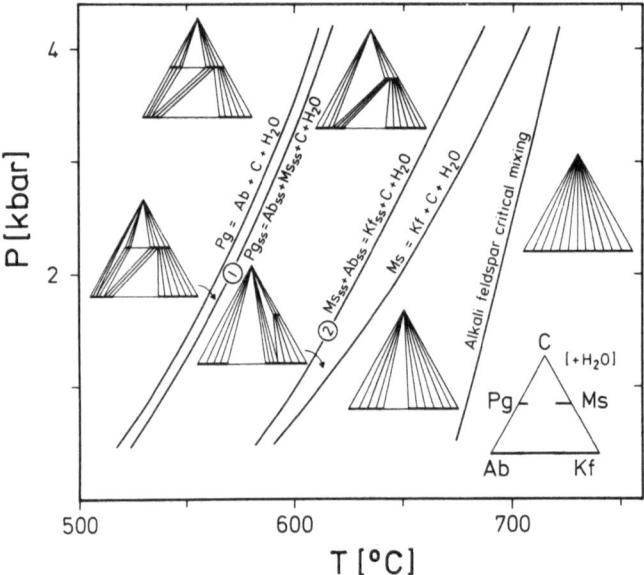

Fig. 9.8 P-T diagram for the paragonite-muscovite pseudobinary. Note that the compositions of the hydrous phases, paragonite and muscovite, are shown in a projection from the H_2O apex onto the H_2O-free base of the $NaAlSi_3O_8$-$KAlSi_3O_8$-Al_2O_3-H_2O quaternary

Similar calculations for a few isobars helped construct the P-T projection illustrated in Fig. 9.8. In addition to the end-member reactions (9.16a,b), the two discontinuous reactions (9.21) and (9.22) also appear as (1) and (2). And finally, the critical mixing curve for the alkali feldspar completes the phase diagram.

9.2.3.3 Computation of a T-X_{CO2} Phase Diagram with a Crystalline Solution

The last example of calculation of phase diagrams will involve a crystalline solution and a binary fluid mixture. Let us consider the reaction

$$Ca[Al_2Si_2O_8] + CaSiO_3 + CaCO_3 = Ca_3Al_2[SiO_4]_3 + CO_2. \qquad (9.23)$$
$$\text{(an in Plag)} \quad \text{(Woll)} \quad \text{(Cc)} \quad \text{(Gr)}$$

As in the foregoing examples, the stoichiometry of the reaction is formulated here in terms of components of the phases participating in the reaction. Let us assume that the only crystalline solution involved is a plagioclase (Plag), a solution of an and ab (albite, $NaAlSi_3O_8$), while wollastonite (Woll), calcite (Cc), and grossular (Gr) remain end-member phases. The second solution, participating in this reaction, is a supercritical H_2O-CO_2 fluid mixture.

The coexistence of five phases (a binary supercritical H_2O-CO_2 fluid mixture, plagioclase, wollastonite, calcite, and grossular) in a six-component (Na_2O-CaO-Al_2O_3-SiO_2-H_2O-CO_2) system implies three degrees of freedom. To compute the univariant T-X_{CO2} curve, we must then specify two intensive variables; these are P and X_{an}. At any P, an isobaric T-X_{CO2} section comprises as many univariant curves as there are X_{an}. That is, our isobaric T-X_{CO2} section will comprise a bunch of X_{an} isopleths.

Considering that wollastonite, calcite, and grossular are end-members, whose activities are unity at all T and P, the condition of equilibrium of reaction (9.23) will be

$$\Delta\mu = 0 = \Delta G^\circ_T + \int_1^P \Delta V_{T,s}(P)dP - RT\ln a_{an} + RT\ln(f^*_{CO_2} \cdot a_{CO_2}). \quad (9.24)$$

In addition to data on ΔG°_T and $\Delta V_s(T,P)$, knowledge of the mixing properties of the plagioclases and the H_2O-CO_2 fluids is necessary to solve (9.24) for X_{CO2} at a given P, X_{an}, and T. For the present purpose, the properties of H_2O-CO_2 fluids, f^*_{CO2} and a_{CO2}, will be taken from Jacobs and Kerrick (1981a). For the plagioclases, the data by Newton et al. (1981) will be utilized. Although the excess mixing properties are given by them in the familiar Margules form, the expressions for the activities may be unfamiliar, because they employed the Al-avoidance model of Kerrick and Darken (1975) to express the ideal mixing term. Following this model, we have

$$a_{ab} = X^2_{ab}(2 - X_{ab})\exp\left[\frac{\{W_{G,ab} + 2X_{ab}(W_{G,an} - W_{G,ab})\}X^2_{an}}{RT}\right], \quad (9.25)$$

$$a_{an} = \frac{X_{an}(1 - X_{an})^2}{4}\exp\left[\frac{\{W_{G,an} + 2X_{an}(W_{G,ab} - W_{G,an})\}X^2_{ab}}{RT}\right], \quad (9.26)$$

with $W_{G,ab} = W_{H,ab} = 28225$ J and $W_{G,an} = W_{H,an} = 8577$ J.

Note that $W_{S,i}$ is zero, as is $W_{V,i}$; therefore, the excess mixing properties of plagioclase are independent of T and P. And finally, ΔG°_T and related data, to be used in this exercise, will be taken from the thermodynamic database of Berman et al. (1985), derived by mathematical programming. These data are listed in Table 9.7. In deriving them, Berman et al. (1985)[1] ignored the T-P-dependences of the molar volumes of the solids. To be consistent with them, we shall replace the volume integral in (9.24) by $\Delta V_{298,s}(P-1)$. This gives us the the final equation,

$$\Delta\mu = 0 = \Delta G^\circ_T + \Delta V^\circ_{298,s}(P-1) - RT\ln a_{an} + RT\ln(f^*_{CO_2} \cdot a_{CO_2}). \quad (9.27)$$

---

[1] The internally consistent thermodynamic dataset of Berman et al. (1985) has been upgraded by Berman (1988). In deriving the latter, V(T,P) was explicitly considered for each mineral.

Table 9.7. Standard state reaction properties for the equilibrium wollastonite-calcite-anorthite-grossular-CO_2

Reaction:	an + Woll + Cc = Gr + CO_2

$\Delta H^\circ_{298} = 40974$ J; $\Delta S^\circ_{298} = 94.921$ J K^{-1}; $\Delta V^\circ_{298,s} = -5.219$ J bar^{-1}

$\Delta C^\circ_P(T)(J\ K^{-1}) = \Delta a + \Delta b \cdot T + \Delta c \cdot T^{-2} + \Delta d \cdot T^{-0.5} + \Delta f \cdot T^{-3} + \Delta g T^{-1}$,

with $\Delta a = -111.5794$, $\Delta b = -0.002$, $\Delta c = -13772160.0$,
       $\Delta d = 3332.259$; $\Delta f = 1985380736.0$, and $\Delta g = -4103.83$

Table 9.8. Numerical examples of computation of X_{an}-isopleths for the grossular-calcite-wollastonite-plagioclase-fluid equilibrium on a T-X_{CO_2} section at 500 bar pressure

T (°C)	X_{an} (J)	X_{CO_2} (J)	ΔG°_T (J)	$\Delta V_s(P-1)$ (J)	$RT\ln f^*_{CO_2}$ (J)	$RT\ln a_{CO_2}$ (J)	$RT\ln a_{an}$	$\Delta\mu$
700	1.0	0.676	-45465.0	-2604.3	51207.7	-3146.6	0.0	-8.2
700	1.0	0.677	-45465.0	-2604.3	51207.7	-3134.7	0.0	3.7
700	0.9	0.579	-45465.0	-2604.3	51207.7	-4383.1	-1243.1	-2.2
700	0.9	0.580	-45465.0	-2604.3	51207.7	-4369.8	-1243.1	11.7
700	0.8	0.532	-45465.0	-2604.3	51207.7	-5057.8	-1909.9	-9.5
700	0.8	0.533	-45465.0	-2604.3	51207.7	-5042.7	-1909.9	5.5
700	0.7	0.509	-45465.0	-2604.3	51207.7	-5408.5	-2268.2	-1.9
700	0.7	0.510	-45465.0	-2604.3	51207.7	-5394.3	-2268.2	12.3
700	0.6	0.488	-45465.0	-2604.3	51207.7	-5743.7	-2599.3	-6.0
700	0.6	0.489	-45465.0	-2604.3	51207.7	-5727.3	-2599.3	10.4
700	0.5	0.452	-45465.0	-2604.3	51207.7	-6353.6	-3207.4	-7.8
700	0.5	0.453	-45465.0	-2604.3	51207.7	-6335.9	-3207.4	9.9
700	0.4	0.387	-45465.0	-2604.3	51207.7	-7587.6	-4439.2	-10.0
700	0.4	0.388	-45465.0	-2604.3	51207.7	-7566.9	-4439.2	10.7
700	0.3	0.290	-45465.0	-2604.3	51207.7	-9877.4	-6733.2	-5.8
700	0.3	0.291	-45465.0	-2604.3	51207.7	-9850.0	-6733.2	21.6
700	0.2	0.174	-45465.0	-2604.3	51207.7	-13937.7	-10769.1	-30.2
700	0.2	0.175	-45465.0	-2604.3	51207.7	-13892.5	-10769.1	15.0
700	0.1	0.069	-45465.0	-2604.3	51207.7	-21333.4	-18174.3	-20.7
700	0.1	0.070	-45465.0	-2604.3	51207.7	-21210.0	-18174.3	102.7

Inserting the appropriate value of a_{an} from (9.26), this equation will have to be solved for X_{CO2} at specified P, X_{an} and T. For P = 500 bar, five isotherms were calculated in the temperature range 500-850°C. Table 9.8 gives an example of computation of a few X_{an}-isopleths at 500 bar and 700°C. Note that a 0.001 bracket for X_{CO2} was established each time, prior to stopping the iteration.

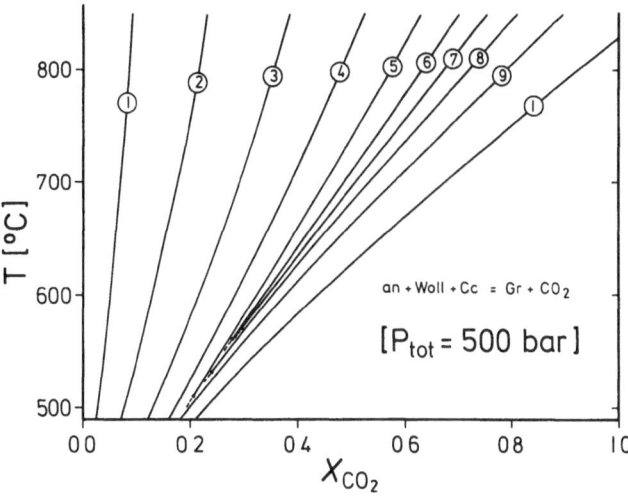

Fig. 9.9 An isobaric (P = 500 bar) T-X_{CO_2} diagram depicting the X_{an}-isopleths of a plagioclase in equilibrium with wollastonite, grossular, calcite, and a binary H_2O-CO_2 fluid mixture

Figure 9.9 displays the T-X_{CO2} diagram. At any temperature, a decrease of X_{an} in the equilibrium assemblage leads to a displacement of the T-X_{CO2} isobar to a more aqueous composition. Note, moreover, that the isopleths in the range 0.5 to 0.8 X_{an} become metastable due to phase separation in plagioclase below 600°C.

9.3 Outlines of Geothermometry and Geobarometry

9.3.1 Scope

As stated earlier, knowledge of the chemical compositions of the coexisting minerals in an equilibrium assemblage helps recover the intensive variables (T, P, a_i, f_i, etc.) prevailing during the equilibration of that mineral assemblage. The terms geothermometry and geobarometry refer to the computation of the equilibrium T and P of an assemblage; the derivation of its equilibrium a_{H2O} and f_{O2} are dubbed geohygrometry and oxygen barometry, respectively. The objective of this section is to focus on the principles underlying such calculations. An in-depth review of the subject is beyond our scope. State-of-the-art reviews and compendiums of geothermometers and geobarometers have been given by Essene (1982,1989), Newton (1983), and Ganguly and Saxena (1987).

Numerous geothermometers and geobarometers have been calibrated during the last two decades. Some of them return good estimates of P and T, if they are not stretched beyond their limits. To get good results, the user must be aware of the thermodynamic basis of calibration of geothermometers and geobarometers, so that he can judge the limits of their applicability. To borrow a phrase from Essene (1982, p.192), there is a lot more to geothermometry and geobarometry than "to plug in the numbers or read off the graphs." Accordingly, in this section we explore the methods of calibration of geothermometers and geobarometers. Any effort toward estimating T and P of equilibration must be matched by a corresponding one to assess their uncertainties. A rather simple procedure to achieve this goal will be outlined below. And finally, two worked examples of evaluation of intensive variables will be given.

9.3.2 Some Fundamental Considerations

9.3.2.1 Identification of Equilibrium Mineral Assemblages

Recognition of an equilibrium mineral assemblage is central to geothermometry and geobarometry. Therefore, we start by exploring the criteria used to identify an equilibrium assemblage. Without going into details, it would suffice to say that both textural and chemical criteria are available to judge how far this fundamental requirement is fulfilled by an observed mineral assemblage.

In metamorphic rocks, with which we shall be concerned here, a mineral assemblage comprises recrystallized and neocrystallized minerals. The process of recrystallization is governed by lattice strain and grain boundary energy, which are small. By contrast, the process of neocrystallization is driven by chemical potential, which can be large in magnitude. The recrystallized phases provide the textural criterion for an equilibrium. An equilibrium texture involving recrystallized minerals with a low surface energy, like quartz or calcite, is characterized by polygonal grains of roughly equal size and shape, with straight grain boundaries meeting at triple points (see Spry 1969, p.160). A chemical equilibrium in an assemblage is *suggested* by (1) unzoned mineral grains in mutual contact lacking every sign of a chemical reaction, and (2) the systematic partitioning of elements between the coexisting phases. Yet another necessary, but not sufficient, criterion is that the assemblage does not violate the phase rule. Based on (1) and (2), overwhelming evidence now exists that in many metamorphic rocks chemical equilibrium was, in general, closely approached (e.g. Jen and Kretz 1981). Given strong evidence for equilibrium, we may apply thermodynamics to compute the intensive variables prevailing during equilibration. What the calculated intensive variables imply in terms of the temporal evolution of a metamorphic rock is, however, an entirely different issue. This must be addressed from the viewpoint of kinetics.

9.3.2.2 Thermodynamic Basis for the Evaluation of the Intensive Variables

Given the compositions of coexisting minerals in an equilibrium assemblage, the calculation of any intensive parameter is based on the application of the thermodynamic criterion of equilibrium. In geothermometry and geobarometry, solid-solid equilibria are of fundamental importance, because they are independent of the fluid fugacities. Accordingly, our discussion focuses on solid-solid equilibria. Choosing the standard state as the pure components with the proper structures at any specified T and P, the condition of equilibrium [Eq.(1.105)] may be written as

$$RT\ln K = -\Delta G^*. \tag{9.28}$$

Recall that K comprises only activity terms a_i, which can be split into the mole fraction X_i and the activitity coefficient γ_i. If we split K in an analogous manner into two parts, K_X and K_γ, (9.28) may be expanded to

$$RT\ln K_X = -\Delta G^* - RT\ln K_\gamma. \tag{9.29}$$

The chemical compositions of coexisting minerals are easily recast to K_X, such that geobarometry and geothermometry boil down to the problem of obtaining data on the quantities on the RHS of (9.29), that is, calibrating $\Delta G^*(P,T)$, and deriving the equations of state of the crystalline solutions involved, from which K_γ is computed. It is also clear that the accuracy of computed P and T will depend upon (1) the accuracy of calibration of $\Delta G^*(P,T)$; (2) the veracity of the equations of state of the crystalline solutions; and on (3) the accuracy of our knowledge of the chemical compositions of the coexisting minerals. At present, the uncertainties due to (1) and (3) can be assessed with some rigor. Unfortunately, (2) very much eludes quantification, particularly for complex solutions, or even a simple crystalline solution having a large number of components.

   As emphasized above, computation of fugacities and activities of various fluid species prevailing during the equilibration of a mineral assemblage also uses Eq.(1.105). Since the underlying principle is identical in both cases, an explicit treatment of the solid-fluid equilibria is deemed unnecessary within the framework of this section. It suffices to point out that solving for a_i (or f_i) of the fluid species i requires that P and T of equilibration be independently known.

9.3.2.3 Choice of Equilibria Appropriate for Geothermometry and Geobarometry

Because a large number of reactions may be formulated between the components of the coexisting phases in a metamorphic rock, and calibration of reaction equilibria by experimental reversals is a tedious and expensive procedure, one may ask how many, and which, reactions to reverse. As pointed out by Thompson (1982, p.34), the number of phase components (or linearly independent end-members of the phase) minus the number of system components equals the number of linearly independent reactions in any system. Thus, the minimum requirement is to calibrate just one set of linearly independent equilibria. Given that, the linearly dependent equilibria can be handled by linear addition (e.g. Aranovich and Podlesskii 1989). Unfortunately, computation of the linearly dependent solid-solid reactions by linear summation of a number of independent reactions is severely plagued by error propagation problems (Sect. 5.5.2).

Given a number of equilibria to choose from, the other problem is which reactions to consider for geobarometry or geothermometry. The criteria for choosing the appropriate ones may be developed by rewriting Eq.(9.28) as follows

$$\ln K = -\frac{\Delta G^*}{RT} = -\frac{\Delta H^*}{RT} + \frac{\Delta S^*}{R} . \tag{9.30}$$

Taking its derivatives with respect to T and P, we have

$$\left(\frac{\partial \ln K}{\partial T}\right)_P = -\left[\frac{\partial(\Delta G^*/RT)}{\partial T}\right]_P = \frac{\Delta H^*}{RT^2}, \quad \text{and} \tag{9.31}$$

$$\left(\frac{\partial \ln K}{\partial P}\right)_T = -\left[\frac{\partial(\Delta G^*/RT)}{\partial P}\right]_T = -\frac{\Delta V}{RT} . \tag{9.32}$$

From (9.31), it is readily apparent that if ΔH^* of a solid-solid reaction is very large, its $\ln K$ term will be strongly temperature-dependent. Such an equilibrium will be a potential geothermometer. In such an equilibrium, a given uncertainty in $\ln K$ will translate to a comparatively small uncertainty in T; thus, the resolution of the geothermometer will be satisfactory. By contrast, solid-solid reactions having a large ΔV will be useful as geobarometers, since their $\ln K$ are highly pressure-dependent [Eq.(9.32)], and thus, a given uncertainty in $\ln K$ corresponds to a small uncertainty in P. The $\ln K$ of certain solid-solid reactions may show intermediate T- and P-dependencies; they are designated as geothermo-barometers. Their utility in deriving P or T will be considerably less.

9.3.3 Setting up Geobarometers and Geothermometers

The approach used to set up geobarometers and geothermometers, and the strategy adopted to calibrate them, are easily appreciated by taking an observed mineral assemblage as an example. Consider the common medium-grade metapelitic assemblage comprising quartz (Q), muscovite (Ms), biotite (Bi), garnet (Gt), sillimanite (Sill) *or* kyanite (Ky), staurolite (St), plagioclase (Plag), and ilmenite (Ilm), with or without rutile (R). Several studies (Guidotti

1974; Mohr and Newton 1983) testify that, based on textural and chemical criteria, it is an equilibrium assemblage. In order to compute the intensive parameters prevailing during its equilibration, we start by writing a set of balanced reactions between the components of those phases. Recalling that solid-solid reactions are fundamental to geothermometry and geobarometry, and restricting ourselves to them, we may, for example, formulate the following four reactions

$$1\ Ca_3Al_2[SiO_4]_3 + 2\ Al_2SiO_5 + SiO_2 = 3\ CaAl_2Si_2O_8, \tag{9.33}$$
$$\text{(gr in Gt)} \qquad \text{(Ky)} \quad \text{(Q)} \quad \text{(an in Plag)}$$

$$1\ Fe_3Al_2[SiO_4]_3 + 3\ TiO_2 = Al_2SiO_5 + 2\ SiO_2 + 3\ FeTiO_3, \tag{9.34}$$
$$\text{(alm in Gt)} \qquad \text{(R)} \quad \text{(Sill)} \quad \text{(Q)} \ \text{(ilm in Ilm)}$$

$$1/3\ Fe_3Al_2[SiO_4]_3 + 1/3\ KMg_3[AlSi_3O_{10}](OH)_2$$
$$\text{(alm in Gt)} \qquad\qquad \text{(phl in Bi)}$$
$$= 1/3\ Mg_3Al_2[SiO_4]_3 + 1/3\ KFe_3[AlSi_3O_{10}](OH)_2, \tag{9.35}$$
$$\text{(py in Gt)} \qquad\qquad \text{(ann in Bi)}$$

$$1/3\ Fe_3Al_2[SiO_4]_3 + MnTiO_3 = 1/3\ Mn_3Al_2[SiO_4]_3 + FeTiO_3. \tag{9.36}$$
$$\text{(alm in Gt)} \quad \text{(pyn in Ilm)} \qquad \text{(spess in Gt)} \quad \text{(ilm in Ilm)}$$

Note that the parenthesized abbreviations appearing in lower case letters designate the phase components, those in upper case letters, the phases. Lack of a specification of components implies that these phases (e.g. Q, Ky, Sill, R) may, for all intents and purposes, be regarded as end-members. Considering that some of the above mentioned phases are multicomponent solutions, many more reactions could have been written. Those given here, however, suffice for our purpose. Note, moreover, that numerous solid-fluid reactions could have been formulated as well. For reasons given earlier, none of them will be explicitly treated in this section.

Of the four reactions given above, the first two belong to the category of net-transfer reactions (Thompson 1982); a net-transfer of matter from one phase to another is involved. They are readily recognized petrographically because of an abrupt change in modal abundance of the phases due to the reaction. A transfer of matter from one phase to another is sometimes accompanied by a change in the coordination number of a cation. Thus, in (9.33), the transfer of Al from an (in Plag) to gr (in Gt) involves an increase of the coordination number of Al from [4] to [6]. Similarly, (9.34) shows an increase in the coordination number of Fe from [6] to [8] when Fe leaves ilm (in Ilm) and enters alm (in Gt). Concurrently, half of Al escaping from Sill to alm (in Gt) increases its coordination number from [4] to [6]. An increase in the coordination number of an atom is always accompanied by a decrease of the molar volume of the phase. As such, these reactions are accompanied by large ΔV, and are thus potentially good geobarometers (see Sect.9.3.2.3). By contrast, (9.35) and (9.36) are exchange reactions, characterized by exchange of cations between a pair of coexisting phases. Typically, the exchange reactions have very small ΔV, and thus, are barely pressure-dependent. If the element fractionation happens to be such that K is much smaller or larger than 1, then, as is seen from Eq.(9.28), its ΔG^* (or $\Delta H^* - T\Delta S^*$) will be large. Such reactions are potential geothermometers. Given a huge amount of observational data on element fractionation collected since the pioneering studies of Kretz (1959) and Mueller (1961), potential geothermometers are easily recognized. Judged by the data on the fractionation of Fe and Mg between Bi and Gt, and of Fe and Mn between Gt and Ilm (Guidotti 1974), both (9.35) and (9.36) are likely to be good geothermometers.

9.3.3.1 Calibration of Geobarometers

Having identified two geobarometers in our metapelite, let us take a quick look at how they are calibrated. Both reactions,

$$1 \ Ca_3Al_2[SiO_4]_3 + 2 \ Al_2SiO_5 + SiO_2 = 3 \ CaAl_2Si_2O_8, \text{ and} \qquad (9.33)$$
$$\text{(gr in Gt)} \qquad \text{(Ky)} \quad \text{(Q)} \qquad \text{(an in Plag)}$$

$$1 \ Fe_3Al_2[SiO_4]_3 + 3 \ TiO_2 = Al_2SiO_5 + 2 \ SiO_2 + 3 \ FeTiO_3, \qquad (9.34)$$
$$\text{(alm in Gt)} \qquad \text{(R)} \quad \text{(Sill)} \qquad \text{(Q)} \ \text{(ilm in Ilm)}$$

involve one solution phase (Gt) on the reactant and one each (Plag or Ilm) on the product sides, the remaining phases being pure end-members. *Calibration of these equilibria for geobarometry implies obtaining standard state thermodynamic data* for these reactions. This may be achieved in two alternative manners, briefly described below.

As indicated in several preceding chapters, reaction reversal experiments with end-member phases often serve as a source of standard state thermodynamic data. Because anorthite and grossular are easily synthesized pure, performing reversal experiments with them pose no problem. Koziol and Newton (1988a) used this method to calibrate the reaction (9.33). Given several reversal brackets for the reaction among the end-member phases (for which $K = 1$ and $RT\ln K = 0$), along with ancillary data on $C°_P(T)$, $V°_{298}$, $\alpha(T)$, and $\beta(P)$ for all phases participating in the reaction, in addition to $\Delta H°_{844}$ and $\Delta S°_{844}$ for the low-quartz/high-quartz phase transition, standard state properties of the reactions, $\Delta H°_{298}$ and $\Delta S°_{298}$, are extracted from

$$-\Delta H°_{298} + T\Delta S°_{298} = \int_{298}^{844} \Delta C°_P(T)dT + \Delta H°_{844} + \int_{844}^{T} \Delta C°_P(T)dT$$

$$-T\left[\int_{298}^{844} \left\{\frac{\Delta C°_P(T)}{T}\right\}dT + \Delta S°_{844} + \int_{844}^{T} \left\{\frac{\Delta C°_P(T)}{T}\right\}dT\right] + \int_{1}^{P} \Delta V_T(P)dP . \qquad (9.13)$$

where T and P refer to each reversal bracketing point.

Prior to Koziol and Newton (1988a), reversal of reaction (9.33) was unsuccessful below 1100 °C due to its immense sluggishness. The geobarometer, denoted henceforth as GASP (for garnet-aluminum silicate-plagioclase-quartz), is, however, applied to rocks at far lower temperatures, on the order of 500-850 °C. This requires it to be extrapolated over a huge temperature range, thereby aggravating its inherent uncertainty. To ameliorate that problem, Koziol and Newton (1988a) strove for direct reversals at lower temperatures. Using a Li_2MoO_4 flux to promote the reaction rate, they eventually succeeded in obtaining tight reversals to 900 °C. To ensure proper extrapolation to yet lower temperatures, they employed an *indirect constraint* at 650 °C, on the basis of a linear combination of the reactions 4 zoisite + quartz = 5 anorthite + grossular + 2 H_2O and 2 zoisite + kyanite + quartz = 4 anorthite + H_2O. Their entire set of data is well represented by the linear relationship,

$$P \text{ (bar)} = 22.80 \text{ T (°C)} - 1093. \tag{9.37}$$

Referring to Eq. (9.13), this linear relation suggests that within the limit of uncertainties of the experimental data, $\Delta C°_P(T)$ of reaction (9.33) is zero and $\Delta H°_{844}$ and $\Delta S°_{844}$ of the high-low phase transition in quartz negligible. It also allows ΔV to be regarded as independent of T and P. Under these conditions, (9.13) may be simplified to

$$-\Delta H°_{<T>} + T\Delta S°_{<T>} = \Delta V(P-1), \tag{9.13a}$$

where $\Delta H°_{<T>}$ and $\Delta S°_{<T>}$ are the average standard state properties for the experimentally explored temperature range, introduced and discussed in detail in Section 5.5. Similarly, ΔV is the average volume difference for the experimental range of T and P. Using a ΔV of 6.210 J bar^{-1} (Koziol and Newton 1988a, p.223), the reversal brackets were fitted to (9.13a) by least squares, giving

$$\Delta H°_{<T>} = 47515 \pm 6014 \text{ J, } \Delta S°_{<T>} = 143.907 \pm 4.583 \text{ J K}^{-1}, \text{ and } \rho_{\Delta H\Delta S} = 0.9914. \tag{9.38}$$

Note that $\Delta H°_{<T>}$ and $\Delta S°_{<T>}$ derived from the experimental reversal data are correlated, the correlation coefficient, $\rho_{\Delta H\Delta S}$, equaling 0.9914. As we shall see later, knowledge of $\rho_{\Delta H\Delta S}$ is pivotal to estimating the uncertainty of a computed P-T curve. Ignoring this term usually inflates the computed uncertainty considerably.

Using the set of standard state reaction properties obtained above, the condition of equilibrium for the GASP assemblage may be extrapolated to multicomponent natural systems via

$$\Delta \mu = 0 = \Delta H°_{<T>} - T\Delta S°_{<T>} + \Delta V(P-1) + RT\ln K. \tag{9.39}$$

Alternatively, if the compositions of coexisting phases have been measured, such that K_X is known, and the equations of state of the solutions are available, giving K_γ at any T and P, *independent knowledge* of T permits solution of (9.39) for P. Because K_γ is, in general, a function of T and P, our estimate of K also depends on T and P. Thus, a solution of (9.39) at a given T can be obtained only by iterations in P. That is how most geobarometers work. In the event of ideal behavior of all solutions involved, $RTlnK_\gamma$ is zero, and a direct solution for P becomes possible for any T,

$$P = \frac{-\Delta H^\circ_{<T>} + T\Delta S^\circ_{<T>} + \Delta V - RTlnK_X}{\Delta V}.$$

(9.40)

It should be borne in mind that the lack of data on equations of state of solutions often leads to application of (9.40) to rocks. In those cases, the results obtained should be regarded as a first approximation, since mineral crystalline solutions scarcely behave ideally.

Having gone through one method of calibrating a geobarometer, let us address the alternative technique, which is a logical extension of the former method. In this, as many *indirect constraints* as possible are utilized to pin down the P-T curve of the reaction. It uses the calorimetric constraints of each phase as well as the reversal brackets of each reaction in which any of those phases participate, and processes them simultaneously by mathematical programming (see Chap. 7). Berman (1988) utilized this technique to calibrate the GASP geobarometer. It suffices to emphasize that the internally consistent thermodynamic dataset obtained by mathematical programming is likely to be somewhat more stringently constrained. That notwithstanding, such data will not be used in this chapter, because their uncertainties remain to be quantified.

The second geobarometer in our metapelite is based on the garnet-rutile-aluminum silicate-ilmenite-quartz assemblage, dubbed GRAIL geobarometer. The basic reaction

$$1\ Fe_3Al_2[SiO_4]_3 + 3\ TiO_2 = Al_2SiO_5 + 2\ SiO_2 + 3\ FeTiO_3,$$ (9.34)
   (alm in Gt)   (R)   (Sill)   (Q) (ilm in Ilm)

was reversibly bracketed by Bohlen et al. (1983), using end-member phases. Their experimental reversals were employed to extract the following standard state properties

$$\Delta H^\circ_{<T>} = 1657 \pm 1053\ J, \Delta S^\circ_{<T>} = 22.507 \pm 1.010\ J\ K^{-1},$$
$$\Delta V = 1.874\ J\ bar^{-1}, and\ \rho_{\Delta H\Delta S} = 0.9910.$$ (9.41)

Berman (1988) also used those brackets as input in mathematical programming to calibrate reaction (9.34). The P-T curve calculated from (9.41) is very similar to those obtained with Berman's (1988) data. Again, we shall use the data in Eq.(9.41) in our subsequent computations, because we need to estimate the uncertainties of the computed equilibrium.

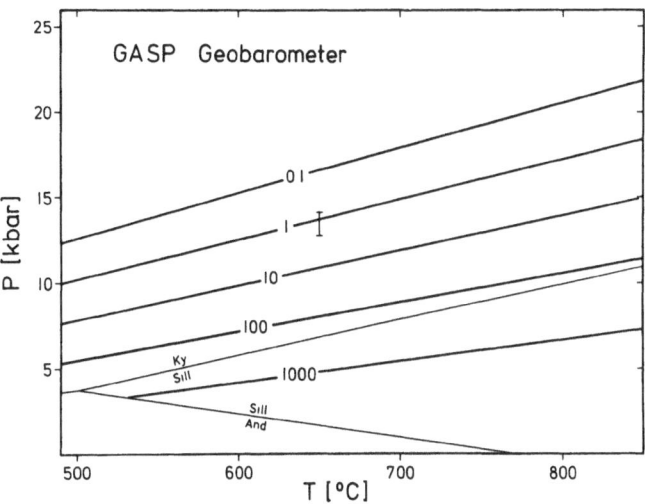

Fig. 9.10 A P-T diagram displaying isopleths of K for the GASP geobarometer. Data from Eq. (9.38) were used to compute the equilibria for K between 0.1 and 100. The reversal bracket indicated at 650°C refers to the indirect constraint discussed by Koziol and Newton (1988a) for the end-member reaction ($K = 1$) involving kyanite. The isopleth for $K = 1000$ involves sillimanite, and was computed from Eq. (9.43). Also indicated on this diagram are the stability fields of the Al_2SiO_5 polymorphs after Holdaway (1971)

Application of both equilibria to natural assemblages requires that they be extrapolated to multicomponent systems. Figure 9.10 displays the effect of such extrapolation by plotting isopleths of K for the GASP assemblage on the P-T diagram. To calculate these isopleths, Eq.(9.39) may be recast to

$$P = \left[\frac{-\Delta H^{\circ}_{<T>} + T\Delta S^{\circ}_{<T>} + \Delta V}{\Delta V} \right] - \left[\frac{RT\ln K}{\Delta V} \right]. \tag{9.42}$$

Note that, in this equation, the first square-bracketed term on the RHS gives the equilibrium pressure for the reaction among the end members, whereas the second one the *displacement* in pressure owing to mixing of components in the solution phases. Using the $\Delta H^{\circ}_{<T>}$, $\Delta S^{\circ}_{<T>}$, and ΔV given in (9.38) [and (9.43), see below], P was obtained for specified T and K, and plotted in Fig. 9.10.

To grasp the significance of Fig. 9.10, recall that $K = 1$ implies a reaction among the pure end-members. Now depending upon the composition of the Plag and the coexisting Gt (Ky and Q are end-member phases), K might assume values smaller or larger than 1. In the former case, the equilibrium is displaced to higher P at any T, and in the latter, to lower P. As is readily apparent from Fig. 9.10, for a K in excess of 100, the GASP assemblage will no longer involve Ky, but Sill. In fact, Ghent (1976, Table 1), who first recognized the barometric potential of GASP, listed a number of rocks in which the numerical value of K_X ranges up to 2679. For all realistic values of K_γ, K for those assemblages would easily exceed 100, and sure enough, they contain Sill, rather than Ky. In other words, the GASP equilibrium must now be computed with Sill. For this purpose, $\Delta H°_{<T>}$, $\Delta S°_{<T>}$, and ΔV of the Ky = Sill reaction (Holdaway 1971) were linearly combined with Eq.(9.38) to calibrate the GASP equilibrium involving sillimanite. The results are

$$\Delta H^\circ_{<T>} = 33619 \pm 6014 \text{ J}, \ \Delta S^\circ_{<T>} = 120.439 \pm 4.583 \text{ J K}^{-1},$$
$$\Delta V = 5.071 \text{ J bar}^{-1}, \text{ and } \rho_{\Delta H \Delta S} = 0.9914. \tag{9.43}$$

These data were used to calculate the $K = 1000$ isopleth of Fig. 9.10. Likewise, the GRAIL reaction was calibrated to include Ky in lieu of Sill,

$$\Delta H^\circ_{<T>} = -5291 \pm 1053 \text{ J}, \ \Delta S^\circ_{<T>} = 10.733 \pm 1.010 \text{ J K}^{-1},$$
$$\Delta V = 1.3405 \text{ J bar}^{-1}, \text{ and } \rho_{\Delta H \Delta S} = 0.9910. \tag{9.44}$$

9.3.3.2 Calibration of Exchange Geothermometers

Calibration of an exchange geothermometer means obtaining the standard state thermodynamic data for that exchange reaction. Once again, an important source of such data are the reaction reversal experiments involving two crystalline solutions, between which two cations (or anions) partition. The nature of those experiments and their thermodynamic evaluation will be briefly examined below.

Consider, for example, the Fe-Mg partitioning between biotite and garnet. The exchange reaction underlying that process is

$$1/3 \ Fe_3Al_2[SiO_4]_3 + 1/3 \ KMg_3[AlSi_3O_{10}](OH)_2$$
$$\text{(alm in Gt)} \qquad \qquad \text{(phl in Bi)}$$
$$= 1/3 \ Mg_3Al_2[SiO_4]_3 + 1/3 \ KFe_3[AlSi_3O_{10}](OH)_2, \tag{9.35}$$
$$\text{(py in Gt)} \qquad \qquad \text{(ann in Bi)}$$

written here on one cation-exchange basis. At equilibrium, we have

$$0 = \Delta G^\circ_T + \int_1^T \Delta V_T(P)dP + RT\ln K . \tag{9.45}$$

Two simplifications are generally introduced at this stage. First, the volume integral is replaced by the product $\Delta V°_{298}(P-1)$. This is justified, because $\alpha(T)$ and $\beta(P)$ of the reactants and products are virtually identical. Second, $\Delta G°_T$ is recast to $\Delta H°_{<T>} - T\Delta S°_{<T>}$, unless a *large* extrapolation outside the experimentally explored range of T is intended. With these simplifications, and the RTln$K$ term split up as before, (9.45) reduces to

$$0 = \Delta H^\circ_{<T>} - T\Delta S^\circ_{<T>} + \Delta V^\circ_{298}(P-1) + RT\ln K_X + RT\ln K_\gamma. \tag{9.46a}$$

Now, in dealing with the exchange reactions, in which a pair of atoms is distributed between a pair of phases, it is *customary* to designate the mole fraction part of the equilibrium constant as K_D, the distribution coefficient. Clearly, K_D is a special case of K_X. Complying with this usage, we rewrite (9.46a) as

$$0 = \Delta H^\circ_{<T>} - T\Delta S^\circ_{<T>} + \Delta V^\circ_{298}(P-1) + RT\ln K_D + RT\ln K_\gamma. \qquad (9.46)$$

Denoting the mole fraction and the activity coefficient of the q-normalized mixing unit of almandine in garnet as X^{Gt}_{Fe} and γ^{Gt}_{Fe}, rather than as $X^{Gt}_{FeAl2/3[SiO4]}$ and $\gamma^{Gt}_{FeAl2/3[SiO4]}$, and likewise for the other components, K_D and K_γ are expressed as

$$K_D = \frac{(X_{Mg}/X_{Fe})^{Gt}}{(X_{Mg}/X_{Fe})^{Bi}}, \quad \text{and} \qquad (9.47)$$

$$K_\gamma = \frac{(\gamma_{Mg}/\gamma_{Fe})^{Gt}}{(\gamma_{Mg}/\gamma_{Fe})^{Bi}}. \qquad (9.48)$$

The objective of these experiments is to reversibly bracket the *compositions* of the coexisting biotite and garnet solutions at a specified set of P and T. In other words, the quantity bracketed by the experiments is K_D. Ferry and Spear (1978) achieved this by equilibrating a garnet and a biotite of known composition at a set of P and T and measuring the composition of the biotite at the end of the experiment. Since the garnet is sluggish to react compared to the biotite, they invariably employed bulk compositions with a molar ratio of garnet:biotite as 98:2. At one T and P, a pair of biotites were equilibrated with a garnet. Figure 9.11 depicts the tie lines of the starting mixtures for their experiments at 799°C and 2.07 kbar by the two dashed straight lines. On equilibration, the tie lines rotate to their final positions shown by the solid straight lines. Because the bulk compositions, shown by encircled dots in Fig. 9.11, were very rich in garnet, the garnets did not change their composition in this process, even though the biotites did. The bracketing tie lines are displayed by the solid straight lines; they correspond to a reversal bracket of

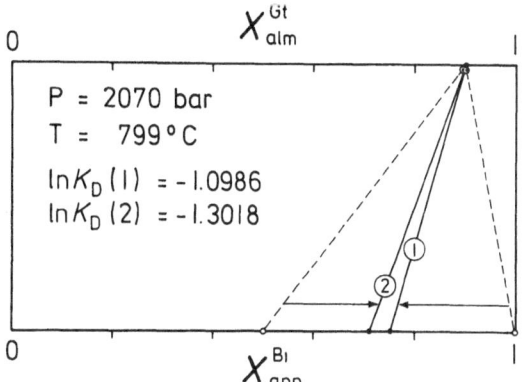

Fig. 9.11 Reversal bracketing of K_D for the Fe-Mg partitioning reaction between garnet and biotite at 2070 bar and 799°C. The tie lines in the starting mixtures are shown by *dashed lines*, the *solid lines* being the final garnet-biotite tie lines. The bulk compositions of the mixtures are indicated by encircled dots

K_D between -1.0986 and -1.3018. In this manner, Ferry and Spear (1978) obtained Fe-Mg partitioning data between biotite and garnet at 2.07 kbar pressure and at temperatures between 550 and 799 °C.

To extract $\Delta H°_{<T>}$ and $\Delta S°_{<T>}$ from such experiments, Eq.(9.46) is rearranged to the following form,

$$-\Delta H°_{<T>} + T\Delta S°_{<T>} = \Delta V°_{298}(P-1) + RT\ln K_D + RT\ln K_\gamma, \qquad (9.49)$$

and $\Delta H°_{<T>}$ and $\Delta S°_{<T>}$ obtained from a least squares fit of the RHS as a function of T. Note that $\Delta V°_{298}$ is known, as is $RT\ln K_D$, not, however, $RT\ln K_\gamma$. To evaluate the RHS completely, knowledge of K_γ is thus essential. In principle, K_γ could be calculated from the equations of state of the garnet and biotite solutions, were they available. Lack of these data forced Ferry and Spear (1978) seek a solution in an indirect manner. They made two sets of experiments with two garnet compositions and found K_D to be independent of the composition of the garnets. They thus concluded that $RT\ln K_\gamma$ is zero within the limits of experimental resolution. This does not necessarily imply that Fe-Mg biotites and Fe-Mg garnets are ideal solutions. All it implies is that possible nonidealities in them offset each other, reducing $RT\ln K_\gamma$ to zero. For this case, $RT\ln K_D$ is equal to $RT\ln K$. Table 9.9 lists Ferry and Spear's (1978) data and evaluates them for the RHS of (9.49). Note that only the set of experiments with $alm_{90}py_{10}$ garnet have been evaluated. Indeed, the reason behind ignoring the runs with $alm_{80}py_{20}$ garnet in Table 9.9 is that those brackets are rather wide, and give no additional information whatsoever, K_D being independent of the composition of the solutions. A least squares fit to the data, weighting each one with the uncertainties indicated by Ferry and

Table 9.9. Evaluation of Fe-Mg exchange equilibrium between garnet and biotite (Ferry and Spear 1978)

T (K)	X^{Gt}_{Fe}	X^{Bi}_{Fe}	$\Delta V°_{298}(P-1)$ (J)	$RT\ln K_D$ (J)	$RT\ln K_\gamma$ (J)	$RT\ln K + \Delta V°_{298}(P-1)$ (J)
823.15	0.900	0.587	164.5	-12631.5	0	-12467
823.15	0.900	0.620	164.5	-11687.2	0	-11523
872.15	0.900	0.608	164.5·	-12750.1	0	-12586
872.15	0.900	0.645	164.5	-11602.8	0	-11438
924.15	0.900	0.661	164.5	-11751.9	0	-11587
924.15	0.900	0.679	164.5	-11126.3	0	-10962
971.15	0.900	0.690	164.5	-11280.8	0	-11116
971.15	0.900	0.704	164.5	-10745.5	0	-10581
1022.15	0.900	0.695	164.5	-11673.7	0	-11509
1011.15	0.900	0.730	164.5	-10110.3	0	-9946
1072.15	0.900	0.710	164.5	-11604.8	0	-11440
1072.15	0.900	0.750	164.5	-9793.2	0	-9629

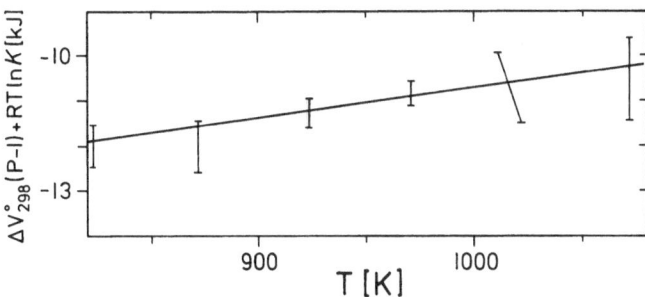

Fig. 9.12 A plot of the RHS of Eq. (9.49) against T for Ferry and Spear's (1978) reversal data on Fe-Mg partitioning between garnet and biotite at 2070 bar and temperatures between 550-799 °C. The result of the least squares fit, reproduced in Eq.(9.50), is shown by a *straight line* passing through the brackets

Spear (1978), gives

$$\Delta H^\circ_{<T>} = 17580 \pm 2136 \text{ J}, \Delta S^\circ_{<T>} = 6.701 \pm 2.287 \text{ J K}^{-1}, \text{ and } \rho_{\Delta H \Delta S} = 0.9960,$$
(9.50)

with ΔV°_{298} known independently as 0.0795 J bar⁻¹. As anticipated, $\Delta H^\circ_{<T>}$ and $\Delta S^\circ_{<T>}$ are similar to those obtained earlier by Ferry and Spear [1978, Eq.(7)]. Additional information obtained here is the value of the correlation coefficient, $\rho_{\Delta H \Delta S}$, to be used later to compute the uncertainty of the geothermometer. Figure 9.12 is a plot of the RHS of (9.49) against T. The straight line through the brackets was recalculated from (9.50). Note that weighting of the input data for least squares analysis forces the curve through the center of the best constrained brackets (see also Fig. 3 in Ferry and Spear 1978).

Given the standard state properties for the exchange reaction, the T of equilibration of the biotite-garnet pairs may be computed by recasting Eq.(9.46) to

$$T = \frac{\Delta H^\circ_{<T>} + \Delta V^\circ_{298}(P - 1)}{\Delta S^\circ_{<T>} - R \ln K}.$$
(9.51)

To the extent that the garnet and the biotite belong to the alm-py and ann-phl binary, $RT \ln K_\gamma = 0$, and $RT \ln K$ reduces to $RT \ln K_D$. Using the data from (9.50), we may thus rewrite (9.51) as

$$T = \frac{17580 + 0.0795(P - 1)}{6.701 - 8.3143 \ln K_D} - 273.15,$$
(9.52)

with T expressed in °C and P in bar. Given independent information on P and data on the compositions of the coexisting garnet-biotite pair, from which K_D is known, a direct solution for T is possible. It is important to remember, however, that garnets and biotites of most metapelites do deviate from those binaries, and therefore, T must be corrected for the $RT\ln K_\gamma$ term. Unfortunately, the mixing properties of the multicomponent garnets and biotites are not yet sufficiently well known to permit a convincing correction for the compositional deviations of both. In absence of such data, Ferry and Spear (1978) suggested that the garnet-biotite Fe-Mg exchange geothermometer be applied to rocks *without* correction, providing the (Ca+Mn)/(Ca+Mn+Fe+Mg) ratio in Gt does not exceed roughly 0.2 and the $(Al^{VI}+Ti)/(Al^{VI}+Ti+Fe+Mg)$ in Bi is not much above 0.15.

Let us now address the Fe-Mn exchange reaction between garnet and ilmenite, experimentally reversed by Pownceby et al. (1987). In terms of the q-normalized mixing units, it can be written as

$$1/3\ Fe_3Al_2[SiO_4]_3 + MnTiO_3 = 1/3\ Mn_3Al_2[SiO_4]_3 + FeTiO_3. \qquad (9.36)$$
$$\text{(alm in Gt)} \quad \text{(pyn in Ilm)} \qquad \text{(spess in Gt)} \quad \text{(ilm in Ilm)}$$

Again, the lower case abbreviations in parentheses designate the components of the phases. In the following, however, we shall use a more condensed form of notations like X^{Ilm}_{Mn}, X^{Ilm}_{Fe}, γ^{Ilm}_{Mn}, and γ^{Ilm}_{Fe} for X^{Ilm}_{pyn}, X^{Ilm}_{ilm}, γ^{Ilm}_{pyn}, and γ^{Ilm}_{ilm}, and so on.

Experimental reversal of K_D between 600-900°C at 2 and 5 kbar pressure revealed a clear composition dependence of K_D, indicating a nonzero $RT\ln K_\gamma$ term for this reaction. Figure 9.13 illustrates the data obtained at 800°C and 5000 bar pressure. For a specified garnet composition, K_D is found to vary substantially with X^{Ilm}_{Mn}.

The derivation of the standard state properties may again make use of Eq.(9.46). However, contrary to the case of Fe-Mg exchange between garnet and biotite, an explicit evaluation of the $RT\ln K_\gamma$ term will be essential now.

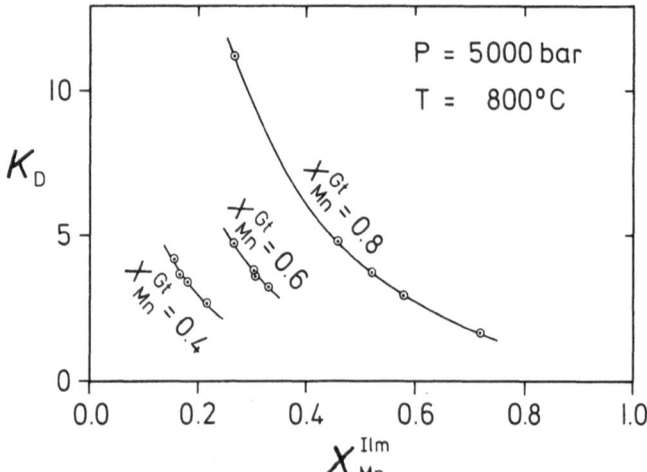

Fig. 9.13 A K_D vs X^{Ilm}_{Mn} plot of the Fe-Mn exchange equilibrium between garnet and ilmenite at 800°C and 5000 bar for three garnet compositions (Pownceby et al. 1987)

Splitting $RT\ln K_\gamma$ into the appropriate $RT\ln\gamma_i$ terms, Eq.(9.46) is rewritten as

$$RT\ln K_D = -\Delta H^\circ_{<T>} + T\Delta S^\circ_{<T>} - \Delta V^\circ_{298}(P-1) - RT[\ln\gamma^{Gt}_{Mn} - \ln\gamma^{Gt}_{Fe}] - RT[\ln\gamma^{Ilm}_{Fe} - \ln\gamma^{Ilm}_{Mn}]. \quad (9.53)$$

Given a large number of reversals of K_D between 600-900°C, and 2-5 kbar pressure (Pownceby et al. 1987, Table 1), it is possible *in principle* to solve for the unkowns on the RHS of (9.53). Such a job is somewhat facilitated by the fact that $(Fe,Mn)TiO_3$ is known to be a symmetric solution (Pownceby et al. 1987, p.121), having a T-P-independent W_G parameter,

$$W^{Ilm}_G = 1757 \pm 180 \text{ J mol}^{-1}. \quad (9.54)$$

The $RT\ln\gamma^{Ilm}_i$ terms of (9.53) may then be obtained from (9.54) via

$$RT\ln\gamma^{Ilm}_{Fe} = W^{Ilm}_G (X^{Ilm}_{Mn})^2, \text{ and } RT\ln\gamma^{Ilm}_{Mn} = W^{Ilm}_G (X^{Ilm}_{Fe})^2. \quad (9.55)$$

Comparable data for Fe-Mn mixing in garnet are lacking; they must be extracted from the Fe-Mn exchange experiments (see below). For this purpose, Pownceby et al. (1987) adopted an analogous mixing model for the garnet, such that,

$$RT\ln\gamma^{Gt}_{Fe} = W^{Gt}_G (X^{Gt}_{Mn})^2, \text{ and } RT\ln\gamma^{Gt}_{Mn} = W^{Gt}_G (X^{Gt}_{Fe})^2. \quad (9.56)$$

Moreover, they observed that the ΔV°_{298} of reaction (9.36) is small enough to be safely ignored. Inserting Eqs.(9.55) and (9.56) into (9.53), and setting $\Delta V^\circ_{298} = 0$, the latter reduces to

$$RT\ln K_D = -\Delta H^\circ_{<T>} + T\Delta S^\circ_{<T>} + W^{Gt}_G (2X^{Gt}_{Mn}-1) - W^{Ilm}_G (2X^{Ilm}_{Mn}-1). \quad (9.57)$$

Using their K_D data, and the W^{Ilm}_G from Eq.(9.54), Pownceby et al. (1987) solved (9.57) by least squares for $\Delta H^\circ_{<T>}$, $\Delta S^\circ_{<T>}$, and W^{Gt}_G. Because the data processing is comparable to that in Section 8.5.4.3, it will not be discussed further. Their results support the postulated solution model for garnet, and yield

$$\Delta H^\circ_{<T>} = -17108 \pm 4217 \text{ J, } \Delta S^\circ_{<T>} = -6.025 \pm 4.100 \text{ J K}^{-1} \text{ and} \quad (9.58)$$

$$W^{Gt}_G = 322 \pm 377 \text{ J mol}^{-1}. \quad (9.59)$$

Since the computation of uncertainty of the geothermometer depends critically upon the use of nonzero correlation coefficients among the extracted parameters, and these were not indicated by Pownceby et al. (1987), we had to fit their experimental data once more to Eq.(9.57). In so doing, W^{Gt}_G was constrained to 322 J mol^{-1}, and W^{Ilm}_G to 1757 J mol^{-1}. This resulted in

$$\Delta H^\circ_{<T>} = -17350 \pm 3749 \text{ J, } \Delta S^\circ_{<T>} = -6.254 \pm 3.656 \text{ J K}^{-1}, \rho_{\Delta H\Delta S} = 0.9962. \quad (9.60)$$

If the compositions of the coexisting ilmenite and garnet do not deviate significantly from the binaries explored experimentally, Eq.(9.57) may be rearranged for geothermometry as

$$T = \frac{\Delta H^\circ_{<T>} + W^{Ilm}_G(2X^{Ilm}_{Mn} - 1) - W^{Gt}_G(2X^{Gt}_{Mn} - 1)}{-RlnK_D + \Delta S^\circ_{<T>}} - 273.15 , \qquad (9.61)$$

with T expressed in °C. Entering the data from Eqs.(9.54), (9.59), and (9.60) into (9.61), we obtain the expression for the garnet-ilmenite Fe-Mn exchange geothermometer:

$$T = \frac{-17350 + 1757(2X^{Ilm}_{Mn} - 1) - 322(2X^{Gt}_{Mn} - 1)}{-8.3143 \ln K_D - 6.254} - 273.15 , \qquad (9.62)$$

where T is given in °C.

9.3.4 Limits of Applicability and Evaluation of Uncertainties in Geothermometry and Geobarometry

Several problems arise in applying experimentally calibrated geothermometers and geobarometers to rocks. These will be treated briefly in this section.

9.3.4.1 Limits of Applicability

As stated earlier, geothermometers and geobarometers are generally calibrated within the framework of a rather simple model system. The mineral assemblages in nature are chemically far more complex. This requires that the user extrapolates the calibrations to the complex systems. This is feasible, given equations of state for the pertinent solutions, whence the $RTlnK_\gamma$ term of Eq. (9.46) may be computed. The knowledge of the standard state properties then permits calculation of T and P for equilibration of the chemically complex mineral assemblage. Unfortunately, many geothermometers or geobarometers involve multicomponent crystalline solutions, whose mixing properties are yet to be convincingly quantified. Such data can barely be utilized with confidence either in geothermometry or in geobarometry. Consider, for example, the biotites. Almost all natural biotites contain Al in excess of those on the phlogopite-annite join; many of them also contain Ti in appreciable amounts. Although the Fe-Mg mixing appears to be near-ideal (Mueller 1972, Wones 1972) on the phlogopite-annite binary, we have yet to assess the property changes due to the incorporation of Ti, or additional Al, from experimental data. Likewise, the mixing properties of Ca, Mg, and Fe in the aluminosilicate garnets are understood (Koziol and

Newton 1988b; Newton et al. 1989), not, however, those of $Mn^{2)}$ in its [8]-site, let alone knowing the properties of garnets with Fe^{3+} and Cr in the [6]-site. In this regard, geothermometers and geobarometers based on solutions with fewer components should be more promising. As an alternative to extrapolating from simple to natural systems, one may experimentally calibrate a geothermometer (or a geobarometer) in more complex systems, as did Pattison and Newton (1989). Such a geothermometer may not permit extrapolation to another bulk composition, however. No matter which approach is chosen, there is still a long way to go. In the worked examples given below, we shall utilize the experimentally calibrated geothermometers and geobarometers, and apply them to rocks, as far as possible within the framework of the calibration.

9.3.4.2 Evaluation of Uncertainties

Ideally, application of geothermometry and geobarometry should be accompanied by a statement regarding the uncertainties of T and P, σ_T and σ_P. Although a rigorous estimation of σ_T and σ_P might be a fairly arduous exercise, satisfactory results are obtained by a *simplified version* of the error propagation calculations outlined earlier for the end-member equilibria (Chaps. 4-6). An extension of that approach to reactions involving crystalline solutions may be attempted as follows.

The general expression for the conditions of equilibrium for a geothermometer or a geobarometer is written as

$$\Delta\mu = 0 = \Delta H^{\circ}_{<T>} - T\Delta S^{\circ}_{<T>} + \Delta V^{\circ}_{298}(P-1) + RT\ln K_D + RT\ln K_\gamma. \tag{9.46}$$

Obtaining σ_T of a geothermometer means solving for the uncertainty in T at any given equilibrium P and compositions of the solutions, expressed in terms of X_i. Likewise, σ_P is obtained by solving for the uncertainty in P at specified equilibrium T and X_i. Therefore, we begin by asking which of the terms on the RHS of (9.46) depend on T, P, and X_i. Evidently, the second and the third terms depend, respectively, upon T and P, the fourth on T and X_i, while the last one is a function of P, T, and X_i. To the extent that $RT\ln K_\gamma$ is amenable to modeling by T-P-independent W_G parameters (as in the garnet-ilmenite thermometer and the GASP geobarometer, see below), it reduces to a function of X_i [cf. Eq.(9.55)]. This enables us to formulate a simple algorithm for calculating σ_T and σ_P. In Section 9.3.5, we shall ignore the minor T-P-dependences of W_G's employed in the GRAIL geobarometer (see later), and apply the simplified algorithm to come up with σ_P. If $RT\ln K_\gamma$ is a sole function of X_i, the variance of $\Delta\mu$ at any P and X_i of equilibrium, $\sigma^2_{\Delta\mu}$, is

---

[2] Reader's attention is drawn to two papers (Koziol 1990, Berman 1990) published after completion of the manuscript for this book. Both explore the properties of spessartine-bearing garnets.

given by (cf. Sect. 4.3.1)

$$\sigma_{\Delta\mu}^2 = \left(\frac{\partial\Delta\mu}{\partial T}\right)^2 \sigma_T^2 = (-\Delta S_{<T>}^\circ + R\ln K_D)^2 \sigma_T^2 . \qquad (9.63)$$

Therefore, given data on $\sigma_{\Delta\mu}$, we have

$$\sigma_T = \frac{\sigma_{\Delta\mu}}{|-\Delta S_{<T>}^\circ + R\ln K_D|} . \qquad (9.64)$$

Similarly, for a geobarometer, at any T and X_i of equilibrium,

$$\sigma_{\Delta\mu}^2 = \left(\frac{\partial\Delta\mu}{\partial P}\right)^2 \sigma_P^2 = (\Delta V_{298}^\circ)^2 \sigma_P^2 , \qquad (9.65)$$

such that

$$\sigma_P = \frac{\sigma_{\Delta\mu}}{|\Delta V_{298}^\circ|} . \qquad (9.66)$$

Let us next address the $\sigma_{\Delta\mu}$ term, and seek to obtain a general expression for it. We shall do this by taking the Fe-Mn exchange reaction between garnet and ilmenite *as an example*. Once this goal is achieved, it could be adapted to the other geothermometers and geobarometers, since $RT\ln K_\gamma$ of many of them are sufficiently well modeled by T-P-independent W_G parameters (or with ones slightly dependent on T and P). The $\Delta\mu$ of the Fe-Mn exchange reaction among garnet and ilmenite is given by

$$\Delta\mu = \Delta H_{<T>}^\circ - T\Delta S_{<T>}^\circ + (P-1)\Delta V_{298}^\circ + RT\ln\frac{X_{Mn}^{Gt}(1-X_{Mn}^{Ilm})}{(1-X_{Mn}^{Gt})X_{Mn}^{Ilm}}$$
$$- W_G^{Gt}(2X_{Mn}^{Gt}-1) + W_G^{Ilm}(2X_{Mn}^{Ilm}-1) , \qquad (9.67)$$

ΔV_{298}° being zero in this case. Regardless of the value of ΔV_{298}°, we may set $\sigma_{\Delta V}$ to zero. Therefore, apart from σ_T and σ_P, which are unknown, and which we wish to solve for, the uncertainties of $\Delta H_{<T>}^\circ$, $\Delta S_{<T>}^\circ$, X_i, W_G^{Gt}, W_G^{Ilm} will contribute to the uncertainty of $\Delta\mu$. Thus, in the context of error propagation

$$\Delta\mu = f(\Delta H_{<T>}^\circ, \Delta S_{<T>}^\circ, X_i, W_G^{Gt}, W_G^{Ilm}), \qquad (9.68)$$

the uncertainties of $\Delta H_{<T>}^\circ$ and $\Delta S_{<T>}^\circ$ being mutually correlated [cf.Eq.(9.60)]. The variance of $\Delta\mu$, $\sigma_{\Delta\mu}^2$, may then be written as

$$\sigma_{\Delta\mu}^2 = \left(\frac{\partial\Delta\mu}{\partial\Delta H}\right)^2 \sigma_{\Delta H}^2 + \left(\frac{\partial\Delta\mu}{\partial\Delta S}\right)^2 \sigma_{\Delta S}^2 + 2\rho_{\Delta H\Delta S}\left(\frac{\partial\Delta\mu}{\partial\Delta H}\right)\left(\frac{\partial\Delta\mu}{\partial\Delta S}\right)\sigma_{\Delta H}\sigma_{\Delta S}$$
$$+ \left(\frac{\partial\Delta\mu}{\partial X_{Mn}^{Gt}}\right)^2 \sigma_{X_{Mn}^{Gt}}^2 + \left(\frac{\partial\Delta\mu}{\partial X_{Mn}^{Ilm}}\right)^2 \sigma_{X_{Mn}^{Ilm}}^2 + \left(\frac{\partial\Delta\mu}{\partial W_G^{Gt}}\right)^2 \sigma_{W_G^{Gt}}^2 + \left(\frac{\partial\Delta\mu}{\partial W_G^{Ilm}}\right)^2 \sigma_{W_G^{Ilm}}^2 ,$$
$$(9.69a)$$

ΔH and ΔS being short-hand symbols for $\Delta H^\circ_{<T>}$ and $\Delta S^\circ_{<T>}$. Taking the derivatives with respect to Eq. (9.67), algebraic manipulation of (9.69a) leads to the final expression for the variance of $\Delta\mu$,

$$\sigma^2_{\Delta\mu} = \sigma^2_{\Delta H} + T^2\sigma^2_{\Delta S} + 2\rho_{\Delta H \Delta S}(-T)\sigma_{\Delta H}\sigma_{\Delta S}$$

$$+ (RT)^2\left[\frac{\sigma^2_{X^{Gt}_{Mn}}}{(X^{Gt}_{Mn})^2(1-X^{Gt}_{Mn})^2} + \frac{\sigma^2_{X^{Ilm}_{Mn}}}{(1-X^{Ilm}_{Mn})^2(X^{Ilm}_{Mn})^2}\right]$$

$$+ [(2X^{Gt}_{Mn} - 1)^2\sigma^2_{W^{Gt}_G} + 4(W^{Gt}_G)^2\sigma^2_{X^{Gt}_{Mn}}]$$

$$+ [(2X^{Ilm}_{Mn} - 1)^2\sigma^2_{W^{Ilm}_G} + 4(W^{Ilm}_G)^2\sigma^2_{X^{Ilm}_{Mn}}]. \qquad (9.69)$$

Given this explicit expression for $\sigma^2_{\Delta\mu}$, Eq. (9.64) may be employed to obtain σ_T for the garnet-ilmenite Fe-Mn exchange geothermometer.

9.3.5 Two Worked Examples of Geothermometry, Geobarometry, and Geohygrometry

Let us apply the geothermometers and geobarometers calibrated in Sect. 9.3.3 to two sets of staurolite-Al_2SiO_5-bearing metapelitic rocks described by Guidotti (1974) and by Mohr and Newton (1983). In both cases, textural and chemical criteria suggest attainment of equilibrium, a necessary prerequisite for geothermometry and geobarometry. Having evaluated their T and P of equilibration, one suitably chosen dehydration equilibrium will be applied to obtain the a_{H2O} of the fluid (geohygrometry) coexisting with the minerals during that episode of equilibration.

It is well known that every metamorphic rock bears the imprint of a temporal evolution in its P-T history, often dubbed as its P-T-t (time) path. Therefore, all P-T-a_{H2O} data obtained from such exercises must be analyzed with regard to their implications. This will be touched upon before closing this section.

9.3.5.1 Staurolite- and Sillimanite-Bearing Metapelitic Rocks, Rangeley Quadrangle, Maine

Guidotti (1974) provided very meticulous data on partitioning of Fe-Mg and Fe-Mn between numerous pairs of coexisting minerals, demonstrating the attainment of equilibrium during metamorphism of staurolite-sillimanite-bearing metapelites. A set of three (Nos. 201, 231, and 247) of his 12 "prime" specimens were culled to obtain P-T-a_{H2O} data. All of them belong to his "lower sillimanite zone".

Application of the garnet-ilmenite Fe-Mn exchange thermometer is based on Eq. (9.62); its uncertainty follows from Eqs. (9.69) and (9.64). Likewise,

Table 9.10. Geothermometry of lower sillimanite zone metapelites from the Rangeley Quadrangle, Maine

	Garnet-ilmenite geothermometry				
Specimen No.	K_D	$\ln K_D$	T(°C)	σ_T(°C)	Grand average
201	5.3657	1.6800	655	49	
231	6.5284	1.8762	587	47	621 ± 68
247	3.8236	1.3412	804[a]	49	

		Garnet-biotite geothermometry				
Specimen No.	P(bar)	K_D	$\ln K_D$	T(°C)	σ_T(°C)	Grand average
201	2000	0.1380	-1.9806	492	22	
	10000			520	21	
231	2000	0.1321	-2.0244	481	26	496 ± 29
	10000			508	25	524 ± 27
247	2000	0.1501	-1.8962	516	26	
	10000			545	25	

[a] Incompatible with the observed coexistence of staurolite+quartz; pruned prior to obtaining the grand average!

the garnet-biotite Fe-Mg exchange geothermometer uses Eq. (9.52) to compute the T of equilibration, its uncertainty estimated in a manner analogous to that of the Gt-Ilm thermometer, noting that all terms due to nonidealities will drop out from Eq. (9.69). Furthermore, to comply with the slow rates of diffusion of ions in garnets compared to those in the other phases, only the compositions of the rims of the garnets were used in the following evaluations of T, reproduced in Table 9.10.

The GASP geobarometer was applied to estimate the pressure of equilibration for the lower sillimanite zone rocks of the Rangeley Quadrangle. Knowledge of K and the standard state properties [see Eq. (9.43)] is needed to solve for P at a given T. The K comprises a^{Plag}_{an} and a^{Gt}_{Ca} (Ca being an abbreviation for the q-normalized grossular component). The appropriate expression for the former is given in Eq. (9.26). The a^{Gt}_{Ca} was taken from Koziol and Newton's (1988b) study of the $(Ca,Mg,Fe)_3Al_2[SiO_4]_3$ garnets. Note that the properties of these garnets were formulated on the basis of experimental data on the compositions of these ternary garnets in equilibrium with anorthite, Al_2SiO_5, and quartz. Consequently, the mixing properties of the garnets are consistent with the standard state data in Eq. (9.43). Table 9.11 reproduces the results of GASP geobarometry. The evaluation of the uncertainties are from (9.66), combined with an expression for $\sigma^2_{\Delta\mu}$ analogous to (9.69), ignoring its nonideality terms for simplicity.

Table 9.11. Examples of evaluation of equilibrium pressure at two specified temperatures for the lower sillimanite zone metapelites of the Rangeley Quadrangle, Maine

Specimen No.	X_{an}^{Gt}	N_{Mg} [a]	GASP geobarometry X_{ca}^{Gt}	T(°C)	P(bar)	σ_P(bar)	Grand average
201	0.205	9.9	0.051	450	2197	569	
				650	6576	430	
231	0.205	10.0	0.055	450	2495	567	2190 ± 657
				650	6948	425	6567 ± 497
247	0.205	10.5	0.047	450	1878	572	
				650	6176	436	

[a] $N_{Mg}^{Gt} \equiv 100 \cdot Mg/Mg+Fe$ (see Koziol and Newton 1988b)

Figure 9.14 depicts the P-T data derived for three staurolite-sillimanite-bearing metapelites from Rangeley Quadrangle, Maine. A remarkable feature of these data is that two discrete temperatures are revealed on the P-T-t path. The Fe-Mn exchange between garnet and ilmenite helps document the higher-temperature episode, while the lower one is discerned by the Fe-Mg exchange garnet-biotite thermometry. The former set of P-T data are in agreement with the stability fields of both staurolite + quartz (indicated by dotted lines) and sillimanite; the second set is clearly located outside the stability fields of either of them. They are best interpreted, respectively, as near-maximum on

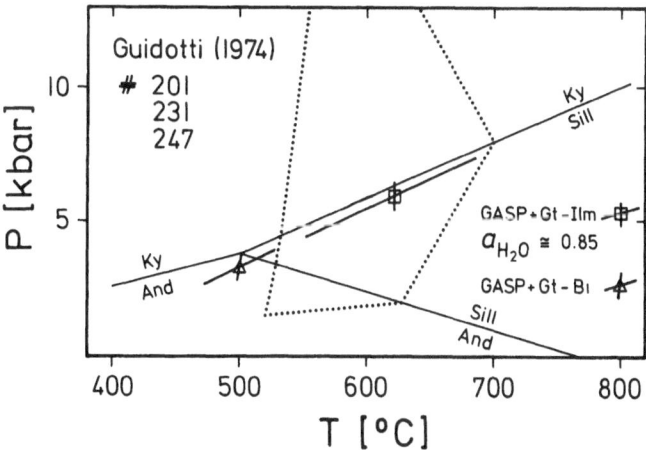

Fig. 9.14 A P-T plot of the geothermometric and geobarometric data for three staurolite-sillimanite-bearing metapelites from the Rangeley Quadrangle, Maine. The specimen numbers are from Guidotti (1974). Indicated by the *dotted lines* is a P-T stability field for the assemblage staurolite + quartz

the P-T-t path and a second point on P-T-t trajectory at which the Fe-Mg exchange between the garnet rim and biotite was eventually frozen in. Ideally, such an interpretation should include an estimation of closure temperature (see Dodson 1973), which is beyond our scope. It should be evident now that the staurolite-sillimanite-quartz assemblage had already become metastable, when biotite and garnet rims were still in the process of exchanging Fe and Mg at around 500°C. Ultimately, this exchange equilibrium was also frozen in, and the entire assemblage became metastable.

Having established the metamorphic P-T conditions, let us try to quantify the a_{H2O} of the fluid prevalent during equilibration in the vicinity of the P-T-t maximum (see above). To this end, any dehydration reaction operating there, and for which thermodynamic data are available, may be considered. One such reaction is

$$NaAl_2[AlSi_3O_{10}](OH)_2 + SiO_2 = Na[AlSi_3O_8] + Al_2SiO_5 + H_2O. \qquad (9.70)$$
$$\text{(pg in Ms)} \qquad \text{(Q)} \quad \text{(ab in Plag)} \quad \text{(Sill)}$$

Note that Q and Sill are end-members, while pg and ab, paragonite and albite components, respectively, reside in the phases Ms and Plag, muscovite and plagioclase. The standard state properties of this reaction can be derived from reversal experiments performed by Holdaway (1971) on the one hand, and by Chatterjee (1972), on the other. The results are

$$\Delta H^\circ_{<T>} = 86575 \text{ J}, \Delta S^\circ_{<T>} = 163.678 \text{ J K}^{-1}, \text{ and } \Delta V^\circ_{298,s} = -0.4278 \text{ J bar}^{-1}. \quad (9.71)$$

Combining these with the appropriate expressions for K, comprising a^{Ms}_{pg} [Eq.(9.20a)] and a^{Plag}_{ab} [Eq.(9.25)], a_{H2O} may be solved for at any given P and T utilizing

$$RT\ln a_{H_2O} = -\Delta H^\circ_{<T>} + T\Delta S^\circ_{<T>} - (P-1)\Delta V^\circ_{298,s} - RT\ln f^*_{H_2O} - RT\ln a^{Plag}_{ab} + RT\ln a^{Ms}_{pg}. \qquad (9.72)$$

The results of derivation of a_{H2O} at the peak P-T condition, 621°C and 5932 bar, are listed in Table 9.12. An activity of H_2O on the order of 0.85 testifies to the presence of species other than H_2O in the metamorphic fluid. The common occurrence of pyrrhotite and graphite in these rocks (see Guidotti 1974, p.478) is suggestive of fluid species such as CO_2, CH_4, S_2, etc.

Table 9.12. Results of the calculation of equilibrium a_{H_2O} in the fluids of the metapelites, Rangeley Quadrangle, Maine

Specimen No.	T(°C)	P(bar)	X^{Ms}_{pg}	X^{Plag}_{ab}	a_{H_2O}
201	621	5932	0.208	0.795	0.879
231	621	5932	0.191	0.795	0.828
247	621	5932	0.197	0.795	0.846

9.3.5.2 Staurolite-Kyanite-Bearing Metapelites from the Great Smoky Mountains, North Carolina

The staurolite-kyanite zone metapelites, described by Mohr and Newton (1983), offer an opportunity to apply the GRAIL barometer, in addition to the other thermometers and barometers already tried out in Sect. 9.3.5.1. Again, three specimens (Nos. 145, 451, 1318) were chosen, each of which permit application of Gt-Bi and Gt-Ilm geothermometers, and both GASP and GRAIL geobarometers. Table 9.13 lists the thermometric data, based on garnet-rim compositions plus ilmenite or biotite.

The evaluation of the P of equilibration for the metapelites with the GASP geobarometer utilized the standard state properties in Eq. (9.38), the source of data on K being the same. The pressures obtained are reproduced in Table 9.14. Application of the GRAIL barometer requires knowledge of the relevant equilibrium constant, K, which again involves two activity terms, a^{Ilm}_{Fe} (Fe for $FeTiO_3$) and a^{Gt}_{Fe} (Fe for $FeAl_{2/3}SiO_4$). The a^{Ilm}_{Fe} was set equal to X^{Ilm}_{Fe}, as justified by the near end-member compositions of the ilmenites in these metapelites. The a^{Gt}_{Fe} was computed from an algorithm given by Newton et al. [1989, Eq. (6), with relevant data from their Table 1]. The evaluation of the uncertainty used Eq. (9.66), coupled with an expression for $\sigma^2_{\Delta\mu}$ analogous to

Table 9.13. Geothermometry of staurolite-kyanite zone metapelites from the Great Smoky Mountains, North Carolina

Specimen No.	K_D	Garnet-ilmenite geothermometry $\ln K_D$	$T(°C)$	$\sigma_T(°C)$	Grand average
145	8.3611	2.1236	513	51	
451	4.4269	1.4877	733[a]	38	583 ± 72
1318	5.3963	1.6857	653	51	

Specimen No.	P(bar)	Garnet-biotite geothermometry K_D	$\ln K_D$	$T(°C)$	$\sigma_T(°C)$	Grand average
145	2000	0.1372	-1.9866	491	24	
	10000			518	22	
451	2000	0.1377	-1.9829	492	24	475 ± 28
	10000			519	22	502 ± 26
1318	2000	0.1140	-2.1715	443	25	
	10000			469	24	

[a] Incompatible with staurolite-quartz coexistence observed; pruned prior to obtaining the grand average.

Table 9.14. Geobarometry of staurolite-kyanite zone metapelites of the Great Smoky Mountains, North Carolina

			GASP geobarometry				
Specimen No.	X_{an}^{Gt}	N_{Mg} [a]	X_{ca}^{Gt}	T(°C)	P(bar)	σ_P(bar)	Grand average
145	0.240	17.8	0.056	450	2030[b]	461	
				650	6263[b]	343	
451	0.250	17.4	0.080	450	3059	454	3412 ± 648
				650	7534	328	7858 ± 478
1318	0.380	15.4	0.138	450	3765	463	
				650	8182	348	

			GRAIL geobarometry				
Sample Nos.	Garnet X_{Fe}	X_{Mg}	X_{ca}	Ilmenite X_{Fe}	T(°C)	P(bar)	σ_P(bar) Grand average
145	0.685	0.148	0.049	0.971	450	6013	324
					650	6341	286
451	0.666	0.140	0.070	0.928	450	6114	328 5950 ± 376
					650	6514	293 6321 ± 335
1318	0.672	0.122	0.127	0.945	450	5724	326
					650	6109	291

[a] $N_{Mg}^{Gt} \equiv 100 \cdot Mg/Mg+Fe$ (see Koziol and Newton 1988b).
[b] Incompatible with stable kyanite; pruned before averaging.

(9.69), disregarding its nonideality terms for simplicity. Table 9.14 reproduces these results.

Figure 9.15 summarizes the P-T data for the three staurolite-kyanite zone metapelites from the Great Smoky Mountains. A feature reminiscent of our previous set of results is the appearance of two discrete temperatures on the P-T-t path. The Fe-Mn exchange between garnet and ilmenite again documents the higher-temperature episode, while the lower one is discerned by the Fe-Mg exchange garnet-biotite thermometry. As before, the higher T-P group agrees with the compatibility of staurolite + quartz (indicated by dotted lines) and of kyanite, while the lower one is outside the quartz + staurolite stability field. Obviously, Fe-Mg exchange between the garnet rims and the matrix biotites continued to be operative down to 500°C. Also interesting, if a bit fortuitous, is the fact that the P-T curves for the two barometers, by mutual "intersection", imply the same temperature as the garnet-ilmenite geothermometer alone. Considering this, the metamorphic P-T peak near 580 ± 72°C and 6280 ± 500 bar is well documented. This is, however, not true for

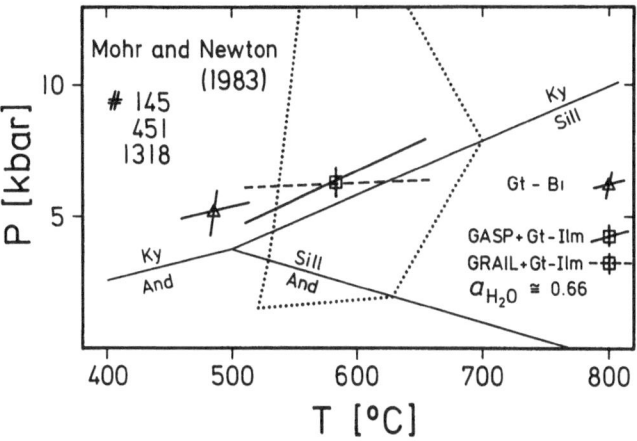

Fig. 9.15 A P-T plot of the geothermometric and geobarometric data for three staurolite-kyanite zone metapelites from the Great Smoky Mountains, North Carolina. The specimen numbers are from Mohr and Newton (1983). The stability field for the quartz + staurolite assemblage is shown by the *dotted lines*

the decompression path. As may be seen from Fig. 9.15, the dP/dT slopes of the two barometers differ somewhat, giving rise to larger uncertainty in specifying the pressure at T around 500°C. Should the P-T-t path have followed the GASP equilibrium, rutile will be expected to be consumed by the GRAIL reaction; no evidence is available for that. By contrast, a near-isobaric cooling path would sustain the GRAIL assemblage, but the plagioclase will tend to become more albitic, for which no evidence exists either. Owing to these problems, the P-T range depicted for the low temperature episode in Fig. 9.15 should be regarded as open to question.

Given the peak metamorphic P-T conditions, evaluation of a_{H2O} may proceed as above [Eq. (9.72)]. The relevant dehydration reaction is

$$NaAl_2[AlSi_3O_{10}](OH)_2 + SiO_2 = Na[AlSi_3O_8] + Al_2SiO_5 + H_2O. \qquad (9.73)$$
$$\text{(pg in Ms)} \qquad \text{(Q)} \quad \text{(ab in Plag)} \quad \text{(Ky)}$$

Kyanite and quartz being end-member phases, we simply need to know the a^{Ms}_{pg} [Eq.(9.20a)] and a^{Plag}_{ab} [Eq.(9.25)], in addition to the standard state properties of reaction (9.73), to solve for a_{H2O}. The latter set of data, extracted from the reversal experiments of Holdaway (1971) and Chatterjee (1972), are

$$\Delta H^{\circ}_{<T>} = 79627 \text{ J}, \Delta S^{\circ}_{<T>} = 151.944 \text{ J K}^{-1}, \text{ and } \Delta V^{\circ}_{298,s} = -0.9973 \text{ J bar}^{-1}. \quad (9.74)$$

Table 9.15 lists the results of evaluation of a_{H2O} at the peak P-T condition, 583°C and 6283 bar. The mean a_{H2O} is 0.66, suggestive of dilution of an aqueous fluid by other gas species. Presence of pyrrhotite and graphite, in addition to ilmenite and rutile, in these metapelites (Mohr and Newton 1983,

Table 9.15. Evaluation of a_{H_2O} in the fluids of staurolite-kyanite zone metapelites, Great Smoky Mountains, North Carolina

Specimen No.	T(°C)	P(bar)	X_{pg}^{Ms}	X_{ab}^{Plag}	a_{H_2O}
145·	583	6283	0.207	0.760	0.649
451	583	6283	0.238	0.750	0.722
1318	583	6283	0.142	0.620	0.608

Table 1), point to CO_2 and CH_4 as possible diluents. The reader may refer to Mohr and Newton (1983, p.125 ff) to appreciate how to solve for the partial pressures of these fluid species. Their results should, however, be regarded as an approximation, because they critically depend on the assumption of ideal mixing in the multicomponent fluid, which is very unlikely at the P and T of interest.

9.3.6 Concluding Remarks

Geothermometry and geobarometry, though still developing, must be regarded as reliable and indispensable tools for petrological research. The rather explicit treatment given here has been at the expense of a broad coverage. The reader's attention is once again drawn to the recent reviews of the subject by Essene (1982, 1989), Newton (1983), and Ganguly and Saxena (1987). A perusal of this literature would help them realize that the basic principles given above are equally applicable to anion-exchange and isotope-exchange reactions, or to solvus thermometry. These are just as worthwhile as the topics discussed in this book; lack of space prohibited them to be included here.

References

Abbott MM, van Ness HC (1972) Thermodynamics. McGraw-Hill, New York

Andersen DJ, Lindsley DH (1981) A valid Margules formulation for an asymmetric ternary solution: revision of the olivine-ilmenite thermometer, with applications. Geochim Cosmochim Acta 45:847-853

Anderson GM (1977a) The accuracy and precision of calculated mineral dehydration equilibria. In: Fraser DG (ed) Thermodynamics in geology. Reidel, Dordrecht, pp 115-136

Anderson GM (1977b) Uncertainties in calculations involving thermodynamic data. In: Greenwood HJ (ed) Short course in application of thermodynamics to petrology and ore deposits. Miner Assoc Canada, Short Course Handbook 2:199-215

Angus S, Armstrong B, de Reuck KM (eds) (1976) International thermodynamic tables of the fluid state - carbon dioxide. IUPAC Chem Data Ser 7, Pergamon, Oxford

Aranovich LYa, Podlesskii KK (1989) Geothermobarometry of high-grade metapelites: simultaneously operating reactions. In: Daly JS, Cliff RA, Yardley BWD (eds) Evolution of metamorphic belts. Geol Soc Spec Publ 43:45-61

Barnes HL, Ernst WG (1963) Ideality and ionization in hydrothermal fluids: the system $MgO-H_2O-NaOH$. Am J Sci 261:129-150

Barrett WT, Wallace WE (1954) Studies of NaCl-KCl solid solutions. I. Heats of formation, lattice spacings, densities, Schottky defects and mutual solubilities. J Am Chem Soc 76:366-373

Bennington KO, Brown RR, Bell HE, Beyer RP (1987) Thermodynamic properties of two manganese silicates, pyroxmangite and fowlerite. US Bur Mines, Rep Invest 9064, Washington D C

Berman RG (1988) Internally-consistent thermodynamic data for minerals in the system $Na_2O-K_2O-CaO-MgO-FeO-Fe_2O_3-Al_2O_3-SiO_2-TiO_2-H_2O-CO_2$. J Petrol 29:445-552

Berman RG (1990) Mixing properties of Ca-Mg-Fe-Mn garnets. Am Miner 75:328-344

Berman RG, Brown TH (1985) Heat capacity of minerals in the system $Na_2O-K_2O-CaO-MgO-FeO-Fe_2O_3-Al_2O_3-SiO_2-TiO_2-H_2O-CO_2$: representation, estimation, and high temperature extrapolation. Contrib Miner Petrol 89:168-183

Berman RG, Brown TH (1986) Heat capacity of minerals in the system Na_2O-K_2O-CaO-MgO-FeO-Fe_2O_3-Al_2O_3-SiO_2-TiO_2-H_2O-CO_2: representation, estimation, and high temperature extrapolation. (Erratum) Contrib Miner Petrol 94:262

Berman RG, Brown TH, Greenwood HJ (1985) An internally consistent thermodynamic data base for minerals in the system Na_2O-K_2O-CaO-MgO-FeO-Fe_2O_3-Al_2O_3-SiO_2-TiO_2-H_2O-CO_2. Atomic Energy Canada Ltd Tech Rep TR-377

Berman RG, Engi M, Greenwood HJ, Brown TH (1986) Derivation of internally consistent thermodynamic data by the technique of mathematical programming: a review with application to the system MgO-SiO_2-H_2O. J Petrol 27:1331-1364

Blander M (1964) Thermodynamic properties of molten salt solutions. In: Blander M (ed) Molten salt chemistry. Wiley, New York, pp 127-237

Bohlen SR, Wall VJ, Boettcher AL (1983) Experimental investigations and geological applications of equilibria in the system FeO-TiO_2-Al_2O_3-SiO_2-H_2O. Am Miner 68:1049-1058

Bottinga Y, Richet P (1981) High pressure and temperature equation of state and calculation of the thermodynamic properties of gaseous carbon dioxide. Am J Sci 281:615-660

Bowers TS, Helgeson HC (1983) Calculation of the thermodynamic and geochemical consequences of non-ideal mixing in the system H_2O-CO_2-$NaCl$ on phase relations in geologic systems: Equation of state for H_2O-CO_2-$NaCl$ fluids at high pressures and temperatures. Geochim Cosmochim Acta 47:1247-1275

Brown TH (1977) Introduction to non-ideal and complex solutions. In: Greenwood HJ (ed) Short course in application of thermodynamics to petrology and ore deposits. Miner Assoc Canada Short Course Handbook 2:126-135

Brown TH, Skinner BJ (1974) Theoretical prediction of equilibrium phase assemblages in multicomponent systems. Am J Sci 274:961-986

Bruckmann-Benke P, Chatterjee ND, Aksyuk AM (1988) Thermodynamic properties of $Zn(Al,Cr)_2O_4$ spinels at high temperatures and pressures. Contrib Miner Petrol 98:91-96

Bunk AJH, Tichelaar GW (1953) Investigations in the system $NaCl$-KCl. Proc Koninkl Nederland Akad Wetenschappen Ser B Phys Sci 56:375-384

Burnham CW, Holloway JR, Davis NF (1969a) Thermodynamic properties of water to $1,000°C$ and 10,000 bars. Geol Soc Am Spec Pap 132:1-96

Burnham CW, Holloway JR, Davis NF (1969b) The specific volume of water in the range 1000 to 8900 bars, $20°$ to $900°C$. Am J Sci 267-A:70-95

Cameron DJ, Unger AE (1970) The measurement of the thermodynamic properties of NiO-MnO solid solutions by a solid electrolyte cell technique. Metall Trans 1:2615-2621

Carnahan NF, Starling KE (1969) Equation of state for nonattracting rigid spheres. J Chem Phys 51:635-636

Carpenter MA, Putnis A, Navrotsky A, McConell JDC (1983) Enthalpy effects associated with Al/Si ordering in anhydrous Mg-cordierite. Geochim Cosmochim Acta 47:899-906

Cemic L, Kleppa OJ (1986) High temperature calorimetry of sulfide systems. I.Thermochemistry of liquid and solid phases of Ni + S. Geochim Cosmochim Acta 50:1633-1641

Charlu TV, Newton RC, Kleppa OJ (1975) Enthalpies of formation at 970 K of compounds in the system $MgO-Al_2O_3-SiO_2$ from high temperature solution calorimetry. Geochim Cosmochim Acta 39:1487-1497

Charlu TV, Newton RC, Kleppa OJ (1978) Enthalpy of formation of some lime silicates by high-temperature solution calorimetry, with discussion of high pressure phase equilibria. Geochim Cosmochim Acta 42:367-375

Chatterjee ND (1970) Synthesis and upper stability of paragonite. Contrib Miner Petrol 27:244-257

Chatterjee ND (1972) The upper stability limit of the assemblage paragonite + quartz and its natural occurrences. Contrib Miner Petrol 34:288-303

Chatterjee ND, Flux S (1986) Thermodynamic mixing properties of muscovite-paragonite crystalline solutions at high temperatures and pressures, and their geological applications. J Petrol 27:677-693

Chatterjee ND, Johannes W (1974) Thermal stability and standard thermodynamic properties of synthetic $2M_1$-muscovite, $KAl_2[AlSi_3O_{10}(OH)_2]$. Contrib Miner Petrol 48:89-114

Chatterjee ND, Leistner H, Terhart L, Abraham K, Klaska R (1982) Thermodynamic mixing properties of corundum-eskolaite, α-$(Al,Cr^{+3})_2O_3$, crystalline solutions at high temperatures and pressures. Am Miner 67:725-735

Chiotti P (1972) New Integration of the Gibbs-Duhem equation and thermodynamics of Pr-Zn alloys. Metallurg Trans 3:2911-2916

Clark SP (1966) Handbook of physical constants. Geol Soc Am Mem 97:1-587

Cohen RE (1986a) Statistical mechanics of coupled solid solutions in the dilute limit. Phys Chem Miner 13:174-182

Cohen RE (1986b) Thermodynamic solution properties of aluminous clinopyroxenes: non-linear ' least squares refinements. Geochim Cosmochim Acta 50:563-575

Cressey G, Schmid R, Wood BJ (1978) Thermodynamic properties of almandine-grossular garnet solid solution. Contrib Miner Petrol 67:397-404

Darken LS, Gurry RW (1953) Physical chemistry of metals. McGraw-Hill, New York

Day HW, Kumin HJ (1980) Thermodynamic analysis of the aluminum silicate triple point. Am J Sci 280:265-287

de Capitani C, Brown TH (1987) The computation of chemical equilibrium in complex systems containing non-ideal solutions. Geochim Cosmochim Acta 51:2639-2652

Delany JM, Helgeson HC (1978) Calculation of the thermodynamic consequences of dehydration in subducting oceanic crust to 100 kb and > 800°C. Am J Sci 278:638-686

Denbigh K (1971) The principles of chemical equilibrium. 3rd edn, Cambridge Univ Press, Cambridge

de Santis R, Breedveld GJF, Prausnitz JM (1974) Thermodynamic properties of aqueous gas mixtures at advanced pressures. Industr Engng Chem Proc Design Dev 13:374-377

Dodson MH (1973) Closure temperature in cooling geochronological and petrological systems. Contrib Miner Petrol 40:259-274

Engi M (1983) Equilibria involving Al-Cr spinel: Mg-Fe exchange with olivine. Experiments, thermodynamic analysis, and consequences for geothermometry. Am J Sci 283:29-71

Essene E (1982) Geologic thermometry and barometry. In: Ferry JM (ed) Characterization of metamorphism through mineral equilibria. Rev Miner 10:153-193

Essene E (1989) The current status of thermobarometry in metamorphic rocks. In: Daly JS, Cliff RA, Yardley BWD (ed) Evolution of metamorphic belts. Geol Soc Spec Publ 43:1-44

Eugster HP (1957) Heterogeneous reactions involving oxidation and reduction at high pressures and temperatures. J Chem Phys 26:1760-1761

Eugster HP, Wones DR (1962) Stability relations of the ferruginous biotite, annite. J Petrol 3:82-125

Evans BW, Johannes W, Oterdoom H, Trommsdorff V (1976) Stability of chrysotile and antigorite in the serpentine multisystem. Schweiz Miner Petrogr Mitt 56:79-93

Fei Y, Saxena SK, Eriksson G (1986) Some binary and ternary silicate solution models. Contrib Miner Petrol 94:221-229

Ferry JM, Spear FS (1978) Experimental calibration of the partitioning of Fe and Mg between biotite and garnet. Contrib Miner Petrol 66:113-117

Flood H, Förland T, Grjotheim K (1954) Über den Zusammenhang zwischen Konzentrationen und Aktivitäten in geschmolzenen Salzmischungen. Z Anorg Allg Chem 276:290-315

Flux S, Chatterjee ND (1986) Experimental reversal of the Na-K exchange reaction between muscovite-paragonite crystalline solutions and a 2 molal aqueous (Na,K)Cl fluid. J Petrol 27:665-676

Franck EU, Tödheide K (1959) Thermische Eigenschaften überkritischer Mischungen von Kohlendioxid und Wasser bis zu 750°C und 2000 atm. Z Phys Chem, Neue Folge, 22:232-245

Froese E (1981) Applications of thermodynamics in the study of mineral deposits. Geol Surv Canada Pap 80-28

Ganguly J (1973) Activity-composition relation of jadeite in omphacite pyroxene: theoretical deductions. Earth Planet Sci Lett 19:145-153

Ganguly J, Ghose S (1979) Aluminous orthopyroxene: Order-disorder, thermodynamic properties, and petrologic implications. Contrib Miner Petrol 69:375-385

Ganguly J, Saxena SK (1987) Mixtures and mineral reactions. Springer, Berlin Heidelberg New York

Garvin D, Parker VB, White HJ (1987) CODATA Thermodynamic tables. Selections for some compounds of calcium and related mixtures: a prototype set of tables. Hemisphere, Washington D C

Gasparik T (1984) Two-pyroxene thermobarometry with new experimental data in the system $CaO-MgO-Al_2O_3-SiO_2$. Contrib Miner Petrol 87:87-97

Gehrig M (1980) Phasengleichgewichte und pVT-Daten ternärer Mischungen aus Wasser, Kohlendioxid und Natriumchlorid bis 3 kbar und 550°C. Thesis Karlsruhe University

Geiger CA, Newton RC, Kleppa OJ (1987) Enthalpy of mixing of synthetic almandine-grossular and almandine-pyrope garnets from high-temperature solution calorimetry. Geochim Cosmochim Acta 51:1955-1963

Ghent ED (1976) Plagioclase-garnet-Al_2SiO_5-quartz: a potential geobarometer-geothermometer. Am Miner 61:710-714

Ghiorso MS, Carmichael ISE, Moret LK (1979) Inverted high temperature quartz. Unit cell parameters and properties of the α-β inversion. Contrib Mineral Petrol 68:307-323

Gill PE, Murray W, Saunders MA, Wright MH (1983) User's guide for SOL/QPSOL: A fortran package for quadratic programming. Systems Optimization Lab, Dep Operations Res, Stanford Univ, Tech Rep 83-7

Gordon TM (1973) Determination of internally consistent thermodynamic data from phase equilibrium experiments. J Geol 81:199-208

Gordon TM, Greenwood HJ (1971) The stability of grossularite in H_2O-CO_2 mixtures. Am Miner 56:1674-1688

Grambling JA (1984) Coexisting paragonite and quartz in sillimanite rocks from New Mexico. Am Miner 69:79-87

Greenwood HJ (1967) Mineral equilibria in the system $MgO-SiO_2-H_2O-CO_2$. In: Abelson PH (ed) Researches in geochemistry. Wiley, New York, 2:542-567

Greenwood HJ (1969) The compressibility of gaseous mixtures of carbon dioxide and water between 0 and 500 bars pressure and 450° and 800° Centigrade. Am J Sci 267-A:191-208

Greenwood HJ (1973) Thermodynamic properties of gaseous mixtures of H_2O and CO_2 between 450° and 800°C and 0 to 500 bars. Am J Sci 273:561-571

Guidotti CV (1974) Transition from staurolite to sillimanite zone, Rangeley Quadrangle, Maine. Geol Soc Am Bull 85:475-490

Haas H (1972) Diaspore-corundum equilibrium determined by epitaxis of dispore on corundum. Am Miner 57:1375-1385

Haas JL, Fisher JR (1976) Simultaneous evaluation and correlation of thermodynamic data. Am J Sci 276:525-545

Haas JL, Robinson GR, Hemingway BS (1981) Thermodynamic tabulations for selected phases in the system $CaO-Al_2O_3-SiO_2-H_2O$ at 101.325 kPa (1 atm) between 273.15 and 1800 K. J Phys Chem Ref Data 10:576-669

Hahn WC, Muan A (1961) Activity measurements in oxide solid solutions: the system NiO-MgO and NiO-MnO in the temperature interval 1100-1300°C. J Phys Chem Solids 19:338-348

Halbach H, Chatterjee ND (1982a) An empirical Redlich-Kwong-type equation of state for water to 1000°C and 200 kbar. Contrib Miner Petrol 79:337-345

Halbach H, Chatterjee ND (1982b) The use of linear parametric programming for determining internally consistent thermodynamic data for minerals. In: Schreyer W (ed) High-pressure researches in geoscience. Schweizerbart, Stuttgart, pp 475-491

Halbach H, Chatterjee ND (1984) An internally consistent set of thermodynamic data for twenty-one $CaO-Al_2O_3-SiO_2-H_2O$ phases by linear parametric programming. Contrib Miner Petrol 88:14-23

Harker RI (1958) The system $MgO-CO_2$-A and the effect of inert pressure on certain types of hydrothermal reaction. Am J Sci 256:128-138

Harker RI, Tuttle OF (1955) Studies in the system $CaO-MgO-CO_2$, part 1. The thermal dissociation of calcite, dolomite and magnesite. Am J Sci 253:209-224

Haselton HT, Hovis GL, Hemingway BS, Robie RA (1983) Calorimetric investigation of the excess entropy of mixing in analbite-sanidine solid solutions: lack of evidence for Na,K short-range order and implications for two-feldspar thermometry. Am Miner 68:398-413

Haselton HT, Newton RC (1980) Thermodynamics of pyrope-grossular garnets and their stabilities at high temperatures and high pressures. J Geophys Res 85(B 12):6973-6982

Haselton HT, Westrum EF (1980) Low-temperature heat capacities of synthetic pyrope, grossular, and $pyrope_{60}grossular_{40}$. Geochim Cosmochim Acta 44:701-709

Helffrich G, Wood B (1989) Subregular model for multicomponent solutions. Am Miner 74:1016-1022

Helgeson HC, Delaney JM, Nesbitt HW, Bird DK (1978) Summary and critique of the thermodynamic properties of rock-forming minerals. Am J Sci 278A:1-229

Hemingway BS, Haas JL, Robinson JR (1982) Thermodynamic properties of selected minerals in the system $Al_2O_3-CaO-SiO_2-H_2O$ at 298.15 K and 1 bar (10^5 Pascals) pressure and at high temperatures. US Geol Surv Bull 1452:1-70

Hemingway BS, Robie RA (1977) Enthalpies of formation of low albite ($NaAlSi_3O_8$), gibbsite ($Al(OH)_3$), and $NaAlO_2$; revised values for $\Delta H^{\circ}_{f,298}$ and $\Delta G^{\circ}_{f,298}$ of some aluminosilicate minerals. US Geol Surv J Res 5:413-429

Hemminger W, Höhne G (1979) Grundlagen der Kalorimetrie. Verlag Chemie, Weinheim New York

Heyen G (1980) Liquid and vapor properties from a cubic equation of state. Eur Feder Chem Engng, Publ Ser 11:9-13

Holdaway MJ (1971) Stability of andalusite and the aluminum silicate phase diagram. Am J Sci 271:97-131

Holdaway MJ (1978) Significance of chloritoid-bearing and staurolite-bearing rocks in the Picuris Range, New Mexico. Geol Soc Am Bull 89:1404-1414

Holland TJB (1979) Experimental determination of the reaction paragonite = jadeite + kyanite + H_2O, and internally consistent thermodynamic data for part of the system Na_2O-Al_2O_3-SiO_2-H_2O, with applications to eclogites and blueschists. Contrib Miner Petrol 68:293-301

Holland TJB (1980) The reaction albite = jadeite + quartz determined experimentally in the range 600° to 1200°C. Am Miner 65:129-134

Holland TJB, Powell R (1985) An internally consistent thermodynamic dataset with uncertainties and correlations: 2. Data and results. J Metamorphic Geol 3:343-370

Holloway JR (1977) Fugacity and activity of molecular species in supercritical fluids. In: Fraser DG (ed) Thermodynamics in geology. Reidel, Dordrecht, pp 161-181

Holmes RD, O'Neil HStC, Arculus RJ (1986) Standard Gibbs free energy of formation for Cu_2O, NiO, CoO, and Fe_xO: High resolution electrochemical measurements using zirconia solid electrolytes from 900-1400 K. Geochim Cosmochim Acta 50:2439-2452

Hovis GL (1988) Enthalpies and volumes related to K-Na mixing and Al-Si order/disorder in alkali feldspars. J Petrol 29:731-763

Hovis GL, Waldbaum DR (1977) A solution calorimetric investigation of K-Na mixing in a sanidine-analbite ion-exchange series. Am Miner 62:680-686

Huckenholz HG (1975) Grossularite, its solidus and liquidus relations in the CaO-Al_2O_3-SiO_2-H_2O system up to 10 kbar. Neues Jahrb Miner Abh 124:1-46

Hultgren R, Desai PD, Hawkins DT, Gleiser M, Kelley KK (1973) Selected values of the thermodynamic properties of the elements. Am Soc Metals, Metals Park, Ohio

Iyanaga S, Kawada Y (1977) Encyclopedic dictionary of mathematics. MIT Press, Cambridge, Massachusetts

Jacob KT (1978) Electrochemical determination of activities in Cr_2O_3-Al_2O_3 solid solution. J Electrochem Soc 125:175-179

Jacobs GK, Kerrick DM (1981a) APL and FORTRAN programs for a new equation of state for H_2O, CO_2, and their mixtures at supercritical conditions. Comput Geosci 7:131-143

Jacobs GK, Kerrick DM (1981b) Methane: an equation of state with application to the ternary system H_2O-CO_2-CH_4. Geochim Cosmochim Acta 45:607-614

Jacobs GK, Kerrick DM (1981c) Devolatilization equilibria in H_2O-CO_2 and H_2O-CO_2-NaCl fluids: an experimental and thermodynamic evaluation at elevated pressures and temperatures. Am Miner 66:1135-1153

Jen LS, Kretz R (1981) Mineral chemistry of some mafic granulites from the Adirondack Region. Can Miner 19:479-491

Johannes W (1969) An experimental investigation of the system MgO-SiO_2-H_2O-CO_2. Am J Sci 267:1083-1104

Johannes W, Metz P (1968) Experimentelle Bestimmungen von Gleichgewichtsbeziehungen im System $MgO-CO_2-H_2O$. Neues Jahrb Miner Mh 1:15-26

Juza J (1966) An equation of state for water and steam, steam tables in the critical region and in the range from 1,000 to 100,000 bars. Rozpr Csek Akad VED Rada Tech VED, Prag, 76:3-121

Kennedy CS, Kennedy GC (1976) The equilibrium boundary between graphite and diamond. J Geophys Res 81:2467-2470

Kerrick DM, Darken LS (1975) Statistical thermodynamic models for ideal oxide and silicate solid solutions, with application to plagioclase. Geochim Cosmochim Acta 39:1431-1442

Kerrick DM, Jacobs GK (1981) A modified Redlich-Kwong equation for H_2O, CO_2, and H_2O-CO_2 mixtures at elevated pressures and temperatures. Am J Sci 281:735-767

King EG, Weller WW (1961) Low-temperature heat capacities and entropies at 298.15°K of some sodium- and calcium-aluminum silicates. US Bur Mines, Rep Invest 5855, Washington D C

King EG, Barany R, Weller WW, Pankratz L (1967) Thermodynamic properties of forsterite and serpentine. US Bur Mines, Rep Invest 6962, Washington D C

Kiseleva IA, Ogorodova LP (1984) High-temperature solution calorimetry for determining the enthalpies of formation for hydroxyl-containing minerals such as talc and tremolite. Geochem Int 21:36-46

Kiseleva IA, Ogorodova LP, Topor ND, Chigareva OG (1979) A thermochemical study of the $CaO-MgO-SiO_2$ system. Geochem Int 16:122-134

Kleppa OJ (1982) Thermochemistry in mineralogy. In: Schreyer W (ed) High-pressure researches in geoscience. Schweizerbart, Stuttgart, pp 461-473

Klotz IM, Rosenberg RM (1972) Chemical thermodynamics, Benjamin, Menlo Park, California, 3rd edn

Koziol A (1990) Activity-composition relationships of binary Ca-Fe and Ca-Mn garnets determined by reversed, displaced equilibrium experiments. Am Miner 75:319-327

Koziol A, Newton RC (1988a) Redetermination of the anorthite breakdown reaction and improvement of the plagioclase-garnet-Al_2SiO_5-quartz geobarometer. Am Miner 73:216-223

Koziol A, Newton RC (1988b) The activity of grossular in ternary (Ca,Fe,Mg) garnet determined by reversed phase equilibrium experiments at 1000° and 900°C. Geol Soc Am Abstr 20:A191

Kretz R (1959) Chemical study of garnet, biotite, and hornblende from gneisses of southwestern Quebec, with emphasis on distribution of elements in coexisting minerals. J Geol 67:371-402

Lee HY, Ganguly J (1988) Equilibrium compositions of coexisting garnet and orthopyroxene: Experimental determinations in the system $FeO-MgO-Al_2O_3-SiO_2$, and applications. J Petrol 29:93-113

Lewis GN, Randall M (1961) Thermodynamics. (revised by Pitzer KS, Brewer L) McGraw-Hill, New York, 2nd edn

Lindsley DH, Grover JE, Davidson PM (1981) The thermodynamics of the $Mg_2Si_2O_6$-$CaMgSi_2O_6$ join: a review. In: Newton RC, Navrotsky A, Wood BJ (eds) Thermodynamics of minerals and melts. Adv Phys Geochem, vol. 1, Springer, Berlin Heidelberg New York, 149-175

Loewenstein W (1953) The distribution of aluminum in the tetrahedra of silicates and aluminates. Am Miner 38:92-96

Luth WC, Fenn PM (1973) Calculation of binary solvi with special reference to sanidine-high albite solvus. Am Miner 58:1009-1015

Maier CG, Kelley KK (1932) An equation for the representation of high-temperature heat content data. J Am Chem Soc 54:3243-3246

Margules M (1895) Über die Zusammensetzungen der gesättigten Dämpfe von Mischungen. Sitzungsber Akad Wiss Wien 104:1243-1278

Meyer SL (1975) Data analysis for scientists and engineers. Wiley, New York

Milton DJ (1986) Chloritoid-sillimanite assemblage from North Carolina. Am Miner 71:891-894

Mohr DW, Newton RC (1983) Kyanite-staurolite metamorphism in sulfidic schists of the Anakeesta Formation, Great Smoky Mountains, North Carolina. Am J Sci 283:97-134

Moser H (1936) Messung der wahren spezifischen Wärme von Silber, Nickel, β-Messing, Quarzkristall und Quarzglas zwischen +50 und 700°C nach einer verfeinerten Methode. Phys Z 37:737-752

Mueller RF (1961) Analysis of relations among Mg, Fe, and Mn in certain metamorphic minerals. Geochim Cosmochim Acta 25:267-296

Mueller RF (1972) Stability of biotite: a discussion. Am Miner 57:300-316

Nacken R (1918) Über die Grenzen der Mischkristallbildung zwischen Kaliumchlorid und Natriumchlorid. Sitzungsber Preuss Akad Wiss Phys Math Klasse 192-200

Navrotsky A (1977) Progress and new directions in high temperature calorimetry. Phys Chem Miner 2:89-104

Navrotsky A (1985) Crystal chemical constraints on the thermochemistry of minerals. In: Kieffer SW, Navrotsky A (eds) Microscopic to macroscopic. Atomic environments to mineral thermodynamics. Rev Miner 14:225-275

Nesterov AR, Rumyantseva YeV (1987) Zincochromite $ZnCr_2O_4$ - new mineral from Karelia. Zap Vses Min Obshch 116:367-371 [in Russian, cited from Am Miner 73:931-932 (1988)]

Newton MS, Kennedy GC (1968) Jadeite, analcime, nepheline, and albite at high temperatures and pressures. Am J Sci 266:728-735

Newton RC (1966a) Kyanite-andalusite equilibrium from 700° to 800°C. Science 153:170-172

Newton RC (1966b) Kyanite-sillimanite equilibrium at 750°C. Science 151:1222-1225

Newton RC (1983) Geobarometry of high-grade metamorphic rocks. Am J Sci 283A:1-28

Newton RC, Charlu TV, Kleppa OJ (1977) Thermochemistry of high pressure garnets and clinopyroxenes in the system $CaO-MgO-Al_2O_3-SiO_2$. Geochim Cosmochim Acta 41:369-377

Newton RC, Charlu TV, Kleppa OJ (1980) Thermochemistry of the high structural state plagioclases. Geochim Cosmochim Acta 44:933-941

Newton RC, Geiger CA, Kleppa OJ (1989) Thermodynamics of $(Fe^{2+},Mg,Ca)_3Al_2Si_3O_{12}$ garnet: a review. (Preprint) Submitted to D.S.Korzhinskii Volume

Newton RC, Wood BJ (1980) Volume behavior of silicate solid solutions. Am Miner 65:733-745

Newton RC, Wood BJ, Kleppa OJ (1981) Thermochemistry of silicate solid solutions. Bull Minér 104:162-171

O'Neill HStC, Navrotsky A (1984) Cation distributions and thermodynamic properties of binary spinel solid solutions. Am Miner 69:733-753

O'Neill MJ (1966) Measurement of specific heat functions by differential scanning calorimetry. Anal Chem 38:1331-1336

Orville PM (1963) Alkali ion exchange between vapor and feldspar phases. Am J Sci 261:201-237

Orville PM (1972) Plagioclase cation exchange equilibria with aqueous chloride solution: results at 700°C and 2000 bars in the presence of quartz. Am J Sci 272:234-272

Pankratz LB, Kelley KK (1963) Thermodynamic data for magnesium oxide (periclase). US Bur Mines, Rep Invest 6295, Washington D C

Pankratz LB, Stuve JM, Gokcen NA (1984) Thermodynamic data for mineral technology. US Bur Mines Bull 677:1-355

Pattison DRM, Newton RC (1989) Reversed experimental calibration of the garnet-clinopyroxene Fe-Mg exchange thermometer. Contrib Miner Petrol 101:87-103

Peng DY, Robinson DB (1976) A new two-constant equation of state. Industr Engng Chem Fundam 15:59-64

Perkins D, Essene E, Wall VJ (1987) THERMO: A computer program for calculation of mixed volatile equilibria. Am Miner 72:446-447

Perkins D, Holland TJB, Newton RC (1981) The Al_2O_3 contents of enstatite in equilibrium with garnet in the system $MgO-Al_2O_3-SiO_2$ at 15-40 kbar and 900-1,600°C. Contrib Miner Petrol 78:99-109

Perkins EH, Brown TH, Berman RG (1986) PTX-System: Three programs for calculation of pressure-temperature-composition phase diagrams. Comput Geosci 12:749-755

Piessens R, de Doncker-Kapenga E, Überhuber CW, Kahaner DK (1983) Quadpack. Springer, Berlin Heidelberg New York

Powell R, Holland TJB (1985) An internally consistent thermodynamic dataset with uncertainties and correlations. 1. Methods and a worked example. J Metamorphic Geol 3:327-342

Powell R, Holland TJB (1988) An internally consistent dataset with uncertainties and correlations: 3. Applications to geobarometry, worked examples and a computer program. J Metamorphic Geol 6:173-204

Pownceby MI, Wall VJ, O'Neill HStC (1987) Fe-Mn partitioning between garnet and ilmenite: experimental calibration and application. Contrib Miner Petrol 97:116-126

Prausnitz JM, Lichtenthaler RN, de Azevedo EG (1986) Molecular thermodynamics of fluid phase equilibria. Prentice-Hall, Englewood Cliffs, New Jersey, 2nd edn

Rammensee W, Fraser DG (1981) Activities in solid and liquid Fe-Ni and Fe-Co alloys determined by Knudsen cell mass spectrometry. Ber Bunsenges Phys Chem 85:588-592

Redlich O, Kister AT (1948) Thermodynamics of nonelectrolyte solutions. Algebraic representation of thermodynamic properties and the classification of solutions. Industr Engng Chem 40:345-348

Redlich O, Kwong JNS (1949) On the thermodynamics of solutions. V. An equation of state. Fugacities of gaseous solutions. Chem Rev 44:233-244

Reid RC, Prausnitz JM, Sherwood TK (1977) The properties of gases and liquids. McGraw-Hill, New York, 3rd edn

Rice MH, Walsh JM (1957) Equation of state of water to 250 kb. J Chem Phys 26:824-830

Richardson SW, Bell PM, Gilbert MC (1968) Kyanite-sillimanite equilibrium between 700° and 1500°C. Am J Sci 266:513-541

Richardson SW, Gilbert MC, Bell PM (1969) Experimental determination of kyanite-andalusite and andalusite-sillimanite equilibria: the aluminum silicate triple point. Am J Sci 267:259-272

Robie RA, Hemingway BS (1972) Calorimeters for heat of solution and low-temperature heat capacity measurements. US Geol Surv Prof Pap 755

Robie RA, Hemingway BS (1984a) Entropies of kyanite, andalusite, and sillimanite: additional constraints on the pressure and temperature of the Al_2SiO_5 triple point. Am Miner 69:298-306

Robie RA, Hemingway BS (1984b) Heat capacities and entropies of phlogopite $(KMg_3[AlSi_3O_{10}](OH)_2)$ and paragonite $(NaAl_2[AlSi_3O_{10}](OH)_2)$ between 5 and 900 K and estimates of enthalpies and Gibbs free energies of formation. Am Miner 69:858-868

Robie RA, Hemingway BS, Fisher JR (1979) Thermodynamic properties of minerals and related substances at 298.15 K and 1 bar (10^5 Pascals) pressure and at higher temperatures. US Geol Surv Bull 1452:1-456 (reprinted with corrections), Washington D C

Robie RA, Hemingway BS, Ito J, Krupka KM (1984) Heat capacity and entropy of Ni_2SiO_4-olivine from 5 to 1000 K and heat capacity of Co_2SiO_4 from 360 to 1000 K. Am Miner 69:1096-1101

Robie RA, Hemingway BS, Takei H (1982) Heat capacities and entropies of Mg_2SiO_4, Mn_2SiO_4, and Co_2SiO_4 between 5 and 380 K. Am Miner 67:470-482

Robin PF (1974) Stress and strain in cryptoperthite lamellae and the coherent solvus of alkali feldspars. Am Miner 59:1299-1318

Robinson GR, Haas JL, Schafer CM, Haselton HT (1982) Thermodynamic and thermophysical properties of selected phases in the $MgO-SiO_2-H_2O-$

CO_2, $CaO-Al_2O_3-SiO_2-H_2O-CO_2$, and $Fe-FeO-Fe_2O_3-SiO_2$ chemical systems, with special emphasis on the properties of basalts and their mineral components. US Geol Surv Open-File Rep 83-79:1-429

Rossini FD (1950) Chemical thermodynamics. Wiley, New York

Sack RO, Ghiorso MS (1989) Importance of considerations of mixing properties in establishing an internally consistent thermodynamic database: thermochemistry of minerals in the system $Mg_2SiO_4-Fe_2SiO_4-SiO_2$. Contrib Miner Petrol. 102:41-68

Sato M (1971) Electrochemical measurements and control of oxygen fugacity and other gaseous fugacities with solid electrolyte sensors. In: Ulmer GC (ed) Research techniques for high pressure and high temperature. Springer, Berlin Heidelberg New York, pp 43-99

Saxena SK, Fei Y (1987a) Fluids at crustal pressure and temperatures. I.Pure species. Contrib Miner Petrol 95:370-375

Saxena SK, Fei Y (1987b) High pressure and high temperature fluid fugacities. Geochim Cosmochim Acta 51:783-792

Saxena SK, Fei Y (1988) Fluid mixtures in the C-H-O system at high pressure and temperature. Geochim Cosmochim Acta 52:505-512

Saxena SK, Ghose S (1971) $Mg^{2+}-Fe^{2+}$ order-disorder and the thermodynamics of the orthopyroxene crystalline solution. Am Miner 56:532-559

Schmidt E (1979) Properties of water and steam in SI units. Springer, Berlin Heidelberg New York

Schramke J, Kerrick DM, Blencoe JG (1982) Experimental determination of the brucite = periclase + water equilibrium with a new volumetric technique. Am Miner 67:269-276

Seetharaman S, Abraham KP (1980) Thermodynamic properties of oxide systems. In: Subbarao EC (ed) Solid electrolytes and their applications. Plenum Press, New York, pp 127-163

Skippen GB (1971) Experimental data for reactions in siliceous marbles. J Geol 79:457-481

Skippen GB (1977) Dehydration and decarbonation equilibria. In: Greenwood HJ (ed) Application of thermodynamics to petrology and ore deposits. Miner Assoc Canada, Short Course Handbook 2:66-83

Slaughter J, Wall VJ, Kerrick DM (1976) APL computer programs for thermodynamic calculations of equilibria in $P-T-X_{CO_2}$ space. Contrib Miner Petrol 54:157-171

Spry A (1969) Metamorphic Textures. Pergamon, London

Spycher NF, Reed MK (1988) Fugacity coefficients of H_2, CO_2, CH_4, and H_2O and of $H_2O-CO_2-CH_4$ mixtures: A virial equation treatment for moderate pressures and temperatures applicable to calculations of hydrothermal boiling. Geochim Cosmochim Acta 52:739-749

Subbarao EC (1980) Solid electrolytes and their applications. Plenum Press, New York

Thompson AB, Perkins EH (1981) Lambda transitions in minerals. In: Newton RC, Navrotsky A, Wood BJ (eds) Thermodynamics of minerals and melts. Adv Phys Geochem, vol 1, Springer, Berlin Heidelberg New York, pp 35-62

Thompson JB (1967) Thermodynamic properties of simple solutions. In: Abelson PH (ed) Researches in geochemistry, vol 2, Wiley, New York, pp. 340-361

Thompson JB (1969) Chemical reactions in crystals. Am Miner 54:341-375

Thompson JB (1982) Reaction space: an algebraic and geometric approach. In: Ferry J (ed) Characterization of metamorphism through mineral equilibria. Rev Miner 10:33-51

Thompson JB, Waldbaum DR (1968) Mixing properties of sanidine crystalline solutions: I. Calculations based on ion-exchange data. Am Miner 53:1965-1999

Thompson JB, Waldbaum DR (1969) Analysis of the two-phase region halite-sylvite in the system NaCl-KCl. Geochim Cosmochim Acta 33:671-690

Torgeson DR, Sahama ThG (1948) A hydrofluoric acid solution calorimeter and the determination of the heats of formation of Mg_2SiO_4, $MgSiO_3$, and $CaSiO_3$. J Am Chem Soc 70:2156-2160

Ulbrich HH, Waldbaum DR (1976) Structural and other contributions to the third-law entropies of silicates. Geochim Cosmochim Acta 40:1-24

van der Waals JD (1881) Die Continuität des gasförmigen und flüssigen Zustandes. Barth, Leipzig

Vidal J (1978) Mixing rules and excess properties in cubic equation of state. Chem Engng Sci 33:787-791

Wagner C (1940) Thermodynamik metallischer Mehrstoffsysteme. In: Masing G (ed) Handbuch der Metallphysik. Bd I, Teil 2. Der metallische Zustand der Materie. Becker und Erler KG, Leipzig

Wagner HM (1959) Linear programming techniques for regression analysis. J Am Stat Assoc 54:206-212

Waldbaum DR, Robie RA (1971) Calorimetric investigation of Na-K mixing and polymorphism in the alkali feldspars. Z Krist 134:381-420

Waldbaum DR, Thompson JB (1968) Mixing properties of sanidine crystalline solutions: II. Calculations based on volume data. Am Miner 53:2000-2017

Waldbaum DR, Thompson JB (1969) Mixing properties of sanidine crystalline solutions: IV. Phase diagrams from equations of state. Am Miner 54:1274-1298

Warner RD, Luth WC (1973) Two-phase data for the join monticellite $(CaMgSiO_4)$-forsterite(Mg_2SiO_4): experimental results and numerical analysis. Am Miner 58:998-1008

Wise SS, Margrave JL, Feder HM, Hubbard WN (1963) Fluorine bomb calorimetry. V. The heats of formation of silicon tetrafluoride and silica. J Phys Chem 67:815-821

Wisniak J, Tamir A (1978) Mixing and excess thermodynamic properties. Elsevier, Amsterdam Oxford New York

Wohl K (1946) Thermodynamic evaluation of binary and ternary liquid systems. Trans Am Inst Chem Eng 42:215-249

Wones DR (1972) Stability of biotite: a reply. Am Miner 57:316-317

Wood BJ, Banno S (1973) Garnet-orthopyroxene and othopyroxene-clinopyroxene relationships in simple and complex systems. Contrib Miner Petrol 42:109-124

Wood BJ, Holland TJB, Newton RC, Kleppa OJ (1980) Thermochemistry of jadeite-diopside pyroxenes. Geochim Cosmochim Acta 44:1363-1371

Wood BJ, Holloway JR (1984) A thermochemical model for subsolidus equilibria in the system CaO-MgO-Al$_2$O$_3$-SiO$_2$. Geochim Cosmochim Acta 48:159-176

Wood BJ, Nicholls J (1978) The thermodynamic properties of reciprocal solid solutions. Contrib Miner Petrol 66:389-400

Wormald CJ, Lancaster NM, Sellars AJ (1986) The excess molar enthalpies of $\{X_{H2O} + (1-X_{CO})\}$(g) and $\{X_{H2O} + (1-X_{CO2})\}$(g) at high temperature and pressure. J Chem Thermodyn 18:135-147

Yund RA (1975) Subsolidus phase relations in the alkali feldspars with emphasis on coherent phases. In: Ribbe PH (ed) Feldspar minerals, Miner Soc Am Short Course Notes 2:Y1-Y28

Zhakirov IV (1984) The P-V-T relations in the H$_2$O-CO$_2$ system at 300 and 400°C up to 1000 bar. Geochem Int 21:13-20

Zen E-an (1977) The phase equilibrium calorimeter, the petrogenetic grid, and a tyranny of numbers. Am Miner 62:189-204

Ziegenbein D, Johannes W (1974) Wollastonitbildung aus Quarz und Calcit bei P_f = 2, 4 u. 6 kb (Abstract). Fortschr Miner, Beiheft 2, 52:77-79

Zyrianov VN, Perchuk LL, Podlesskii KK (1978): Nepheline-alkali feldspar equilibria: I. Experimental data and thermodynamic calculations. J Petrol 19:1-44

Subject Index